DATE DUE			
Feb21 80			

FLAVOR OF FOODS AND BEVERAGES

Chemistry and Technology

Academic Press Rapid Manuscript Reproduction

Proceedings of a conference sponsored by the Agricultural and Food Chemistry Division of the American Chemical Society, held in Athens, Greece, June 27–29, 1978

FLAVOR OF FOODS AND BEVERAGES

Chemistry and Technology

EDITED BY

GEORGE CHARALAMBOUS

Technical Center
Anheuser Busch, Inc.
St. Louis, Missouri

GEORGE E. INGLETT

U.S. Department of Agriculture
Federal Research, North Central Region
Peoria, Illinois

ACADEMIC PRESS **New York San Francisco London** **1978**
A Subsidiary of Harcourt Brace Jovanovich, Publishers

ACADEMIC PRESS, INC.
111 Fifth Avenue, New York, New York 10003

United Kingdom Edition published by
ACADEMIC PRESS, INC. (LONDON) LTD.
24/28 Oval Road, London NW1 7DX

Library of Congress Cataloging in Publication Data
Main entry under title:

Flavor of foods and beverages.

Proceedings of a symposium held June 27–28, 1977, Athens, Greece.
1. Flavor—Congresses. 2. Flavoring essences—Congresses. I. Charalambous, George, Date II. Inglett, G.E., Date
TP372.5.F55 664'.5 78-11177
ISBN 0-12-169060-1

PRINTED IN THE UNITED STATES OF AMERICA

CONTENTS

List of Contributors *ix*
Preface *xiii*

1. Flavor Modification 1
   C. E. Eriksson

2. New Developments in Meat Aroma Research 15
   Ivon Flament, Bruno Willhalm, and Günther Ohloff

3. Flavor Problems in the Application of Soy Protein Materials as Meat Substitutes 33
   Godefridus A. M. van den Ouweland and Leonard Schutte

4. Flavoring Vegetable Protein Meat Analogs 43
   R. F. Heinze, M. B. Ingle, and J. F. Reynolds

5. Flavor Development of White and Provincial Greek Bread. Traditional and Rural Breadmaking Process 57
   E. C. Voudouris, M. E. Komaitis, and B. Pattakou

6. Some Recent Advances in the Knowledge of Cheese Flavor 65
   Jacques Adda, Sylviane Roger, and Jean-Pierre Dumont

7. Flavor Aspects of Chocolate 75
   Dominique Reymond and Walter Rostagno

8. Cocoa Substitution 81
   Willi Grab, Walter Brugger, and Derek Taylor

9. Biomolecular Studies of a Class of Sweet Compounds 91
   Lloyd N. Ferguson, Rose W. Bragg, Yvonne Chow, Susan Howell, and Manque Winters

10. Intense Sweetness of Natural Origin 97
G. E. Inglett

11. Flavor Interactions with Food Ingredients from Headspace Analysis Measurements 113
Fouad Z. Saleeb and John G. Pickup

12. Occurrence of Amadori and Heyns Rearrangement Products in Processed Food and Their Role in Flavor Formation 131
Godefridus A. M. van den Ouweland, Hein G. Peer, and Sing Boen Tjan

13. Formation of Lactones and Terpenoids by Microorganisms 145
Roland Tressl, Martin Apetz, Ronald Arrieta, and Klaus Günter Grünewald

14. The Chemistry of the Black-Caramel Color Substances of Several Human Foods 169
J. P. J. Casier, A. Ahmadi Zenouz, and G. M. J. De Paepe

15. Characterization of Synthetic Substances in Food Flavors by Isotopic Analysis 199
Jacques Bricout and Joseph Koziet

16. Analysis of Air Odorants in the Environment of a Confections Manufacturing Plant 209
M. M. Hussein and D. A. M. Mackay

17. Cat and Human Taste Responses to L-α-Amino Acid Solutions 231
James C. Boudreau

18. Aroma Analysis of Virgin Olive Oil by Head Space (Volatiles) and Extraction (Polyphenols) Techniques 247
G. Montedoro, M. Bertuccioli, and F. Anichini

19. The Volatile Fraction of Orange Juice. Methods for Extraction and Study of Composition 283
José Alberola and Luis J. Izquierdo

20. The Aroma of Various Teas 305
Tei Yamanishi

21. Relationship between Physical and Chemical Analysis and Taste-Testing Results with Beers 329
Manfred Moll, That Vinh, and Roland Flayeux

22. The Aroma Composition of Distilled Beverages and Perceived Aroma of Whisky 339
Paula Jounela-Eriksson

23. Wine Flavor 355
Pascal Ribéreau-Gayon

24. Alteration in a Wine Distillate during Ageing 381
S. van Straten, G. Jonk, and L. van Gemert

25. Possibilities of Characterizing Wine Quality and Vine Varieties by Means of Capillary Chromatography 391
A. Rapp, H. Hastrich, L. Engel, and W. Knipser

Index *419*

LIST OF CONTRIBUTORS

Numbers in parentheses indicate the pages on which the authors' contributions begin.

Jacques Adda (65), Institut National de la Recherche Agronomique, F-78350 Jouy-en-Josas, France

José Alberola (283), Instituto de Agroquímica y Technología de Alimentos, Jaime Roig 11, Valencia 10, Spain

F. Anichini (247), Istituto Industrie Agrarie, Università Studi Perugia, I-06100 Perugia, San Costanzo, Italy

Martin Apetz (145), Lebensmitteltechnologie und Biotechnologie, Technische Universität Berlin, FB 13—Chem.-techn. Analyse, D-1000 Berlin 65, Seestrasse 13, Germany

Ronald Arrieta (145), Lebensmitteltechnologie und Biotechnologie, Technische Universität Berlin, FB 13—Chem.-techn. Analyse, D-1000 Berlin 65, Seestrasse 13, Germany

M. Bertuccioli (247), Istituto Industrie Agrarie, Università Studi Perugia, I-06100 Perugia, San Costanzo, Italy

James C. Boudreau (231), Sensory Sciences Center, University of Texas, Houston, Texas 77030

Rose W. Bragg (91), California State University, Los Angeles, California 90032

Walter Brugger (81), Givaudan Dübendorf Limited, Ueberlandstrasse 138, CH-8600 Dübendorf, Switzerland

Jacques Bricout (199), Institut de Recherches Appliquées aux Boissons, 120 avenue du Maréchal Foch, F-94003 Créteil Cedex, France

J. P. J. Casier (169), Katholieke Universiteit Leuven, Laboratory of Applied Carbohydrate and Cereal Chemistry, 36 Prinses Lydialaan, B-3030 Heverlee, Belgium

Yvonne Chow (91), California State University, Los Angeles, California 90032

G. M. J. De Paepe (169), Katholieke Universiteit Leuven, Laboratory of Applied Carbohydrate and Cereal Chemistry, 36 Prinses Lydialaan, B-3030 Heverlee, Belgium

Jean-Pierre Dumont (65), Institut National de la Recherche Agronomique, F-78350 Jouy-en-Josas, France

L. Engel (391), Bundesforschungsanstalt für Rebenzuchtung Geilweilerhof, 6741 Siebeldingen, Federal Republic of Germany

C. E. Eriksson (1), SIK, The Swedish Food Institute, Fack, S-400 23 Göteborg, Sweden

Lloyd N. Ferguson (91), California State University, Los Angeles, California 90032

Ivon Flament (15), Firmenich S.A., Research Laboratories, CH-1211 Geneva 8, Switzerland

Roland Flayeux (329), Centre de Recherches et Développement TEPRAL, 2, rue Gabriel Bour, F-54250 Champigneulles, France

Willi Grab (81), Givaudan Dübendorf Limited, Ueberlandstrasse 138, CH-8600 Dübendorf, Switzerland

Klaus Günter Grünewald (145), Lebensmitteltechnologie und Biotechnologie, Technische Universität Berlin, FB 13—Chem.-techn. Analyse, D-1000 Berlin 65, Seestrasse 13, Germany

H. Hastrich (391), Bundesforschungsanstalt für Rebenzuchtung Geilweilerhof, 6741 Siebeldingen, Federal Republic of Germany

R. F. Heinze (43), Miles Laboratories, Inc., 1127 Myrtle Street, Elkhart, Indiana 46514

Susan Howell (91), California State University, Los Angeles, California 90032

M. M. Hussein (209), Life Savers, Inc., N. Main Street, Port Chester, New York 10573

M. B. Ingle (43), Miles Laboratories, Inc., 1127 Myrtle Street, Elkhart, Indiana 46514

G. E. Inglett (97), Northern Regional Research Center, SEC, 1815 N. University Street, Peoria, Illinois 61604

Luis J. Izquierdo (283), Instituto de Agroquímica y Tecnología de Alimentos, Jaime Roig 11, Valencia 10, Spain

G. Jonk (381), Central Institute for Nutrition and Food Research TNO, 48 Utrechtsweg, Zeist, The Netherlands

Paula Jounela-Eriksson (339), Research Laboratories of the State Alcohol Monopoly (ALKO), Box 350, SF-00101 Helsinki 10, Finland

W. Knipser (391), Bundesforschungsanstalt für Rebenzuchtung Geilweilerhof, 6741 Siebeldingen, Federal Republic of Germany

M. E. Komaitis (57), Department of Food Chemistry, National University, Athens, Greece

Joseph Koziet (199), Institut de Recherches Appliquées aux Boissons, 120 avenue du Maréchal Foch, F-94003 Créteil Cedex, France

D. A. M. Mackay (209), Life Savers, Inc., N. Main Street, Port Chester, New York 10573

Manfred Moll (329), Centre de Recherches et Développement TEPRAL, 2, rue Gabriel Bour, F-54250 Champigneulles, France

G. Montedoro (247), Istituto Industrie Agrarie, Università Studi Perugia, I-06100 Perugia, San Costanzo, Italy

Günther Ohloff (15), Firmenich, S.A. Research Laboratories, CH-1211, Geneva 8, Switzerland

B. Pattakou (57), Cereal Institute, Thessaloniki, Greece

Hein G. Peer (131), Unilever Research Vlaardingen/Duiven, Postbos 7, Zevenaar, The Netherlands

John G. Pickup (113), Physical Chemistry Laboratory, Technical Center, General Foods Corporation, White Plains, New York 10625

A. Rapp (391), Bundesforschungsanstalt für Rebenzuchtung Geilweilerhof, 6741 Siebeldingen, Federal Republic of Germany

Dominique Reymond (75), Nestlé Technical Assistance Company (NESTEC), C.P. 1009, CH-1000 Lausanne, Switzerland

J. F. Reynolds (43), Miles Laboratories, Inc., 1127 Myrtle Street, Elkhart, Indiana 46514

Pascal Ribéreau-Gayon (355), Institut d'Œnologie, Université de Bordeaux II, 351 cours de la Libération, F-33405, Talence Cedex, France

Sylviane Roger (65), Institut National de la Recherche Agronomique, F-78350 Jouy-en-Josas, France

Walter Rostagno (75), Nestlé Technical Assistance Company, C.P. 1009, CH-1000 Lausanne, Switzerland

Fouad Z. Saleeb (113), Physical Chemistry Laboratory, Technical Center, General Foods Corporation, White Plains, New York 10625

Leonard Schutte (33), Unimills, B.V., Lindtsedijk 8, Zwijndrecht, The Netherlands

Derek Taylor (81), Givaudan Dübendorf Limited, Ueberlandstrasse 138, CH-8600 Dübendorf, Switzerland

Sing Boen Tjan (131), Unilever Research Vlaardingen/Duiven, Postbos 7, Zevenaar, The Netherlands

Roland Tressl (145), Lebensmitteltechnologie und Biotechnologie, Technische Universität Berlin, FB 13—Chem. techn. Analyse, Seestrasse 13, D-1000 Berlin 65, Germany

Godefridus A. M. van den Oueweland (33, 131), Unilever Research Vlaardingen/Duiven, Postbos 7, Zevenaar, The Netherlands

L. van Gemert (381), Central Institute for Nutrition and Food Research TNO, 48 Utrechtsweg, Zeist, The Netherlands

S. van Straten (381), Central Institute for Nutrition and Food Research TNO, 48 Utrechtsweg, Zeist, The Netherlands

That Vinh (329), Centre de Recherches et Développement TEPRAL, 2, rue Gabriel Bour, F-54250 Champigneulles, France

E. C. Voudouris (57), Department of Food Chemistry, National University, Navarinou 13a, Athens, Greece

Bruno Willhalm (15), Firmenich, S.A., Research Laboratories, CH-1211 Geneva 8, Switzerland

Manque Winters (91), California State University, Los Angeles, California 90032

Tei Yamanishi (305), Department of Food and Nutrition, Ochanomizu University, Ohtsuka, Bunkyo-ku, Tokyo 112, Japan

A. Ahmadi Zenouz (169), Katholieke Universiteit Leuven, Laboratory of Applied Carbohydrate and Cereal Chemistry, 36 Prinses Lydialaan, B-3030 Heverlee, Belgium

PREFACE

The flavor of foods and beverages is of perpetual importance to growers, processors, manufacturers, brewers, distillers, and, ultimately, the consumer. Flavor is the most important quality of foods and beverages for determining consumer acceptability. Regardless of how nutritious a product may be, it will not be consumed unless the product has acceptable taste and aroma.

Rapid changes in flavor chemistry and technology require frequent updating of recent progress. This book is the proceedings of an international conference to bring existing information together on the flavor of foods and beverages. The conference was sponsored by the Agricultural and Food Chemistry Division of the American Chemical Society and was held in Athens, Greece, June 27–29, 1978.

Topics of the conference covered such wide ranging subjects as flavor of meat, meat analogs, chocolate and cocoa substitutes, cheese aroma, beer, wine, and spirits, baked goods, confections, tea, citrus and other fruits, olive oil, and sweeteners. New analytical methodology on taste and aroma was emphasized by several speakers. Flavor production, stability, and composition were covered by various other speakers.

This volume should be a useful book for students, chemists, technologists, and executives who are involved with any facet of producing foods and beverages.

We wish to thank all who attended and participated in the conference, and particularly the speakers. The publisher's help and advice is also greatly appreciated.

FLAVOR OF FOODS AND BEVERAGES

Chemistry and Technology

FLAVOR MODIFICATION

C.E. Eriksson

SIK-The Swedish Food Institute
Göteborg
Sweden

I. INTRODUCTION

The flavor of various foods can undergo great changes during handling of the raw material, the processing into a product or the storage. Primary flavor compounds can be lost or secondary ones formed, sometimes to the benefit of the final product as for instance in ageing processes, often, however, such changes lead to loss in sensory quality of food.

The concept of flavor modification has been attributed to two different changes; firstly, the ones caused by environmental and technical influence leading to unintentional changes and secondly, different chemical, physical and technical operations to deliberately produce a changed flavor in a product for certain reasons. Such reasons can be to regenerate lost flavor or to remove or neutralize an impairing flavor that was formed in a product or to produce a new flavor sensation. In the present paper only deliberate flavor modification will be dealt with.

Such flavor modification can be performed in a number of ways, for instance through:

- *de novo* synthesis of flavor compounds
- inhibition of the formation of particular flavor compounds
- removal av particular flavor compounds
- masking of flavor
- conversion of one flavor compound into another

In this presentation examples will be given on small but important flavor changes in foods and how these changes can be performed or counteracted. Illustration will be made to each

ISBN 0-12-169060-1

of the ways of doing so, as listed above. The examples chosen are regeneration of flavor in processed vegetable products, heat induced flavor, lipid oxidation and its inhibition by untraditional means, bitterness caused by protein hydrolysis, by certain compounds in citrus products and by lipid oxidation and finally interconversion of flavor compounds. Much work has been devoted to these problems in our institute.

II. *DE NOVO* SYNTHESIS OF FLAVOR COMPOUNDS

More than twenty years ago the idea was introduced that fresh flavor lost during processing of food could be restored by added enzymes with the ability to produce new fresh flavor from the precursor compounds still present in the processed food material according to the scheme below

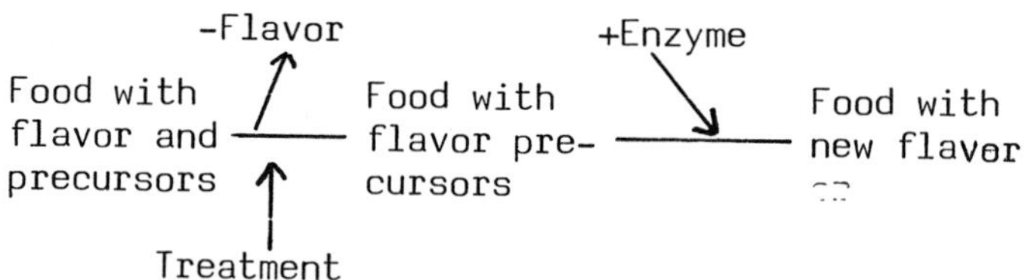

This possibility was demonstrated with dehydrated watercress and cabbage, whose fresh flavor had unintentionally been removed in the drying process. On rehydration of the dry watercress or cabbage in water containing an enzyme preparation obtained from white or black mustard seeds, the characteristic flavor of both watercress and cabbage was rapidly regained (1).

Watercress and cabbage like several other vegetables in the *Cruciferae* family get much of their flavor from volatile allylisothiocyanate which is produced from the non-volatile precursor sinigrin which is a thioglycoside containing sulfate. Watercress and cabbage both contain the enzyme myrosinase which converts the precursor into allylisothiocyanate, glucose and bisulfate. The heat treatment during the drying of watercress and cabbage causes inactivation of the myrosinase and therefore watercress and cabbage on rehydration in pure water does not develop fresh flavor. Mustard seed, however, is not heat treated, and contains a high concentration of myrosinase and was therefore found to be a suitable enzym source. These results led to great expectations in the field of flavor regeneration.

The same principle was later applied in order to improve the flavor of other food material for instance carrots (2,3), tomato juice (4), citrus juice (5), and banana (6). These experiments showed that it is generally difficult to reproduce a correct flavor by this technique. The reasons for that is the chemical complexity of the flavor of these materials in comparison with watercress and cabbage whose flavor is mainly derived from one type of precursor. At that time the metabolic pathways for the formation of flavor compounds in vegetables were little known. In this respect the situation has changed dramatically during the past 20 years. Most flavor biosynthesis is today known to originate from the metabolism of amino acids, nucleic acids, carbohydrates, lipids, terpenes, and cinnamic acid. For example in banana 3-methylbutanal, 3-methylbutanol and 3-methylbutanoate are formed from amino acids through a sequence of reactions involving enzymatic transamination, decarboxylation, reduction and oxidation. The 3-methylbutanol can then be alkylated into 3-methylbutyl esters and the 3-methylbutanoate acylated via the CoA ester, into 3-methylbutyrate esters. Both types of esters are produced in banana and contribute to its flavor (7). Another example showing the complexity of flavor biosynthesis is the formation of the bitter substances in hops, lupulone and humulone from acetyl-CoA involving carboxylase, transacylase and ligase enzymes (7,8).

The development of sensitive analytical tools such as combined gas chromatography-mass spectrometry has shown that the volatile fraction of most food materials is very complex. Although such vast information of the chemical composition of the volatile fraction of foods now exists little is yet known about the chemistry of the flavor of the same material; in other words, which set-up of volatiles in certain concentration proportions will produce the correct flavor of a certain food.

To sum up the prerequisites for *de novo* synthesis of flavor often suffer from complexity of two kinds.

- several volatiles in certain proportions are required
- the formation of each flavor compound can be complex.

Hence, the possibility of *de novo* synthesis of flavor is the best when the flavor is determined by a single or a few volatile compounds which are derived directly from closely related non-volatile precursors. This is the case with many *Allium* species where S-1-propenyl, S-2-propenyl- and S-methyl-L-cysteine sulfoxides are the principal precursors of flavor (9).

Another group of closely related flavor precursors is the unsaturated fatty acids which are known to give rise to a great number of volatile compounds, most of which are potential flavor compounds. The oxidation of unsaturated fatty acids is often looked upon as a reaction leading to off-flavor formation. The following example, however, illustrates a case where positive aroma might be derived from lipid breakdown.

The formation of C_6 and C_9 aliphatic straight chain saturated and unsaturated aldehydes and alcohols from lipids in tomato and cucumber occurs on disintegration of the tissues through a series of enzymic oxidation, cleavage and isomerisation reactions (10), according to the following, simplified scheme:

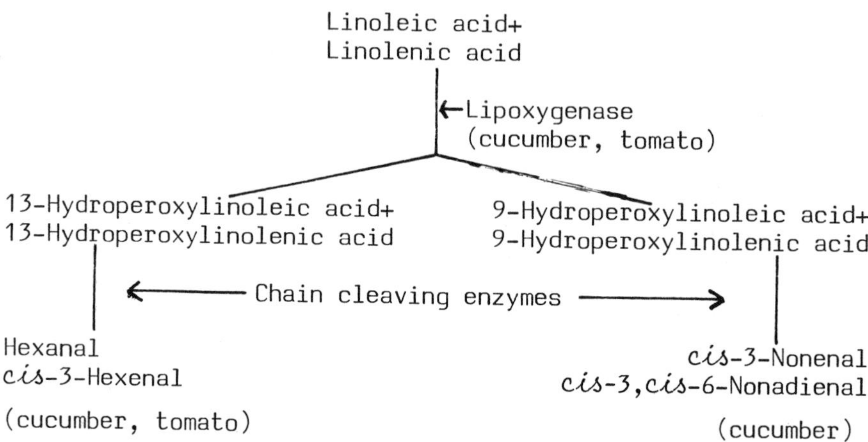

It is interesting to note that such different fruits produce these odorous compounds basically according to the same reaction scheme and that they also differ at certain important points. Thus the precursors of hexanal and *cis*-3-hexenal are the 13-hydroperoxylinoleic and -linolenic acid respectively produced from linoleic and linolenic acid by the action of lipoxygenase in the presence of oxygen. The precursors of *cis*-3-nonenal and *cis*-3, *cis*-6-nonadienal are 9-hydroperoxylinoleic and -linolenic acid respectively produced analogically to the 13-hydroperoxyisomers. *cis*-3-Hexenal, *cis*-3-nonenal and *cis*-3,*cis*-6-nonadienal can easily isomerize into *trans*-2-hexenal, *trans*-2-nonenal and *trans*-2,*cis*-6-nonadienal respectively. Both the tomato and cucumber lipoxygenase favor the formation of the 9-isomers (95%) over the 13-isomers (5%). In the following reaction, however, the chain cleavage of the hydroperoxy acids, cucumber enzymes cleave both the 9- and 13-hydroperoxy acids whilst the tomato enzymes only cleave the 13-isomer (11, 12). With this background it might be interesting to study the possibilities of reproducing tomato and cucumber flavor in processed tomato and cucumber respectively by the use of proper enzyme preparations.

Today's knowledge of flavor biosynthesis and chemistry and the access to novel analytical tools seem to have given new incentive to enzymatic flavor regeneration work. Two examples will be mentioned.

Tomato powder was used together with crude enzyme preparations from different parts of the tomato plant, stems, leaves, green fruit, yellow fruit and ripe fruit which all produced a tomato flavor character in the product, however, somewhat different from that of fresh tomato and also depending on the enzyme preparation used (13). The authors followed the changes in selected flavor compounds during the reaction and found for instance that the concentration of a number of aliphatic straight and branched chain alcohols as well as of terpene alcohols increased during the reaction while the concentration of other compounds remained unchanged.

Another element involved in the regeneration of flavor from different precursors is the enzyme(s). An enzyme preparation for this purpose should contain the specific enzymes for flavor production and as little as possible of other enzymes, at least such enzymes should do as little harm as possible. The enzyme preparation should also be cheap which means that the source of the enzyme must either be a material of low alternative value or be produced by microorganisms and should in no case need much purification. In the case of microbial enzymes only certain microorganisms can presently be allowed as enzyme sources. There is also a technical difficulty involved in using the enzymes namely the possibility to add them under sterile conditions. Some interest has been taken in producing microbial enzyme preparations for flavor production. An enzyme preparation isolated from *Oospora suaveloens* was found to be capable of producing fruit flavors when added to strawberry, banana, peach, pineapple, apple, and melon purees and grape juice and the results seemed to be promising, since this enzyme preparation was able to produce a mixture of aliphatic alcohols and esters typical of fruits (Table I). It was also shown that the amino acids valine, leucine, isoleucine and threonine were precursors of branched chain esters (14).

III. HEAT INDUCED FLAVOR CHANGES

Among the many types of objectionable flavors produced during

TABLE I. Volatile compounds produced by cultures of *Oospora suaveloens* (14)

Alcohols	*Iso*-butyl esters
Ethyl alcohol	*Iso*-butyl acetate
Iso-propyl alcohol	*Iso*-butyl n-propionate
Iso-butyl alcohol	*Iso*-butyl n-butyrate
Iso-amyl alcohol	*Iso*-butyl *iso*-butyrate
Ethyl esters	*Iso*-butyl 2-methylbutyrate
Ethyl acetate	*Iso*-amyl esters
Ethyl n-propionate	*Iso*-amyl acetate
Ethyl n-butyrate	*Iso*-amyl n-propionate
Ethyl-*iso*-butyrate	*Iso*-amyl *iso*-butyrate
Ethyl 2-methyl butyrate	*Iso*-amyl 2-metylbutyrate
Ethyl n-valerate	*Iso*-amyl *iso*-valerate
Ethyl *iso*-valerate	
Ethyl caproate	

food handling the ones causing cooked flavor and retort flavor have been investigated by several groups. Heat induced flavor occurs in many kinds of heat treated foods, in this survey milk, black currant juice and canned meat products have been selected as examples where objectionable flavor and the inhibition of its formation has been studied.

In a recent review on enzymatic modification of milk flavor (15) reference was made to sulfhydryl oxidase and the possibility to use this enzyme to prevent sulfhydryl compound from being accumulated in HTST-sterilized milk and thus cause a "cooked flavor". The enzyme oxidizes the sulfhydryl compounds formed to disulfides. This reaction is further commented on by Menz in the chapter about milk flavor.

Dimethylsulfide concentration was shown to increase considerably in both black currant juice and mash during heating. This was also the case with aliphatic aldehydes while the concentration of terpene hydrocarbons decreased (16). The flavor quality of the product underwent changes during heating and unpleasant odor notes among them "cooked odor" and "pungency" increased while pleasant ones like "fruity", "floral" and "fragrant" decreased. The rate of all changes, both increments and decrements increased as a function of temperature (17). By computerized psychophysical and statistical analysis the negative sensory changes were found to correlate well with the increase in dimethylsulfide and aliphatic aldehydes and with the decrease in the terpene hydrocarbons (18).

TABLE II. Mean panel values of intensities of positive and negative flavor notes of black currant as influenced by heat treatment (intensity scale 0 - 10). (17).

Odor notes	Panel values after	
	75°C, 15 min.	55°C, 30 min.
Positive		
Fruity, citrus	2.5	2.6
Fruity, other	4.1	4.3
Floral	2.5	2.7
Fragrant	3.4	3.3
Negative		
Cooked odor	3.4	2.8
Burnt, smoky	1.0	0.6
Sickly	2.3	1.9
Sharp, pungent	3.5	3.1

TABLE III. Influence of arginine added to a beef formulation on the development of aldehydes and "retort" and "burnt" flavor during sterilization of canned beef (21).

Compound	Control (no additive)	Arginine added (1.5%)
Butanal	2.2 ppb	1.2 ppb
Pentanal	3.7 ppb	1.9 ppb
Hexanal	6.1 ppb	3.6 ppb
2-Methyl propanal	47 ppb	30 ppb
2-Methyl butanal	43 ppb	24 ppb
3-Methyl butanal	56 ppb	33 ppb
Retort flavor*)	5.0 ppb	3.0 ppb
Burnt flavor*)	5.0 ppb	3.0 ppb

*) evaluated on a 0-10 point scale.

This knowledge might be used to modify flavor of black currant juice by selecting the process parameters carefully particularly the heating temperature, the treatment time which should be as short as possible, rapid initial temperature rise and final drop. The influence of heat treatment on the intensity of "positive" and "negative" odor notes in black currant is shown in Table II. Another way of modifying the "cooked flavor" is to selectively remove objectionable flavor compounds such as dimethylsulfide and aliphatic aldehydes by performing the heat treatment under vacuum in a cooking device allowing certain types of compounds to be added back to the heat treated juice and others to be removed from it (19).

"Retort flavor" of canned beef most probably depends on the formation of volatile sulfur compounds such as hydrogensulfide, methylmercaptan and dimethylsulfide and of aldehydes like 2-methylpropanal, 2-methylbutanal, and 3-methylbutanal. The same compounds, particularly 2-methylpropanal and dimethylsulfide, also contribute to the odor note "burnt" (20). The flavor of canned beef could be modified by the choice of heating and packaging technique and by chemical means. Both aseptic canning and HTST-sterilization particularly when performed in thin layers resulted in less formation of the undesired compounds. Addition of small amounts of free arginine and lysine to the can content before sterilization decreased the concentration of objectionable aldehydes whilst addition of disodium fumarate or maleinate caused markedly less formation of the sulfur compounds and improvement of the beef aroma. Combinations of both types of additions caused even better improvement (21) (see Table III).

IV. NONTRADITIONAL INHIBITION OF LIPID OXIDATION

Lipid oxidation gives rise to a large number of volatile compounds each one capable of flavor contribution sometimes at very low concentrations. Volatile lipid oxidation products at superthreshold concentration also cover a wide range of odor quality notes. It is still very little known how different volatile lipid oxidation products sensorically interact with each other or with volatile compounds derived from other sources. Such interaction can be for instance synergism, additive effects or masking. Volatile lipid oxidation products often possess high chemical reactivity and can thus react with other flavor compounds *e.g.* sulfur compounds (22) for instance on hot-keeping of food in catering. Hence, volatile lipid oxidation products can contribute to flavor in at least three ways: direct flavor contribution, removal of other flavor compound, and inducing new flavor.

Consequently, inhibition of lipid oxidation can modify flavor of foods in several ways. Traditionally lipids are protected from being oxidized by the use of synthethic antioxidants. On the other hand unavoidable off-flavor consequences of lipid oxidation are sometimes masked by the addition of other flavoring compounds. Investigations during recent years on natural ways of protecting lipids from oxidation have pointed at several possibilities for instance through formulation of recipes and processing whereby browning reaction products are formed which can protect lipids from oxidation and also contribute with own flavor. "Warmed over flavor" formation in cooked meats could thus be avoided if Maillard reaction components like glycine + glucose, lysine + glucose, leucine + glucose and glycine + lactose were added to the meat before heat treatment. At the same time a much lesser degree of lipid oxidation occurred as measured by the TBA reaction. The use of some non-meat protein products had similar effects (23).

Several authors have reported antioxidative effect of the browning reaction (24, 25, 26). In our laboratory we have studied the antioxidative effect of browning reaction products in some more detail. By polarographic measurement of oxygen consumption in linoleic acid emulsion and gas chromatographic measurement of the formation of volatile aldehydes in the headspace over oxidizing linoleic acid emulsion we found that certain combinations of amino acids + sugars, for instance histidine + glucose, and arginine + xylose, produced more efficient antioxidative products than other combinations. Some combinations also produced prooxidative material. When histidine and glucose were added to doughs containing lard before baking cookies from them we found that the acceptability time of the cookies was doubled as compared both with cookies containing no addition or a mixture of synthetic antioxidants. We also found that reaction products both of the dipeptides histidine-glycine and glycine-histidine as well as of enzymatic protein hydrolysates possessed antioxidative power. Glycine + glucose reaction products alone did not show any such activity (27).

Protein hydrolysate (autolyzed yeast protein) also showed antioxidative effect alone or synergistically with phenolic antioxidants like BHA when tested in freeze-dried food models. Also some spices and herbs acted as antioxidants in the same model system (28). The antioxidative effect of certain spices and herbs may be attributed to their content of flavonoid compounds(29). The application of Maillard reaction products, protein hydrolysates, spices, and herbs are thus involved in flavor modification of foods by utilization of two combined effects, the inhibition of rancid flavor formation and the introduction of additional flavor since all these materials also contribute with their own flavor.

IV. BITTERNESS

Protein hydrolysates can contribute to flavor in two ways either due to their content of many precursors of flavor compounds which are produced *e.g.* on heating of the protein hydrolysates or by their content of bitter substances consisting of hydrophobic peptides. Protein hydrolysates from different sources differ in bitterness, depending on the hydrophobicity of the intact protein and the degree of hydrolysis. The bitterness of protein hydrolysates obtained from casein, soy protein, gliadin, glutenin, zein, and of peptides with known structure has been extensively reviewed (30, 31, 32, 33). Bitterness of protein hydrolysates can be reduced or removed in several ways. Enzymic soy protein hydrolysates were debittered through the action of acid carboxypeptidases from *Aspergillus* (34). Milk protein hydrolysates were debittered through the action of pig kidney homogenate on a laboratory (35) and a large scale (36). From fish protein hydrolysates the bitter peptides could be extracted with *sec*-butanol or removed by hydrophobic interaction chromatography (37). The extracted bitter peptides could also be debittered by applying plastein reactions to them.

The bitterness of protein hydrolysates has also been reduced by performing the enzymatic hydrolysis in the presence of certain amino acids, *e.g.* glutamic and aspartic acid, alanine, glycine, serine, proline, hydroxyproline, and threonine (38). Masking of the bitterness in pepsin hydrolyzed fish protein could be accomplished with peptides rich in glutamic acid, particularly L-glutamyl-L-glutamic acid. This peptide showed a general bitter-masking effect since other bitter substances as brucine, caffeine and magnesium chloride were also debittered (39), indicating that the modifying effect occurred at the receptor or higher level.

Other well known examples of debittering are the enzymic conversion of the bitter principles in citrus fruits. Basically there are two bitter components,the naringin (7-rhamnosiodoglucosides of 4',5,7-trihydroxy flavonone) in grapefruit and limonin in grapefruit and navel orange juice. The bitterness of these substances can be significantly decreased by the aid of enzymes.

Naringin can be hydrolyzed to the non-bitter mixture of rhamnose and prunin (naringenin-7, β-glucoside) by an enzyme preparation called naringinase which is a mixture of a rhamnosidase and a β-glucosidase. It is the rhamnosidase that acts as the major debittering agent since prunin is much less bitter than naringin. The glucosidase can hydrolyse the prunin further with less effect on the bitterness. Limonin is converted to its non-bitter 17-dehydro derivative by the action of limonoate dehydrogenase and $NAD(P)^+$. The bitterness of citrus products and the mechanism of enzymatic debittering has recently been reviewed (40, 41).

Even lipids can form bitter substances. Oxidation of linoleic acid by a soy bean protein fraction containing lipoxygenase and peroxidase led to the formation of 9,12,13-trihydroxyoctadec-10-en acid (I) and 9,10,13-trihydroxyoctadec-11-en acid (II). The bitterness of these compounds is determined to a great deal by the double bond since hydrogenation of the double bond resulted in a threefold increase of the taste threshold (42, 43). Lipid oxidation protection will most certainly inhibit the formation of this kind of bitterness.

$$CH_3-(CH_2)_4-\underset{OH}{CH}-\underset{OH}{CH}-CH=CH-\underset{OH}{CH}-(CH_2)_7-COOH \quad \text{(I)}$$

$$-\underset{OH}{CH}-CH=CH-\underset{OH}{CH}-\underset{OH}{CH}- \quad \text{(II)}$$

V. INTERCONVERSION OF FLAVOR COMPOUNDS

One example has already been mentioned above, namely the conversion of sulfhydrylcontaining volatile compounds responsible for cooked flavor of milk into disulfides possessing another odor note (15). The conversion of methylmercaptan into dimethyldisulfide will be followed by a drop in odor intensity since the odor threshold in air is 0.99 ppb for methylmercaptan and 7.6 ppb for dimethyldisulfide (44).

In our laboratory we studied the conversion of aliphatic aldehydes derived from lipid oxidation into the corresponding alcohols by use of the enzyme alcohol dehydrogenase and the coenzyme NADH according to the general reaction:

$$\text{Aldehyde} + \text{NADH} + \text{H}^+ \underset{}{\overset{\text{ADH}}{\rightleftharpoons}} \text{Alcohol} + \text{NAD}^+$$

The equilibrium in this system depends on the pair of aldehyde and alcohol involved, while the NADH/NAD$^+$ ratio and pH can be used to control the degree of interconversion (45). It is easy to convert most of the common aldehydes into alcohols. Since these kinds of aldehydes were shown to have odor threshold 20 - 250 times lower than that of the corresponding alcohols (46) and also to differ somewhat in odor quality (47) this type of interconversion is followed by both quantitative and qualitative sensory effects. These ideas were applied on cow's milk containing increased proportions of polyunsaturated fatty acids obtained by feeding the cows with a lipid protected diet containing high proportions of polyunsaturated fatty acids. Such milk rapidly develop typical lipid oxidation derived aldehydes for instance hexanal and so called "oxidized flavor". By treating whole milk and skimmilk with the enzyme--coenzyme combination the hexanal and other aldehydes did not accumulate in the milk or could the aldehydes already formed be removed (47). Instead hexanol and other alcohols were produced but had less effect on the oxidized flavor of the milk depending on its much higher odor threshold (46).

Intact strawberries were shown to adsorb volatile aliphatic alcohols and aldehydes from the atmosphere around them and to convert both the alcohols and the aldehydes into carboxylic esters of the acids naturally present in strawberry. By applying small amounts of both alcohols and carboxylic acids normally not present in strawberry novel sensory sensations could be introduced (48, 49, 50). The aldehydes were first reduced to alcohols by alcohol dehydrogenase present in strawberry before esters were formed (51). These findings may open interesting possibilities to introduce flavor in strawberry and maybe other fruits merely by introducing small amounts of properly selected alcohols and aldehydes into the atmosphere of stored fruit.

REFERENCES

1. Hewitt, E.J., Mackay, D.A.M., Konigsbacher, K.S., and Hasselstrom, T., Food Technol. 10:487 (1956).
2. Schwimmer, S., J. Food Sci. 28:460 (1963).
3. Heatherbell, D.A., and Wrolstad, R.E., J. Agr. Food Chem. 19:281 (1971).

4. Yu, M.-H., Olsen, L.E., and Salunkhe, D.L., Phytochemistry 7:561 (1968).
5. Bailey, S.D., Bazinet, M.L., Driscoll, J.L., and McCarthy, A.I., J. Food Sci. 26:163 (1961).
6. Hultin, H.O., and Proctor, B.E., Food Technol. 16:108 (1962).
7. Drawert, F., in "Proc. Int. Symp. Aroma Research" (H. Maarse and P.J. Groenen, eds.) p. 13, Pudoc, Wageningen, 1975.
8. De Keukeleire, D., and Verzele, M., in "Industrial Aspects of Biochemistry" (B. Spencer, ed.) vol. 30, p. 261 North Holland/American Elsevier, Amsterdam, 1974.
9. Freeman, G.G., and Whenham, R.J., J. Sci. Fd. Agric. 26:1869 (1975).
10. Gaillard, T., in "Proc. Plant Lipid Symp.", p. 12, Tokyo, 1976.
11. Gaillard, T., Phillips, D.R., and Reynolds, J., Biochim. Biophys. Acta, 441:181 (1976).
12. Gaillard, T., and Matthew, J.A., Phytochemistry 16:339 (1977).
13. Gremli, H., and Wild, J., in "Proc. IV Int. Congress Food Sci. and Technol.", vol. I, p. 158 Valencia, 1976.
14. Hattori, S., Yamaguchi, Y., and Kanisawa, T., in "Proc. IV Int. Congress Food Sci. and Technol.", vol. I, p. 143, Valencia, 1976.
15. Shipe, W.F., in "Enzymes in Food and Beverage Processing" (R.L. Ory and A.J. St. Angelo, eds.) p. 57, ACS Symposium Series No. 47, American Chemical Society, Washington, D.C. 1977.
16. von Sydow, E., and Karlsson, G., Lebensm.- Wiss. u. Technol. 4:54 (1971).
17. von Sydow, E., and Karlsson, G., Lebensm.- Wiss. u. Technol. 4:152 (1971).
18. Karlsson-Ekström, G., and von Sydow, E., Lebensm.- Wiss. u. Technol. 6:86 (1973).
19. Karlsson-Ekström, G., and von Sydow, E., Lebensm.- Wiss. u. Technol. 6:165 (1973).
20. Persson, T., and von Sydow, E., J. Food Sci. 39:357 (1974).
21. Persson, T., and von Sydow, E., J. Food Sci. 39:406 (1974).
22. Badings, H.T., Maarse, H., Kleipool, R.J.C., Tas, A.C., Neeter, R., and ten Noever de Brauw, in "Proc. Int. Symp. Aroma Research", (H. Maarse and P.J. Groenen, eds.) p. 63, Pudoc, Wageningen, 1975.
23. Sato, K., Hegarty, G.R., and Herring, H.K., J. Food Sci. 38:398 (1973).
24. Maleki, M., Fette, Seifen, Anstrichmitteln 75:103 (1973).
25. Yamaguchi, N., Fujimaki, M., Nippon, Shokuhin Kogyo Gakkai Shi 21:13 (1974).

26. Eichner, K., in "Water Relations in Foods" (R.B. Duckworth ed.) p. 417, Academic Press, London, 1975.
27. Lingnert, H., and Eriksson, C., in "Proc. 9th Nordic Lipid Symposium" (R. Marcuse ed.) Lipidforum, SIK, Göteborg, 1978 (in press).
28. Bishov, S.J., Nasuoka, Y., and Kapsalis, J.G., J. Food Processing and Preservation 1:153 (1977).
29. Pratt, D.E., in "Phenolic, Sulfur and Nitrogen Compounds in Food Flavors" (G. Charalambous and I. Katz, eds.) p.1, ACS Symposium Series No. 26, American Chemical Society, Washington, D.C., 1976.
30. Petritschek, A., Lynen, F., and Belitz, H.-D., Lebensm.-Wiss. u. Technol. 5:47 (1972).
31. Petritschek, A., Lynen, F., and Belitz, H.-D., Lebensm.-Wiss. u. Technol. 5:77 (1972).
32. Belitz, H.-D., Z. Ernährungswiss. Suppl. 16:150 (1973).
33. Guigoz, Y., and Solms, J., Chemical Senses and Flavor 2:71 (1976).
34. Arai, S., Noguchi, M., Kurosawa, S., Kato, H., and Fujimaki, M., J. Food Sci. 35:392 (1970).
35. Clegg, K.M., and McMillan, A.D., J. Fd. Technol. 9:21 (1974).
36. Clegg, K.M., Smith, G., and Walker, A.L., J. Fd. Technol. 9:425 (1974).
37. Lalasidis, G., and Sjöberg, L.-B., J. Agr. Food Chem. (1978) (in press).
38. Fukumoto, J., and Okada, S., Kokai 40:995 (1973).
39. Noguchi, M., Yamashita, M., Arai, S., and Fujimaki, M., J. Food Sci. 40:367 (1975).
40. Chandler, B.V., and Nicol, K.J., CSIRO Fd. Res. Q. 35:79 (1975).
41. Bruemmer, J.H., Baker, R.A., and Roe, B., in "Enzymes in Food and Beverage Processing" (R.L. Ory and A.J. St. Angelo, eds.) p. 1, ACS Symposium Series No. 47, American Chemical Society, Washington, D.C., 1977.
42. Baur, C., Grosch, W., Wieser, H., and Jugel, H., Z. Lebensm. Unters.-Fosch. 164:171 (1977).
43. Baur, C., and Grosch, W., Z. Lebensm. Unters.-Fosch. 165:82 (1977).
44. Wilby, F.V., J. Air Pollution Control Association, 19:96 (1969).
45. Eriksson, C.E., J. Food Sci. 35:525 (1968).
46. Eriksson, C.E., Lundgren, B., and Vallentin, K., Chemical Senses and Flavor 2:3 (1976).
47. Eriksson, C.E., Qvist, I., and Vallentin, K., in "Enzymes in Food and Beverage Processing" (R.L. Ory and A.J. St. Angelo, eds.) p. 132, ACS Symposium Series No. 47, American Chemical Society, Washington, D.C. 1977.

48. Yamashita, I., Nemoto, Y., and Yoshikawa, S., Agr. Biol. Chem. 39:2303 (1975).
49. Yamashita, I., Nemoto, Y., and Yoshikawa, S., Phytochemistry 15:1633 (1976).
50. Yamashita. I., Iino, K., Nemoto, Y., and Yoshikawa, S., J. Agr. Food Chem. 25:1165 (1977).
51. Yamashita, I., Nemoto, Y., and Yoshikawa, S., Agr. Biol. Chem. 40:2231 (1976).

NEW DEVELOPMENTS IN MEAT AROMA RESEARCH

Ivon Flament
Bruno Willhalm
Günther Ohloff

Firmenich SA
Research Laboratories
Geneva, Switzerland

I. INTRODUCTION

Meat is usually processed in two ways: roasting at 150 - 160°, and cooking in an aqueous medium at 100°. The first method, which results in the formation of an aroma of the roasted or grilled type, can advantageously be simplified by subjecting only the aroma precursors to heat and not the entire raw meat with its protein and lipid content. Thus, there is obtained by dry heating a basic grilled meat aroma which is a very powerful and characteristic mixture. In order to investigate the aroma of cooked meat it is much simpler to use a commercial extract and isolate the aroma by steam distillation and solvent extraction. Two types of aromas prepared in this manner were analyzed, not for the purpose of a comparative study, but rather for identifying in each the largest possible number of characteristic components. Both extracts contain several hundred components, most of which are common to both. The isolation and identification of some of these products, selected on the grounds of their original structural or organoleptic features, will be described.

ISBN 0-12-169060-1

II. SEPARATIONS AND IDENTIFICATIONS

A. Analysis of a Meat Aroma Prepared by Thermolysis of the Water-Soluble Precursors

1. Preparation of the Extract. In 1949 Bouthilet (1) observed that aroma precursors could be extracted from chicken muscle by polar solvents. This observation was confirmed shortly afterwards by Pippen et al. (2-3), but it was only at the beginning of the sixties that more thorough analytical investigations were made with various types of meat by Hornstein et al. (4-6), Batzer et al. (7-8), Jacobson and Koehler (9), then by Zaika et al. (10-11), Mabrouk et al. (12) and Macy et al. (13). A survey of these results was published by Solms (14) and by Hornstein (15). All these investigations confirm that the precursors of meat aroma are water-soluble, that they consist of reducing sugars and amino acids, that the composition of lean bovine, porcine and ovine meat is very similar, and finally that the fibrillary and sarcoplasmic proteins do not directly contribute to the aroma which develops as a result of heating. We therefore extracted the water-soluble precursors according to fig. 1. One of the first operations consists in

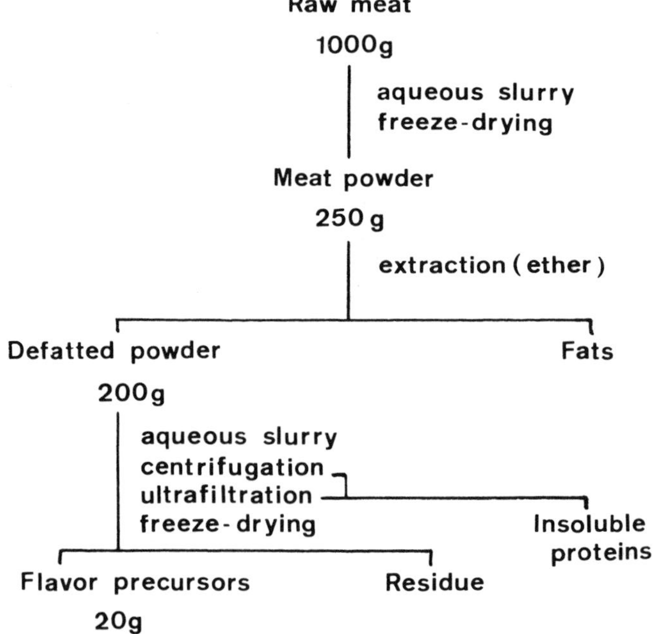

FIGURE 1. Preparation of flavor precursors.

the removal of any trace of lipid. It is known (16-19) that decomposition of fats results in the formation of a wide variety of products, in particular lactones, aldehydes and aliphatic esters which, although being characteristic of an animal species, are not necessarily present in the "basic meat aroma" which we intend to prepare. Likewise, in order to avoid the presence of proteins, polypeptides and other components having too high a molecular weight, we carried out a double ultrafiltration on a membrane before the final freeze-drying. The precursors obtained in a yield of 2%, based on the starting fresh meat, form a white, almost odourless powder having a sweetish and at the same time slightly bitter taste.

The precursor powder is heated in vacuo to 160° (fig. 2), and the volatile products which form are condensed in a series of cooled traps. After removal of the water, the oily phase is distilled and divided into two fractions, one containing the relatively volatile products (A, b.p.$_{11}$ < 160°) and the other the heavier components (B, b.p.$_{0.01}$ < 150°). The flavour of the crude extract has a pleasant character typical of grilled steak, this note being found in both fractions A and B.

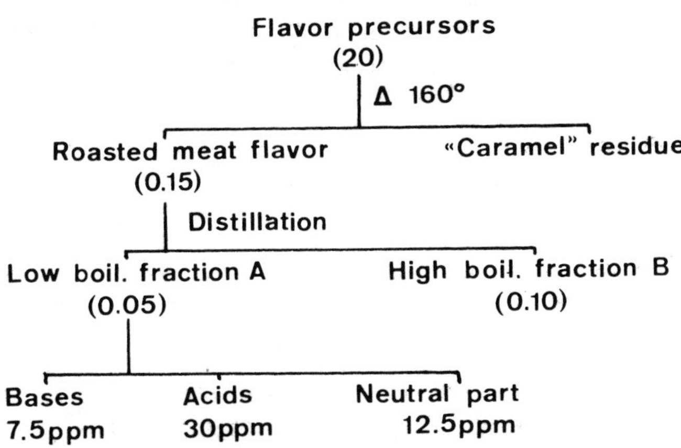

FIGURE 2. Preparation of roasted meat flavor by thermolysis of precursors.

2. Fractionation of the Extract. The extract obtained by thermolysis is relatively unstable: even if stored at 0° in the absence of air, it becomes more and more viscous and its colour turns from a pale yellow to a dark brown within a few weeks. For this reason we did not treat the entire 4 kilos of precursor powder (obtained from 200 kg of fresh meat) at once,

but carried out a series of analyses by treating 500 g portions of meat precursors using a different separation procedure each time. However, all methods involved one common step: chemical fractionation into neutral, acid and basic components (fig. 2). The separation techniques used in this investigation comprise vapour-phase chromatography on filled columns or on capillary glass columns, and adsorption chromatography of various types (dry column chromatography, liquid chromatography, thin-layer chromatography). The identification techniques used are mass spectrometry, nuclear magnetic resonance spectrometry and IR and UV spectrophotometry. Except for the last steps, the fractions obtained in the various separation steps were evaluated organoleptically, and their components were always characterized by olfaction at the outlet of the chromatograph.

3. Structure of the Identified Products.

a. Cyclohexenones (fig. 3). Among the pyrolysis products of the precursors, four methylated cyclohexenones were identified, of which only 2-methyl-2-cyclohexen-1-one 1 had been previously identified in coffee aroma (20). The 4-methylated isomer 2 and the 2,6-dimethylated homolog 3 were identified

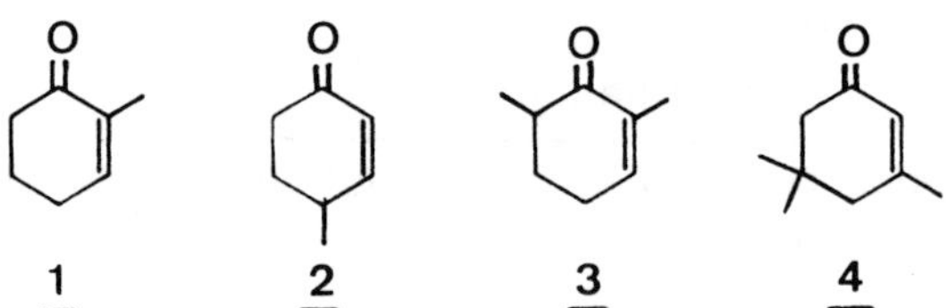

FIGURE 3. Cyclohexenones.

for the first time in an aroma. Isophorone (3,5,5-trimethyl-2-cyclohexen-1-one 4) had already been identified in cranberries (21), macadamia nuts (22), roasted filberts (23), saffron and mushrooms (24). The mono- and dimethylated cyclohexenones can be prepared easily by Birch reduction of the corresponding anisoles (25). These unsaturated ketones have interesting organoleptic properties; if combined with an alkylpyrazine (and possibly with a 2-alkylthiazolidine), they allow the organoleptic characteristics of certain foodstuffs to be enhanced, improved or modified. Such compositions impart to the foodstuffs in which they are incorporated a taste having a roasted, grilled, sometimes earthy note, or a more pronounced taste of meat, cereals, hazelnut, walnut or cocoa (26).

b. Lactyl lactates (fig. 4). Products 6 and 7, which were identified for the first time in an aroma, make up almost 5% of the neutral fraction. Their presence can be easily explained by the conversion of lactic acid, one of the main components of fresh meat. Besides, their synthesis is effected by alcoholysis of lactide 5, a condensation product of lactic acid (27).

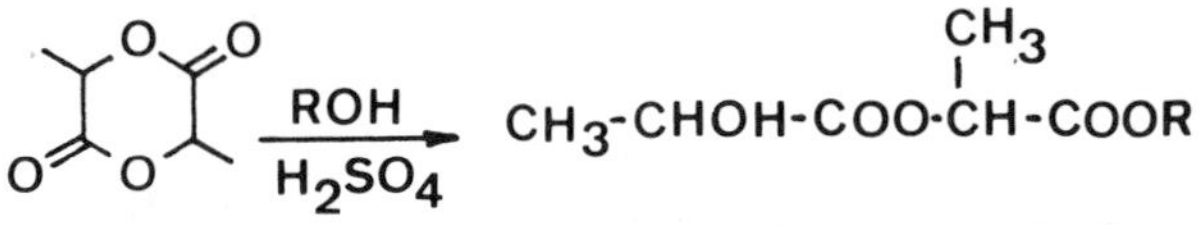

$$\mathbf{5} \xrightarrow[H_2SO_4]{ROH} CH_3\text{-}CHOH\text{-}COO\text{-}CH(CH_3)\text{-}COOR$$

$R = C_2H_5$ (6) or C_4H_9 (7)

FIGURE 4. Formation of lactyl lactates by alcoholysis of lactide.

c. Lactones (fig. 5). Anhydromevalolactone 8, which was already identified in tobacco aroma (28), is a dehydration product of mevalolactone which, in turn, is formed by dehydration of mevalonic acid, an important precursor in the biosynthesis of cholesterol (29). Pantolactone 9, which is a decomposition product of pantothenic acid and which was also

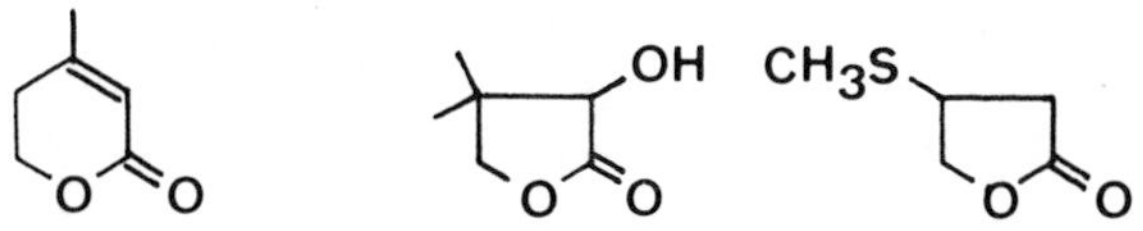

8 9 10

FIGURE 5. Structure of some lactones.

identified in a Spanish sherry (30), is relatively abundant in the neutral fraction. β-Methylthio-γ-butyrolactone (4,5-dihydro-4-methylthio-2(3H)-furanone) 10 is a new and particularly original component. This compound is obtained by the addition of methanethiol to 2-butenolide, a component of numerous foodstuffs which is particularly abundant in meat aroma. The reference is obtained in the same manner, the addition being catalyzed by triethylamine. This lactone 10 has a characteristic sulfur-like note of onions.

d. Five-membered nitrogen-containing heterocycles (fig. 6). Dimethyl- and ethyl-methyl-maleinimide 11 and 12 were also identified for the first time in a meat aroma. Compound 12, which was already detected in a tobacco aroma (28), is relatively abundant in certain fractions. 2-Methylthiobenzothiazole 13 is also an original product; it has a relatively characteristic fatty and smoky odour.

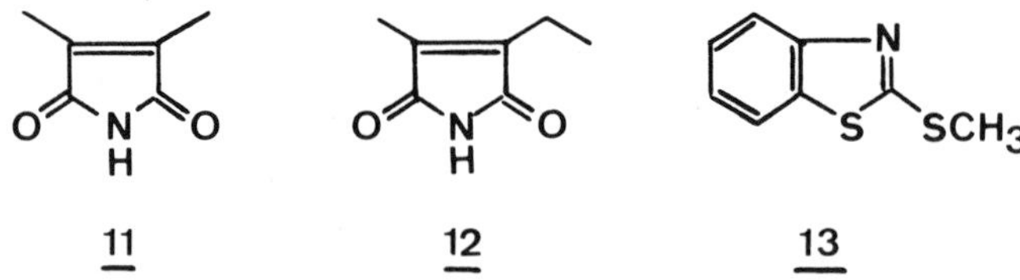

FIGURE 6. Maleinimides and benzothiazole.

Finally, our particular attention was drawn to another new product the mass spectrum of which, however, gave little information about its structure, showing the main fragment of m/e 58 and the molecular ion at m/e 129. The NMR spectrum is very simple; it comprises a doublet at 1.59 ppm ($J = 7$ cps) showing the presence of a methyl group attached to a CH (quadruplet at 4.92 ppm), and a singlet at 3.08 ppm corresponding to a methyl group attached to a nitrogen atom. These spectral data allowed the new compound to be identified as 3,5-dimethyl-1,3-oxazolidine-2,4-dione 14. This probably results from the condensation between a lactate and urea. The method of synthesis used (31) follows a similar pathway (fig. 7).

e. Condensed six-membered bicyclic nitrogen-containing heterocycles (fig. 7 and 8). Among the basic components, we have identified 4,6-dimethylpyrimidine 15, the mass spectrum of which is sufficiently different from those of its isomers, the dimethylpyrazines, to prevent any confusion. The reference product is easily prepared by catalytic dehalogenation of 2-chloro-4,6-dimethylpyrimidine (32). Numerous alkylpyrazines 16 are also present in the aroma of grilled meat (33). In order to facilitate their identification we have synthesized and characterized by spectroscopy the 64 mono- and multi-sub-

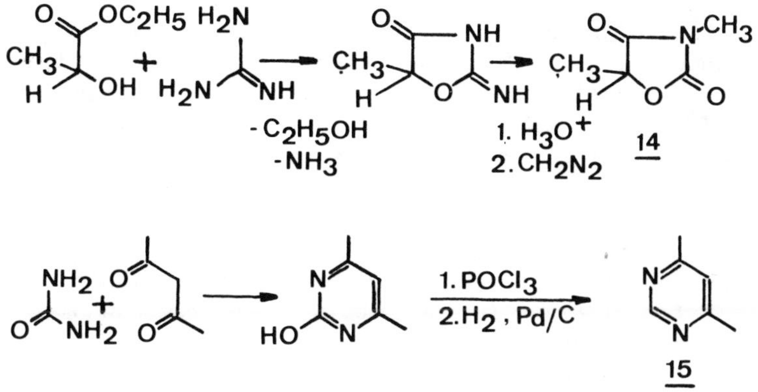

FIGURE 7. Synthesis of 3,5-dimethyl-1,3-oxazolidin-2,4-dione 14 and of 4,6-dimethylpyrimidine 15.

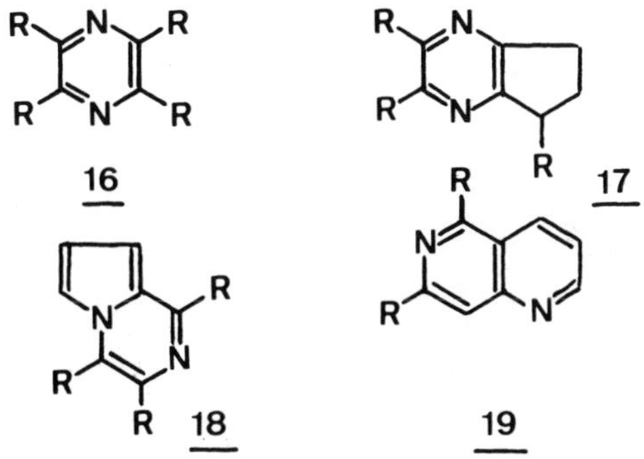

FIGURE 8. Six-membered ring and condensed bicyclic nitrogen compounds (R = H or alkyl).

stituted homologs in the series extending from methylpyrazine (molecular weight 94) to the isomers of pentylpyrazine (molecular weight 150) (34). It is superfluous to recall the organoleptic properties of this important class of heterocyclic compounds, some lower homologs having been already identified in 1926 in a coffee aroma (35) and numerous members also having been discovered in foodstuffs (24) (36). Several alkylated homologs of 6,7-dihydro-5H-cyclopenta|b|pyrazine 17 were also identified in the course of this analysis (37). Their synthesis and their organoleptic properties have been described before (38). A further class of new compounds are the alkyl-pyrrolo|1,2-a|pyrazines 18, the structure of which we have recently elucidated by NMR spectroscopy (39). It is very likely that certain members of this class have been confused with other diazaindenes or even with diazanaphthalenes; the mass spectra of the pyrrolo|1,2-a|pyrazines are practically identical with those of the benzimidazoles, indazoles, clyclopenta|b|pyrazines and dihydroquinoxalines.

1,6-Naphthyridine 19 (R = H) was also identified by its NMR spectrum, the data from the mass spectrum alone being insufficient for determining its structure. Some of its methylated homologs also occur in nature; as a rule, their synthesis is carried out by the method of Skraup (40-41). These heterocyclic compounds have a very characteristic odour and taste (42).

4. Remarks. The novelty of the method of preparing meat aroma by thermolysis of a precursor powder has involved considerable experimental difficulty. The preparation of this powder is time-consuming and demanding, owing to the nature of the components involved (sugars, amino acids, nucleosides, etc.). In particular, hundreds of litres of very dilute aqueous extract must be subjected to slow and carefully controlled freeze-drying, taking into account that this operation has to be repeated three times every time the method is carried out (fig. 1). As can be seen (fig. 2), the treatment of 2 kg of precursor powder (100 kg of raw meat) yields only 5 g of fraction A, a complex mixture which must be subjected to multiple separations before the most characteristic components can be isolated and identified with certainty. In spite of the application of microtechniques, more than 200 components isolated in the course of the present investigation remain to be identified, and these identifications are by no means easy. In view of the fact that a large number of these components are present in beef concentrates commercially available in large quantities, we are continuing our analytical investiga-

tions using aroma from boiled meat.

B. Analysis of a Commercial Beef Extract

1. Preparation and Fractionation of the Extract. To elucidate the structure of further components of meat aroma, we decided to prepare an aroma concentrate of the "boiled beef" type by extracting a concentrated beef extract (origin: Argentina and Brazil). This extract was subjected to steam distillation, then extracted with a solvent and distilled (fig. 9). Because of the complexity of these fractions, lengthy separation techniques were needed. Thus, a chromatogram of fraction B2 obtained on a capillary glass column shows more than 200 individual peaks, and one of its sub-fractions, isolated by preparative gas chromatography in a region relatively "poor" in components, contains nearly one hundred components. Nevertheless, owing to the availability of the starting product (up to now 1 ton of meat extract has been treated) it is often possible to isolate certain components in sufficient quantity and quality to complete the MS identification by an NMR spectrum. A selection of a few new products identified in the course of this analysis follows.

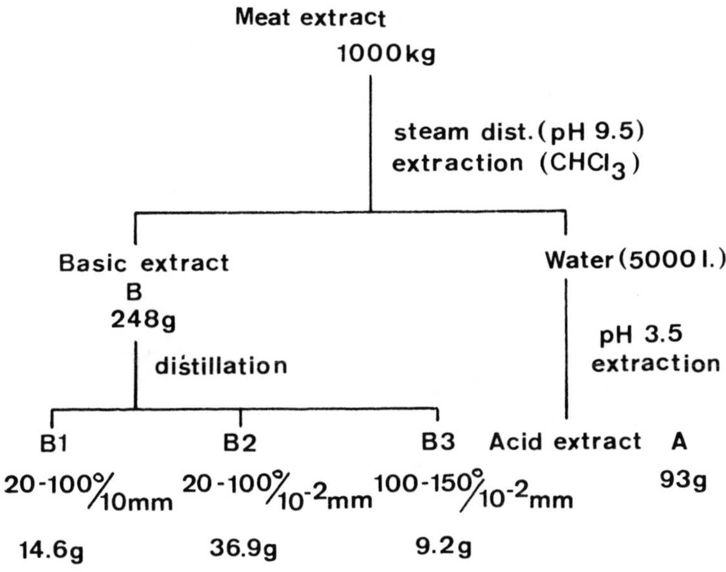

FIGURE 9. Isolation of flavor fraction from a commercial meat extract

2. Structure of Identified Products.

a. Sulfur-containing cyclic compounds (fig. 10). 1,2,4-Trithiolane has been isolated only from a red alga (43), but its 3,5-dimethylated homolog was identified for the first time in 1968 in a boiled beef aroma by Herz and Chang (44-45). Since then, it has also been identified in roasted filberts (23), mushrooms (46), pork meat (47), dry red beans (48), grape leaves (49) and potatoes (50). In addition we have identified two higher homologs (both in *cis* and *trans* forms): 3-methyl-5-ethyl-1,2,4-trithiolane 20 and 3-methyl-5-isopropyl-1,2,4-trithiolane 21. These compounds were synthesized according to the method of Asinger (51) by treating an equimolecular mixture of acetaldehyde and propionaldehyde (or isobutyraldehyde) with hydrogen sulfide in the presence of sulfur and diisobutylamine. The asymmetrical trithiolanes were separated from the symmetrical compounds by distillation. Both consist of a mixture of *cis* (a) and *trans* (b) isomers as found in the natural extract. In the further course of this analysis we have isolated a compound, the mass spectrum of which is very similar to that of 3,5-dimethyl-1,2,4-trithiolane and which could have been mistaken for the latter if the NMR spectrum had not shown the presence of two vicinal methylene groups (multiplet between 2.7 and 3.6 ppm) in addition to a CH_3-CH< group. Hence, the compound in question is 3-methyl-1,2,4-trithiane 22 which has been reported to be present only

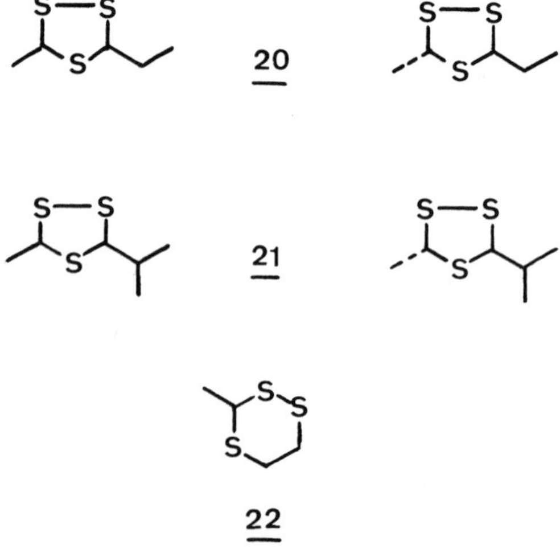

FIGURE 10. 1,2,4-Trithiolanes and 1,2,4-trithiane

among the decomposition products of cysteine used as a reagent in non-enzymatic model browing reactions (52-53).

The synthesis of compound 22 is carried out by reacting acetaldehyde, hydrogen sulfide and 1,2-dimercaptoethane. Trithiane 22 has very interesting organoleptic properties; its effect is already noticeable at concentrations of the order of 0.05 to 0.5 ppm (42).

b. 4,5-Dimethyl-4-cyclopentene-1,3-dione 23 (fig. 11). This is a simple new compound the structure of which, however, was not deducible from an analysis of the mass spectrum (MW 124 (base peak) and fragments at m/e 54, 39 and 53). The NMR spectrum also is strikingly simple: 2 singlets at 2.00 and 2.84 ppm (ratio of intensities 3 : 1). This highly symmetrical compound, therefore, exists mainly in the diketonic form 23. Its synthesis followed a sequence of known reactions (54-56).

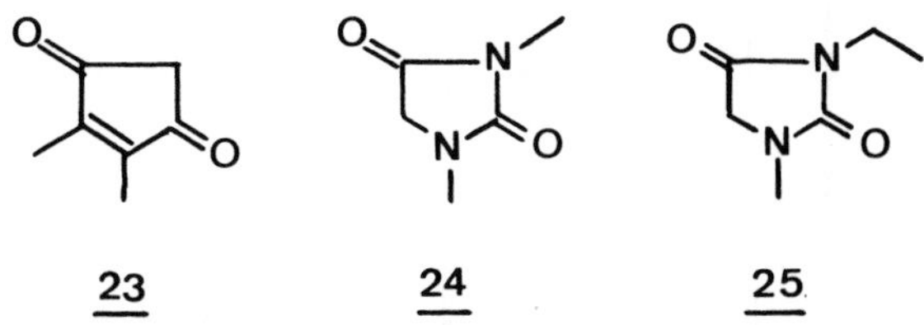

FIGURE 11. 4,5-Dimethyl-4-cyclopenten-1,3-dione and hydantoins.

c. Hydantoins (fig. 11). These compounds were also identified for the first time in an aroma. They consist of 1,3-dimethylhydantoin 24 and 1-methyl-3-ethylhydantoin 25 which probably are formed by degradation of uric acid (57). As in the preceding cases, their structure was elucidated on the base of their NMR spectra. Biltz and Lemberg (58), who synthesized dimethylhydantoin in 1923, describe this compound as having a peculiar acrid taste ("eigentümlichen brennenden Geschmack"). Its homolog 25 is prepared by ethylation of 1-methylhydantoin; only its isomer, 1-ethyl-3-methylhydantoin, is mentioned in the literature (59).

d. 4-Acetyl-2-methylpyrimidine 26 (fig. 12). Once again, the mass spectrum of this compound, which is very typical of meat aroma, has a striking resemblance with that of an isomer, 2-acetyl-3-methylpyrazine (fig. 13a), already identified in the aroma of tobacco (28), white bread and coffee (24). The

$\xrightarrow{POCl_3}$ $\xrightarrow{1. (CH_3)_3N \quad 2. KCN}$ $\xrightarrow{1. CH_3MgI \quad 2. H_3O^+}$ 26

FIGURE 12. Synthesis of 4-acetyl-2-methylpyrimidine.

NMR spectrum excluded this structure and showed that a pyrimidine was involved; the aromatic protons are located at 7.69 and 8.87 ppm (doublets, J = 6 cps), whereas they appear at 8.46 and 8.60 ppm (doublets, J = 2 cps) in the case of pyrazine (60) (fig. 13b). The synthesis of 26 was carried out according to the sequence shown in fig. 12. 4-Acetyl-2-methylpyrimidine has a very interesting grilled note and is detectable at a dose of 0.5 ppm (42).

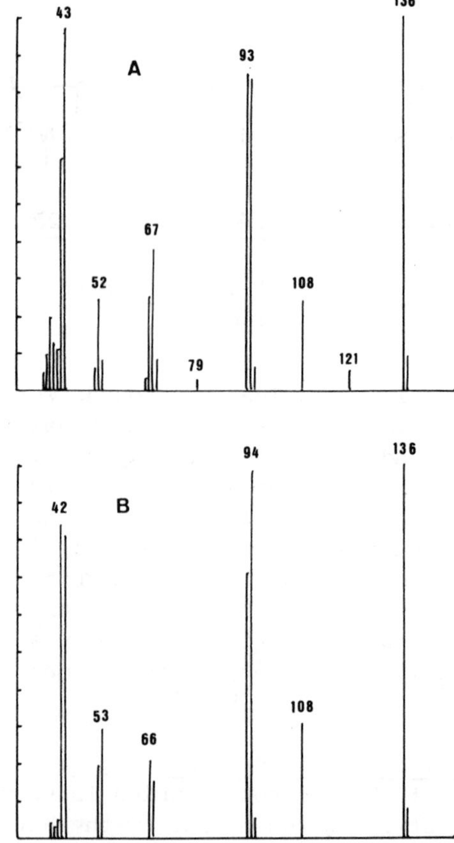

FIGURE 13a. Mass spectra of 2-acetyl-3-methylpyrazine (A) and of 4-acetyl-2-methylpyrimidine (B)

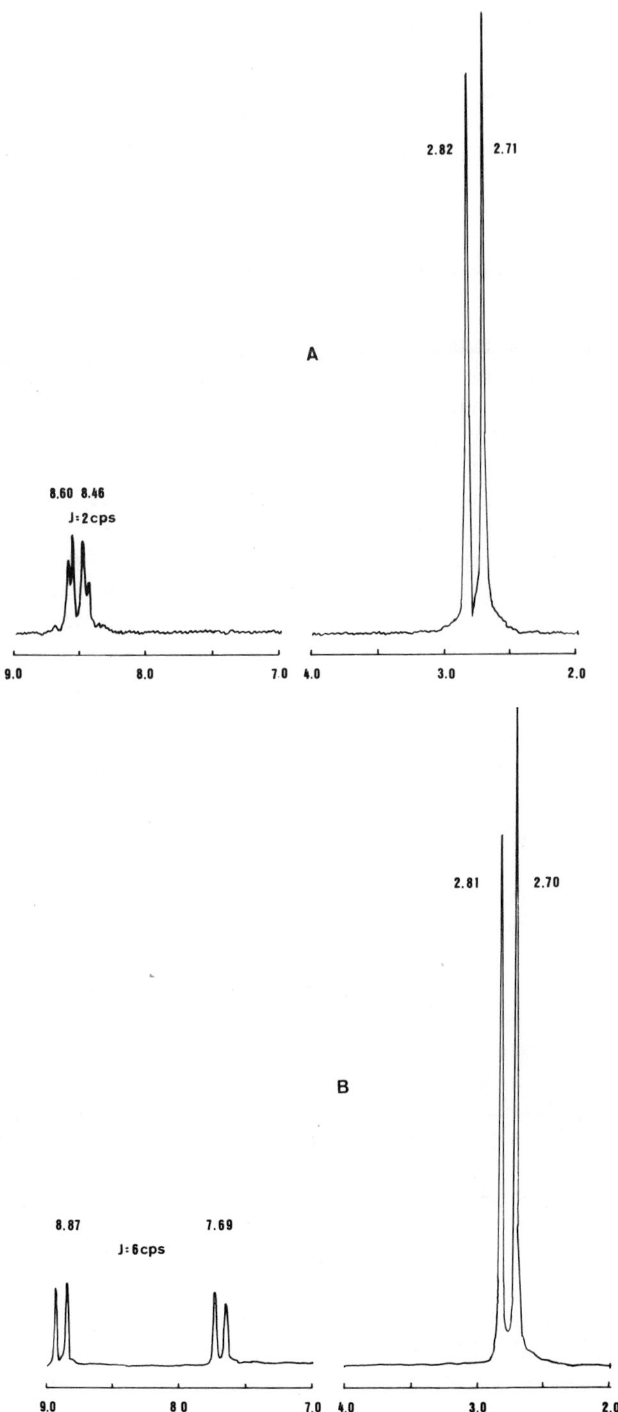

FIGURE 13b. NMR spectra of 2-acetyl-3-methylpyrazine (A) and of 4-acetyl-2-methylpyrimidine (B)

Acknowledgements. The authors are indebted to Dr. P. Dietrich and Mr. A. Smith, flavourists, who have evaluated the various fractions and synthetic products organoleptically. They also express their gratitude to Miss M. Kohler and Messrs R. Aschiero, P. Sonnay, E. Pittet and R. Tonna for their technical assistance.

III. REMARKS AND CONCLUSIONS

It is obvious that within the framework of the present paper we can only present a few fragmentary results of the analytical work in the field of meat aroma which we have been following since 1970. Up to now, about 600 components are mentioned in the literature or have been identified, and we can estimate that 400 further products have been isolated or detected, the structures of which are under investigation. Before concluding, some general remarks should be made on the technical possibilities and the present tendancies in the analysis of aromas as well as on the presentation of the results.

a. Development of techniques and number of identified components. Since 1975 we observe a distinct reduction in the number of publications and at the same time in the number of components identified in aromas, whereas the expansion was continuous during the preceding decade (61). It is commonplace to connect the advent of gas chromatography with the development of the analysis of aromas, although about 5 years passed between the actual commercialization of this technique (1960) and the first crop of results (1965). Although chromatography on glass capillary columns (GCGC or GC2) has replaced gas chromatography on packed columns during the last years and in spite of the considerably higher resolving capacity of these capillary columns, the number of identifications does not seem to have increased proportionally. Although the combination GCGC-MS has in turn replaced the former techniques, this often resulted merely in an accumulation of hundreds of spectra from which it is very difficult to extract some results. In order to take up and to maintain the rythm of identifications performed during the period 1965-75, it will be necessary to proceed simultaneously in two distinct directions.

The first consists in developing microtechniques and above all in acquiring analytical instruments of high performance such as those in which the technology of Fourier transformation is applied to NMR and IR spectrometers. Modern spectrometers (360 MHz) indeed allow high quality spectra to be recorded on samples of a few µg. We have already shown above the importance of NMR spectrometry in the elucidation of new structures.

The second direction to be followed consists in increasing the quantities of starting material, i.e. in passing from a few kilograms to several hundreds of kilograms or even several tons in certain cases in order to have at one's disposal at the end of the separation sufficient quantities of each of the most typical components so that the above-mentioned microtechniques can be applied efficiently. We are far from having penetrated the secret of all aromas, and numerous characteristic and powerful components remain to be identified.

b. Presentation of analytical results. In apparent contradiction with the preceding remarks, a considerable lengthening of the lists of products mentioned in publications has been observed during recent years. However, very often these lists are tedious repetitions in which the really new structures are lost among numerous known components. It is desirable that the authors of such papers limit the contents of the lists to products identified for the first time in the aroma under investigation and that in all cases these identifications be confirmed without any ambiguity by comparison with a synthetic product. Much too often structures are proposed or presented as certain whereas no reference data or sometimes at most 3 or 4 MS fragments published in an earlier paper are cited.

In conclusion, it seems that we are entering a new phase of analytical research applied on aromas. The time of facility is definitly terminated: it is not sufficient any more to inject any summarily prepared extract to give the proof of originality. As pointed out by Rykens and Boelens (61), most of the common aromas (about 150) have been analyzed, and among these 50 have been analyzed with particular care. Each of these aromas comprises 300 to 800 components, and a total of 2600 individual components are known today (24). Considering this sum of results, it is easier to understand what effort is necessary to pierce through the labyrinth exhibited, say, by the aromas of grilled or roasted products. The analyst enters step by step into new corridors, and at each turn he discovers new passages hidden by the walls. In order not to lose his way, he must be able to identify the sites he reaches and to mark out his path, and when the moment comes where he is compelled to stop due to lack of food, perhaps he will have acquired sufficient knowledge to establish an account of his travel and to reconstitute the aroma labyrinth in which he has been travelling.

REFERENCES

1. Bouthilet, R.J., *Food Technol. (Chicago)* 3:118 (1949).
2. Pippen, E.L., Campbell, A.A., and Streeter, I.V., *J. Agric. Food Chem.* 2:364 (1954).
3. Pippen, E.L., and Klose, A.A., *Poult. Sci.* 34:1139 (1955).
4. Hornstein, I., Crowe, P.F., and Sulzbacher, W.L., *J. Agric. Food Chem.* 8:65 (1960).
5. Hornstein, I., and Crowe, P.F., *J. Agric. Food Chem.* 8:494 (1960).
6. Hornstein, I., and Crowe, P.F., *J. Agric. Food Chem.* 11:147 (1963).
7. Batzer, O.F., Santoro, A.T., Tan, M.C., Landmann, W.A., and Schweigert, B.S., *J. Agric. Food Chem.* 8:498 (1960).
8. Batzer, O.F., Santoro, A.T., and Landmann, W.A., *J. Agric. Food Chem.* 10:94 (1962).
9. Jacobson, M., and Koehler, H.H., *J. Agric. Food Chem.* 11:336 (1963).
10. Zaika, L.L., Wasserman, A.E., Monk, C.A., and Salay, J., *J. Food Sci.* 33:53 (1968).
11. Zaika, L.L., *J. Agric. Food Chem.*17:893 (1969).
12. Mabrouk, A.F., Jarboe, J.K., and O'Connor, E.M., *J. Agric. Food Chem.* 17:5 (1969).
13. Macy, R.L., Naumann, H.D., and Bailey, M.E., *J. Food Sci.* 35:78 (1970).
14. Solms, J., *Fleischwirtschaft* 48:287 (1968).
15. Hornstein, I., in "The Science of Meat and Meat Products" (J.F. Price and B.S. Schweigert, ed.) p. 348. W.H. Freeman and Co., San Francisco, 2nd Ed., 1971.
16. Watanabe, K., and Sato, Y., *Agric. Biol. Chem.* 33:242 (1969).
17. Watanabe, K., and Sato, Y., *Agric. Biol. Chem.* 32:1318 (1968).
18. Watanabe, K., and Sato, Y., *Agric. Biol. Chem.* 32:191 (1968).
19. Watanabe, K., and Sato, Y., *Agric. Biol. Chem.* 35:756 (1971).
20. Bondarovich, H.A., Friedel, P., Krampl, V., Renner, J.A., Shephard, F.W., and Gianturco, M.A., *J. Agric. Food Chem.* 15:1093 (1967).

21. Anjou, K., and von Sydow, E., *Acta Chem. Scand.* 21:2076 (1967).
22. Crain, W.O., and Tang, C.S., *J. Food Sci.* 40:207 (1975).
23. Kinlin, T.E., Muralidhara, R., Pittet, A.O., Sanderson, A., and Walradt, J.P., *J. Agric. Food Chem.*20:1021 (1972).
24. Weurman, C., in "Lists of Volatile Compounds in Food" (Central Institute for Nutrition and Food Research, TNO, Zeist, Metherlands).
25. Birch, A.J., *J. Chem. Soc.* 430 (1944).
26. Flament, I. (Firmenich SA), Fr. Demande 2.201.839, *Chem. Abstr.* 82:155239 (1975). See also Swiss 568 024 and 568 722 (1975).
27. Claborn, H.V., U.S. 2.371.281, *Chem. Abstr.* 39:4089 (1945).
28. Demole, E., and Berthet, D., *Helv. Chim. Acta* 55:1866 (1972).
29. Wagner, A.F., and Folkers, K., *Endeavour* 177 (1971).
30. Webb, A.D., Kepner, R.E., and Maggiora, L., *Amer. J. Enol. Vitic.* 18:190 (1967) and 19:116 (1968).
31. Traube, W., and Ascher, R., *Ber. Dtsch. Chem. Ges.* 46:2077 (1913).
32. Angerstein, S., *Ber. Dtsch. Chem. Ges.* 34:3957 (1901).
33. Flament, I., and Ohloff, G., *Helv. Chim. Acta* 54:1911 (1971).
34. Flament, I. (Firmenich SA), Swiss 540016 (1973), Prior. date March 9, 1971.
35. Reichstein, T., and Staudinger, H., Brit. 260960 (1926), *Chem. Zentralbl.* 1927 I, 2613. See also *Angew. Chem.* 62:292 (1950).
36. Maga, J.A., and Sizer, C.E., *J. Agric. Food Chem.* 21:22 (1973).
37. Flament, I., Kohler, M., and Aschiero, R., *Helv. Chim. Acta* 59:2308 (1976).
38. Flament, I., Sonnay, P., and Ohloff, G., *Helv. Chim. Acta* 56:610 (1973).
39. Flament, I., Sonnay, P., and Ohloff, G., *Helv. Chim. Acta* 60:1872 (1977).
40. Paudler, W.W., and Kress, T.J., *J. Heterocycl. Chem.* 4:284 (1967).
41. Hamada, Y., and Takeuchi, I., *Chem. Pharm. Bull. (Tokyo)* 19:1857 (1971), *Chem. Abstr.* 75:151701d (1971).
42. Flament, I. (Firmenich SA), Pat. pend.
43. Wratten, S.J., and Faulkner, D.J., *J. Org. Chem.* 41:2465 (1976).
44. Herz, K.O., Ph. D. Thesis, Rutgers State University, 1968 (Univ. Micr. Ord. No 68-14248).

45. Chang, S.S., Hirai, C., Reddy, B.R., Herz, K.O., Kato, A., and Sipma, G., *Chem. Ind. (London)* 1639 (1968).
46. Thomas, A.F., *J. Agric. Food Chem.* 21:955 (1973).
47. Swain, J.W., Ph. D. Thesis, Univ. Missouri, Columbia 1972 (Univ. Micr. Ord. No 73-21836).
48. Buttery, R.G., Seifert, R.M., and Ling, L.C., *J. Agric. Food Chem.* 23:516 (1975).
49. Wildenradt, H.L., Christensen, E.N., Stackler, B., Caputi, A., Slinkard, K., and Scutt, K., *Amer. J. Enol. Vitic.* 26:148 (1975).
50. Buttery, R.G., Seifert, R.M., and Ling, L.C., *J. Agric. Food Chem.* 18:538 (1970).
51. Asinger, F., Thiel, M., and Lipfert, G., *Liebigs Ann. Chem.* 627:195 (1959).
52. Mulders, E.J., *Z. Lebensm. Unters. Forsch.* 152:193 (1973).
53. Ledl, F., *Z. Lebensm. Unters. Forsch.* 161:125 (1976).
54. Anteunis, M., and Compernolle, F., *Bull. Soc. Chim. Belg.* 76:482 (1967).
55. Van Wÿnsberghe, L., and Vandewalle, M., *Bull. Soc. Chim. Belg.* 79:699 (1970).
56. Vandewalle, M., Van Wÿnsberghe, L., and Witvrouwen, G., *Bull. Soc. Chim. Belg.* 80:39 (1971).
57. Brandenberger, H., *Helv. Chim. Acta* 37:641 (1954).
58. Biltz, H., and Lemberg, R., *Liebigs Ann. Chem.* 432:137 (1923).
59. Biltz, H., and Slotta, K., *J. Prakt. Chem.* |*2*| 113:233 (1926).
60. Wolt, J., *J. Org. Chem.* 40:1178 (1975).
61. Rÿkens, F., and Boelens, H., in "Proc. int. Symp. Aroma Research, Zeist", p. 203. Pudoc, Wageningen 1975.

FLAVOR PROBLEMS IN THE APPLICATION OF SOY PROTEIN MATERIALS AS MEAT SUBSTITUTES

Godefridus A.M. van den Ouweland

Unilever Research Duiven
Zevenaar, The Netherlands

Leonard Schutte

Unimills
Zwijndrecht, The Netherlands

I. INTRODUCTION

A person of good taste is someone who enjoys the good qualities of life. Similarly, a good quality food can only be enjoyed if it has a good taste. And although food, of course, is necessary to sustain life, the pleasure we derive from eating determines for a great part the quality of life.

In other words, flavors are very important in our appreciation of food. Therefore, when new food ingredients are being offered, it is not just their nutritional aspect, functionality and price that determines their applicability: flavor plays a major role.

In the last decade novel protein ingredients have gradually been introduced on the market place as a partial relief for the ever rising demand for protein foods like meat. The best known products are textured soy proteins, but also other soy products like soy flour, concentrates and isolates have been offered.

Nutritional value and price of these materials, it is true, are very favourable, their functionality adequate for incorporation into food, often mixed with minced or comminuted meat. However, the flavor problems have been neglected or underestimated, certainly by the soy protein manufacturers, but also by the flavor industry.

ISBN 0-12-169060-1

There exist two major problems. Problem one is the presence of off-flavors inherent to the soy. Problem two is the absence of an attractive positive, e.g. meat-like, flavor. We shall deal with the off-flavors first.

II. OFF-FLAVORS

Various types of off-flavors have been distinguished in soy protein products. Most of them are developed during processing, which means that their precursors are present in the bean. Identification of the off-flavor compounds and understanding their mode of formation from their precursors may lead to effective methods to obtain bland materials. We can distinguish the following off-flavors.

A. Bitter Flavors

According to Sessa et al.[1] oxidized phosphatidylcholines are responsible for the bitter taste. In hexane-defatted flakes three types of phosphatidylcholines were found to be present, which upon autoxidation yielded a bitter taste. It is not clear whether lipoxygenase stimulates this autoxidation. The identity of the bitter principles has not yet been established. There exists, however, a distinct relationship between bitterness or a pungent aftertaste and extractable phosphorus.

B. Sweet Taste

A significant amount of sucrose (12%) in defatted soy flour can give rise to peculiar flavor impressions when a savoury meat flavor is applied in products containing such flour or extruded defatted soy.

C. Green, Grassy Odors

Volatile carbonyl compounds having a green, grassy aroma have been found in soy protein products. The compounds are formed by lipoxygenase iso-enzymes in soybeans from linoleic and linolenic acid, both present is soybeans[2]. A range of saturated and unsaturated carbonyl compounds was isolated from incubation experiments with the polyunsaturated fatty acids[3,4] (see Table 1).

TABLE 1. Carbonyl Compounds Formed from Incubation Experiments with Lipoxygenase Iso-enzymes (Green Grassy Odors

from linoleic acid	from linolenic acid
pentanal	acetaldehyde
hexanal	propanal
heptanal	2trans-pentenal
2trans-octenal	2trans-hexenal
2trans,4trans-nonadienal	2trans,4cis-heptadienal
2trans,4cis -decadienal	2trans,6cis-nonadienal
2trans,4trans-decadienal	3,5-octadien-2-one
	2,4,6-nonatrienal

D. Cooked Soybean Odor

Upon heating, a distinctly unpleasant cooked soybean odor is developed. This off-flavor is frequently encountered in extruded soy flour and in sterilized soy protein products. The compounds found to be characteristic of this off-flavor are *p*-vinylphenol and *p*-vinylguaiacol[5]. The volatiles are generated by heat from the precursors *p*-coumaric acid and ferulic acid, respectively. These phenolic compounds are derived from the lignin which is typical of plant materials.

E. Burnt Flavor

Under excessive heating and extrusion with high temperatures a burnt flavor may be formed. This is the result of the well-known Maillard reaction between the amino acids in the soy proteins and the carbohydrates, particularly the oligosaccharides present in defatted soy flour. Apart from a great amount of hydrocarbons, over 70 compounds, amongst which aldehydes, ketones, furan derivatives and sulfur compounds, have been identified from a soy protein isolate[6] and 15 pyrazines could be isolated from regular and toasted soy[7]. Although essentially the same compounds are found in the unheated and heated material (except for sulfur compounds), it is interesting to note that the concentration of almost all volatiles increases on heating.

F. Artifacts

Residues of solvents being used for protein purification may give rise to off-flavors like a sweet fusel note or catty odors arising from the reaction of mesityl oxide from traces of acetone and hydrogen sulfide from sulfur-containing amino acids.

III. METHODS TO PREVENT FORMATION OF OFF-FLAVORS

The enzymic formation of off-flavors can be prevented by destruction of the enzyme system in the bean by a heat treatment. A dry heat treatment at 100°C for 28 min and an open steam treatment for 5 min proved to be successful[8,9]. These procedures should, however, be preceded by careful selection of intact beans, since in broken beans the enzymes may have been active before the heat treatment.

The typical cooked soybean flavor formed during extrusion or heat sterilization from phenolic compounds in the bean may be prevented by texturing at moderate temperatures. This procedure is, however, questionable since also the destruction of anti-nutritional factors like trypsin inhibitor is incomplete under these curcumstances. The product must, therefore, be heated at a later stage of the food preparation, which logically may lead to the formation of the cooked soybean odor.

IV. REMOVAL OF OFF-FLAVORS

Since prevention of the off-flavors can only be partially effective and is in many cases impracticable, removal of both the off-flavor compounds and their precursors by proper refining techniques offers a wider scope. The most effective way to remove off-flavors from defatted soy flour is extraction with an alcohol such as ethanol and 2-propanol. This was already realized as early as 1948 by Beckel et al.[10]. Careful extraction with these solvents can lead to bland concentrates with sufficient functionality to be still extrudable[11]. In soy isolates extraction of the protein can be performed under conditions (e.g. treatment with a 1 mol l^{-1} NaCl solution at a pH of 4.5[12]), whereby the off-flavors and their precursors are largely removed. When such isolates are textured, e.g. by wet spinning, residual off-flavors will largely be leached out, thus yielding bland protein fibres.

It has been claimed that proteolytic enzymes can be used to facilitate the removal of off-flavors from soybean products[13]. Off-flavor compounds that are partially bound to the protein are released more easily than without enzymic incubation. An extraction step still seems to be required to remove the off-flavors. In general, refining processes to remove off-flavor will be necessary to yield materials that can be used at high levels in food stuffs. A flow sheet setting out the

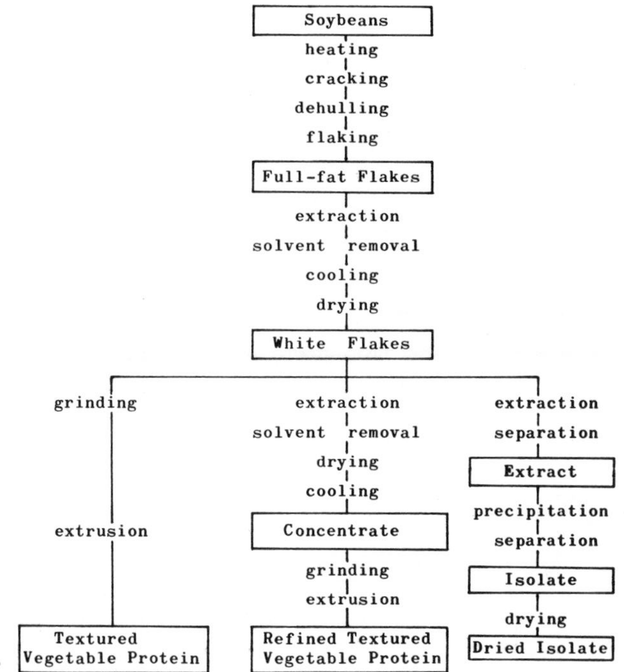

SCHEME 1. Refining processes of soybeans

processing of soybeans to obtain soy protein materials is given in Scheme 1.

The first stage is to clean and dehull the beans, and then crush them between rollers from which they emerge as a mass of flakes. The oil is then extracted with a hydrocarbon solvent and the defatted flakes can be used in the process. These flakes can be further improved by a refining step in which they are treated with aqueous alcohol which removes factors responsible for the cooked bean odor, sweet taste, the burnt flavor and bitter taste. The protein concentrate so obtained can be textured, yielding a bland tasting, refined, textured soy protein, containing about 70% protein.

V. INTERACTION AND RELEASE OF FLAVORS IN SOY PROTEIN

After having produced a bland-tasting soy protein, it is then the not easy task of the flavorist to create a flavor in soy protein-containing foods. If, for instance, in meat products 50% or more of the meat is replaced by soy protein, it is obvious that for the drop in flavor in these products has to be compensated by addition of appropriate flavors. It is known that when model flavor compounds are added to soy pro-

TABLE 2. Flavor Threshold Values of Meat Flavor Component

Compound	Threshold value (ppm)			
	in water	in o/w emulsion	in textured soy protein	in meat
2-Heptanone	0.08	1	1	5
1-Octen-3-ol	0.05	0.7	0.5	2.3
1,5-Dodecalactone	0.06	3	3	6.5
Benzaldehyde	0.05	0.4	0.6	2.0
4-Hydroxy-2,5-dimethyl-[2H]-furan-3-one	0.03	0.2	0.1	5.7

tein in an aqueous medium, interactions of a number of these compounds with the soy protein occur, even if no heat treatment is involved. A study into these interactions, which resulted in a suppression of the flavor impact, was conducted by Gremli[14] and Solms[15].

In extended meat products we have a meat/vegetable protein mixture into which flavor, color and fat are incorporated during preparation. The fat then often exists in these products as a component of a fat/water emulsion. In Table 2 flavor threshold values are given for some meat flavor components in four different media.

As shown, the threshold values in an o/w emulsion are for all flavor components higher than in water. This is in agreement with studies carried out by McNulty[16], who established the relationship between partition coefficient and taste threshold values in o/w emulsions. The threshold values of the components in an o/w emulsion are of the same order of magnitude as those found for the textured soy protein into which this oil is incorporated. In bland-tasting freeze-dried meat pieces rehydrated with the same emulsion the threshold values are in all cases higher than in the textured soy protein.

An explanation for this interesting phenomenon seems to be that during mastication of textured soy protein the emulsion is pressed out already during the first bites, while in the meat system the release of the emulsion is much more gradual.

The problem of the quick release of water from textured soy protein presents a challenge to those involved in flavoring vegetable protein products and it seems necessary to develop meat flavor blends very much adapted to the quick release of water from the material, i.e. by incorporation of mixtures with a strong and lasting taste. An even greater challenge is presented to those involved in texturing vegetable proteins. It seems that approximation of the structure-dependent eating characteristics of meat will be necessary.

Another serious flavor problem arises when the food, into

which the textured soy protein is incorporated, has to be sterilized to improve its storage stability. Often spices and condiments are incorporated into a meat flavor to accentuate or to round off the meaty note. In the Figures 1a and 1b the dramatic changes of the volatile part of onion powder and garlic powder are demonstrated when these materials are sterilized in the presence of textured soy protein. Especially the sulfur-containing compounds so characteristic of these materials have mainly disappeared and new compounds have been formed.

These flavor problems can only be solved by very close co-operation of the food technologist with the creative flavor chemist.

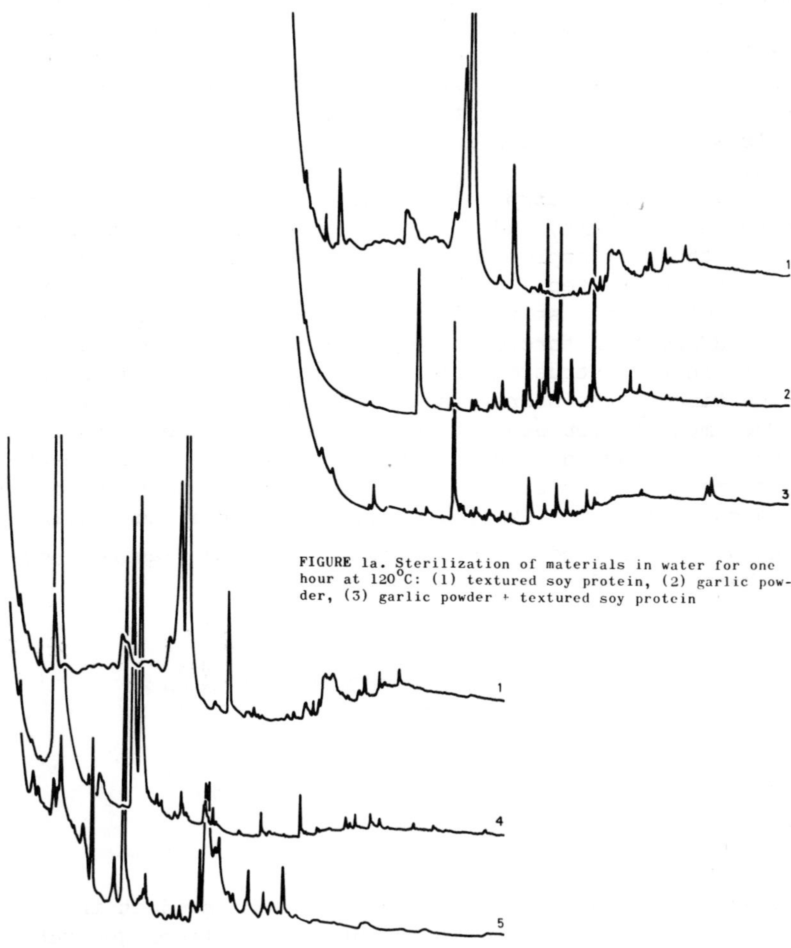

FIGURE 1a. Sterilization of materials in water for one hour at 120°C: (1) textured soy protein, (2) garlic powder, (3) garlic powder + textured soy protein

FIGURE 1b. Sterilization of materials in water for one hour at 120°C: (1) textured soy protein, (4) onion powder, (5) onion powder + textured soy protein

VI. UTILIZATION OF EXTRUSION PROCESS FOR MEAT FLAVOR DEVELOPMENT

The origin of meat flavor in natural meats lies in a series of complex reactions taking place during the cooking of meat. From studies concerning the source of meat flavor it was concluded that the reaction of reducing sugars with amino acids play a dominant role and that the contribution of cysteine in this reaction is rather unique. Since the early 1960's this knowledge has been applied in the manufacturing of meat-like flavors on an industrial scale[17]. A range of hexoses and pentoses is utilized in this reaction and since there is similarity in composition between the free amino acids found in meat and those found in hydrolysed vegetable proteins and yeast autolysates,combinations of the latter materials are often applied in the meat flavor-generating reaction.

If we want to incorporate the meat flavor system into the textured soy protein, and there are certainly advantages in it, then the most appropriate stage in the process to do so is the extrusion step. In this step the soy protein concentrate is mixed to a stiff slurry with water and this slurry is heated and then forced through a nozzle. The temperature employed in the extrusion process is about 150°C and the residence time of the slurry in the extruder is about 1 minute.

When meat flavor-generating reactions must be operative during the extrusion step, a proper choice of reaction partners seems unavoidable. The early steps in the amino acid reducing sugar reaction (Maillard reaction) are a sugar-amine condensation followed by a transformation into 1-amino-1-deoxy-2-ketose (Amadori rearrangement):

$$\begin{array}{c} HC{=}O \\ | \\ CHOH \\ | \\ (CHOH)_n \\ | \\ CH_2OH \end{array} \xrightarrow{\begin{array}{c} NH_2 \\ | \\ R\text{-}CH\text{-}COOH \end{array}} \begin{array}{l} HC\!\!-\!\!N(H)\text{-}C(H)(R)\text{-}COOH \\ |\;\backslash OH \\ CHOH \\ | \\ (CHOH)_n \\ | \\ CH_2OH \end{array} \xrightarrow{-H_2O} \begin{array}{l} H_2C\text{-}N(H)\text{-}C(H)(R)\text{-}COOH \\ | \\ C{=}O \\ | \\ (CHOH)_n \\ | \\ CH_2OH \end{array}$$

Application of the Amadori rearrangement product in the extrusion step does have more potentials in flavor generation than using two compounds of which the product is made up. A further feature of the Maillard reaction is that fission of the carbohydrate chain occurs resulting in very reactive in-

TABLE 3. Carbonyl Intermediates from Carbohydrate Chain Fission

Compound	Formula
glycolaldehyde	$HO\text{-}CH_2\text{-}CHO$
glyoxal	$CHO\text{-}CHO$
pyruvaldehyde	$CH_3\text{-}CO\text{-}CHO$
dihydroxyacetone	$HO\text{-}CH_2\text{-}CO\text{-}CH_2\text{-}OH$
glyceraldehyde	$HO\text{-}CH_2\text{-}CHOH\text{-}CHO$
acetoin	$CH_3\text{-}CHOH\text{-}CO\text{-}CH_3$
diacetyl	$CH_3\text{-}CO\text{-}CO\text{-}CH_3$
hydroxydiacetyl	$CH_3\text{-}CO\text{-}CO\text{-}CH_2\text{-}OH$

termediates. A number of these intermediates found in Maillard reaction products is given in Table 3.

A well balanced mixture of these carbonyl intermediates is more adapted to the extrusion process conditions in providing a meat flavor, in their reaction with hydrolysed vegetable proteins and yeast autolysates.

A combination of the above-discussed reaction partners gives to the textured soy protein a meaty background. If, however, a material with a more meat-like flavor character is wanted, then extension of the foregoing reactions with two other meat flavor-generating reactions will be necessary.

The first one is an interesting example of the way in which amino acids can act as flavor precursors for the formation of 1-methylthioethanethiol. The formation of this compound has been described by Schutte and Koenders[18] and the net result of this reaction is as follows:

$$CH_3CHO + CH_3SH + H_2S \rightarrow CH_3\text{-}\underset{\displaystyle H}{\overset{\displaystyle SH}{\overset{|}{\underset{|}{C}}}}\text{-}S\text{-}CH_3$$

The other meat flavor-generating reaction is based on the reaction of hydroxyfuranones with hydrogen sulfide sources, such as cysteine[19]:

$$\text{(4-hydroxy-3(2H)-furanone; HO, O, } R_1\text{, } R_2\text{)} + H_2S \rightarrow \text{MEAT FLAVOR}$$

The meat flavor derived from this reaction much more closely resembles the authentic flavor of roasted meat.

Appropriately selected precursors[20] can combine these two meat flavor-generating reactions:

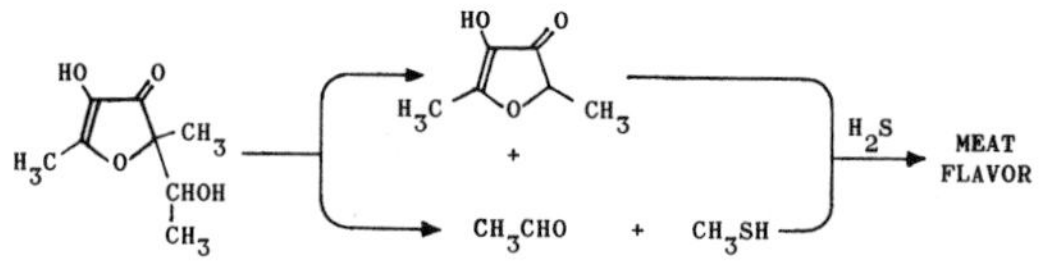

Combination, in te extrusion process, of these reactions with the Maillard-type reactions discussed earlier provides the textured soy protein with a flavor which simulates more closely the flavor of natural meats.

REFERENCES

1. Sessa, D.J., Warner, K. and Rackis, J.J., J. Agr. Fd. Chem. 24: 16 (1976).
2. Wolf, W., J. Agr. Fd. Chem. 23: 136 (1975).
3. Grosch, W. and Laskawy, G., J. Agr. Fd. Chem. 23:791(1975).
4. Fischer, K. and Grosch, W., Z. Lebensm. Unters. Forsch. 165: 137 (1977).
5. Greuell, E.H.M., J. Amer. Oil Chemists' Soc. 51: 98 (1974).
6. Qvist, I.H. and Sydow, E.C.F. van, J. Agr. Fd. Chem. 22: 1077 (1974).
7. Maga, J.A., Lebensm. Wiss. Technol. 10: 100 (1977).
8. Mustakas, G.C., Albrecht, W.J., McGee, J.E., Black, L.T. and Bookwalter, G.N., J. Amer. Oil Chemists' Soc. 46: 623 (1969).
9. Mustakas, G.C., J. Amer. Oil Chemists' Soc. 48: 815 (1971).
10. Beckel, A.C., Belter, P.A. and Smith, A.K., J. Amer. Oil Chemists' Soc. 25: 10 (1948).
11. Cowan, J.C., Rackis, J.J. and Wolf, W.J., J. Amer. Oil Chemists' Soc. 50: 425 (1973).
12. Anderson, R.L. and Warner, K., J. Fd. Sci. 41:293 (1976).
13. Noguchi, M., Arai, S., Kato, H. and Fujimaki M., J. Fd. Sci. 35: 211 (1970).
14. Gremli, H.A., J. Amer. Oil Chemists' Soc. 51: 95A (1974).
15. Beyeler, M. and Solms, J., Lebensm. Wiss. Technol. 7: 217 (1974).
16. McNulty, P.B., Fourth Int. Congr. Food Sci. Technol., Madrid, 22-27 September 1974.
17. Morton, I.D., Akroyd, P. and May, C.G., Brit. Pat. 836694 (June 9, 1960).
18. Schutte, L. and Koenders, E.B., J. Agr. Fd. Chem. 20: 181 (1972).
19. Ouweland, G.A.M. van den, and Peer, H.G., J. Agr. Fd. Chem. 23: 501 (1975).
20. Ouweland, G.A.M. van den, U.S. Pat. 4 070 381 (Jan.24,1978).

FLAVORING VEGETABLE PROTEIN MEAT ANALOGS

R. F. Heinze
M. B. Ingle
J. F. Reynolds

Miles Laboratories, Inc.
Elkhart, Indiana
U.S.A.

Meat Analogs can be defined as products nutritionally equivalent to their biological counterparts, however, they do not contain meat protein or meat by-products. They are developed to have the appearance, texture and flavor of their meat counterparts. The proteins in meat analogs are derived from vegetable, dairy and other non-meat sources. On the otherhand, meat extenders are normally vegetable proteins added to processed meats to supply nutritional and functional properties.

Worthington Foods was the modern day pioneer in the fabrication of meat analogs with the introduction in June 1962 of SOYA-MEAT[R], a canned chicken analog. These meat analogs were based on the Robert Boyer patent which employed the use of spun soy fiber from soy isolate to impart a meaty texture.

Consumers are motivated to purchase fabricated foods for many reasons but the principal factors are nutrition, health, convenience, economy and variety. The finished meat analog must meet the consumer's expectations in taste, aroma, texture and appearance.

In the early 1970's, government and industry sources indicated highly optimistic projections for sales of meat analogs. This optimistic forecast was based on significant improvement by 1985 in the flavor, texture and overall nutritional composition of these products. However, sales of this magnitude never materialized.

ISBN 0-12-169060-1

TABLE 1 Comparative Data of the Early and Current Forecasts

	Meat Analog Sales - Millions of Dollars	
Year	1970 Forecast	1977 Forecast
1966	3	3
1970	10	15
1977	1000	25
1980	1800	40
1985	3200	125

There are many factors which influence the sales forecast for meat analogs. The poor flavor and texture of soy protein based analogs along with the inherent flatulence factor were major drawbacks in early fabricated food products. Significant improvements in the texture, appearance, nutrition, and flavor have occurred over the last ten years and further improvements are being made. However, improved meat flavor for meat analogs would definitely increase consumer acceptance. Consumers expect meat analogs to taste like their meat counterparts. When this criteria is met, sales of meat analogs will become a significant factor in the food processing industry.

The flavoring of meat analogs include the following topics: A) development of characteristic meat flavors; B) properties of meat analogs as compared to meat and; C) the application of meat flavors in meat analogs concerning ingredients and processing. Flavoring meat analogs involves two related major problems, namely, developing characteristic meat flavors and the application of these flavors in the analogs.

A. DEVELOPMENT OF CHARACTERISTIC MEAT FLAVORS

The flavor of meat is attributed to a complex mixture of compounds produced by heating a heterogeneous system containing non-odorous precursors which can develop volatile compounds, non-volatile compounds with taste properties, potentiators and synergists (Dwivedi 1975). An approach to the development of meat flavors involves the following investigations.

1. The isolation and identificiation of volatile flavor compounds found in processed meat.

2. The identification of flavor precursors in meat and their role in flavor development.
3. The role of biological processes in the development of meat flavor.
4. The study of model systems that produce desirable meat type flavors.
5. Combinations of above.

These are very complex investigations due to physical and chemical properties, size and chemistry of the compounds. New analytical techniques have been developed to pursue these investigations. In the last 20 years, there has been a tremendous advance in the understanding of meat flavor technology. To date, the majority of investigations by government, universities and industry have been concerned with the basic understanding of the flavor of beef, pork and poultry.

Dwivedi (1975) summarized the volatile compounds identified in Beef and Pork. In beef, he lists 6 acids, 31 aldehydes, 3 esters, 1 ether, 2 pyroles, 25 alcohols, 23 ketones, 19 hydrocarbons, 12 benzene compounds, 11 lactones, 8 furans, 53 sulfur compounds, and 37 nitrogen compounds. The compounds listed from pork are similar in number and type. Chang (1976) characterized lactones, furanoids, sulfur compounds and pyrazines as compounds which may be important contributors to meat flavor.

One would think there would be more characteristic meat flavors available from this type of investigation. However, as compared to fruit, dairy and nut type flavors, meat flavor has not been found to contain a single predominant characterizing volatile compound. In addition, since the flavor of meat is the result of thermal processing, the typical flavor varies depending upon the degree and length of heating. This also adds to the complexity of flavor characterization. Just the number of volatile compounds identified poses a development problem.

Mabrouk (1976) lists the aqueous meat flavor precursors which encompass fifteen classes of organic compounds (Table 2).

TABLE 2 Aqueous Meat Flavor Precursors

Free Nucleotides	Glycopeptides	Amino Acids
Nucleotide Sugaramine	Nucleic Acids	Free Sugars
Sugar Phosphate	Organic Acids	Sugaramine
Peptide-bound Nucleotides	Peptides	Amines
Nucleotide Acetylsugaramine	Nucleotide Sugar	

His work indicates that the overall impression of cooked meat is the result of these organic precursors. Of the twenty-three amino compounds identified and quantified in beef extract, he concluded that Methionine and Cysteic Acid are the most important amino acids contributing to meat flavor. It is generally agreed that the development of volatiles responsible for typical meaty flavor is the result of non-enzymatic browning reactions. These volatile flavors are produced mainly from reactions of reducing sugars with amines, amino acids, peptides and proteins (Dwivedi 1975).

It is uncertain what significant effect lipids have on the development of meat aroma, however, it appears to be important in identifying overall characteristic meat flavor. Flavor components are formed as a result of thermal oxidation of lipid constituents.

Many of the meat flavor precursors found in meat are the result of biochemical reactions catalyzed by enzymes. Dwivedi (1975) reviewed anti- and post-mortem enzymatic factors relating to the development of meat flavor precursors. Post-mortem tissue proteins degrade to release amino acids by the action of microbial proteases. In addition, other precursors such as peptides, nucleotides and carbohydrates also result from biochemical reactions. These precursors are important taste compounds and also develop volatile compounds upon heat processing.

Cultures and lipolytic enzymes have been used for decades for developing the characteristic flavor of sausage. Free fatty acids liberated from sausage lipids by lipase provide the substrate for the development of carbonyl compounds (Demeyer, 1974). These free fatty acids and carbonyl compounds are important in characterizing dry sausage.

Studies of model systems containing precursors identified in meat, and systems containing compounds that may produce desirable meaty aromas and flavor have been investigated. Shibamoto et al (1977) studied the reaction of D-Glucose, Hydrogen Sulfide and Ammonia. Fifty-one compounds were identified, the main constituents being thiophenes, furans, pyrazines and thiazoles. Twenty-four of these compounds were also identified in heated beef. By varying the levels and ratios of reactants, the reaction products will vary in concentration and composition. This example demonstrates the utility of this approach in developing meat flavors.

Many patents have been issued covering the use of certain chemicals in meat flavor formulations. In addition, patents covering flavor formulations utilizing maillard and other reactions producing meaty compounds have been issued over the last 15 years (May, 1960).

TABLE 3 Types of Meat Flavors Used in Meat Analogs

Beef	Meat Fat	Salami
Pork	Liver	Bologna
Poultry	Hamburger	Pepperoni
Bacon	Meat Loaf	Stewed Meat
Pork Sausage	Corned Beef	Fish
Frankfurter	Ethnic Sausages	Seafood

Even though meat has been intensely investigated using modern technology, characteristic meat flavors have not been developed to satisfaction for meat analogs. One of the reasons may be that the majority of the investigations concern beef, pork and poultry, whereas, meat analogs utilize other types of meat flavors as found in Table 3. Many of these meats are combinations of various cuts and types of meats, fats, and spices. In addition, many "home made" recipes add to the complication of characterizing these processed meats. One of the major flavor differences between meat and meat analogs is the development of characteristic meat aroma upon consumer processing. This is caused by the level and type of flavor added and the inherent chemical properties of meat analogs.

To date, the development of characteristic meat flavor includes the combined use of volatile compounds, maillard and other selected reactions, precursors and enhancers, and biological processes. Meat flavors for analogs must not only supply the characterizing flavor of the meat, they must also mask the flavor of the analog ingredients. Analogs that contain spicy flavors such as sausage and hamburger are somewhat easy to flavor as compared to analogs that contain delicate flavors like ham.

B. PROPERTIES OF MEAT ANALOGS AS COMPARED TO MEAT

Part of the problem in flavoring meat analogs is due to the interaction of meat flavors with analog ingredients and the development of off flavors from analog ingredients. Table 4 shows ingredients commonly found in meat analogs to supply nutritional and functional properties. Many of these ingredients can affect added flavor and develop off flavors during processing. Vegetable proteins are important relative to flavor degradation. In addition, other ingredients such as vitamins and starch also affect flavor but not to the degree of

TABLE 4 Typical Ingredients Found in Meat Analogs

Textured vegetable protein (wheat gluten, soy concentrate and isolate), egg whites, soybean and/or corn oil, partially hydrogenated soybean and cottonseed oil, flavorings (artificial flavors, salt, monosodium glutamate, spices, onion powder), sodium caseinate, wheat starch, sodium phosphates (tripolyphosphate, pyrophosphate, hexa-metaphosphate, monophosphate), modified corn starch, caramel color, sugar, vitamins (niacinamide, thiamine mononitrate (B_1), pyridoxine hydrochloride (B_6), riboflavin (B_2), B_{12}), iron (as ferrous sulfate).

soy protein. The various fractions of soy protein such as soy flour, concentrate and isolate and other vegetable proteins such as wheat gluten all affect added flavor differently.

Soy protein fractions are used in meat analogs in either the extruded form, wet spun form or as a flour. Table 5 gives the basic composition of Soy Flour, Concentrate and Isolate. The values given are average values and may vary according to soybean variety and methods used during processing.

The composition of the soy fractions vary in flavor and potential to develop off flavors during processing. The carbohydrates, vitamins and minerals are responsible in part to develop off flavors during heat processing. Thirteen vitamins ranging up to 3.8 mg/100 g and eighteen minerals ranging up to 2360 mg/100 g are found in soy flour (Kellor 1974). Carbohydrate includes Sucrose, Stachyose, Raffinose, Arabinogalactin and Polysaccharides (Horan 1974). One of the major problems in flavoring meat analogs containing soy protein is to mask the typical green, beany, bitter flavor associated with soy proteins. The degree of processing affects the overall flavor of the soy fractions. In general, soy flour has more undesirable flavor than concentrate which has more undesirable flavor than isolate. Also, the use of soy isolate produces a more

TABLE 5 Proximate Composition of Soy Flour, Concentrate and Isolate

Protein	55	72	96
Fat	1	1	.1
Carbohydrates	38	22	.4
Ash	6	5	3.5

bland analog after processing than using flour or concentrate. Since the isolate has less fat, carbohydrates and minerals, it reacts less to added flavor during processing and develops less off flavor due to its composition. The fraction of soy protein used in meat analogs plays an important role in the resultant flavor of the product.

Many investigators have tried to identify the chemicals responsible for the typical off flavor associated with soy. Mattick and Hand (1969) reported the typical soy flavor was associated with ethyl vinyl ketone. However, this compound only represents part of the flavor. Other investigators studied the flavor properties of the classes of chemicals identified in soy and arrived at varied conclusions. Maga (1973) gives an excellent review of the chemicals identified in various soy fractions and lists acids, alcohols, aldehydes, amines, ketones and phenolic acids. Table 6 shows chemicals from these classes of compounds that would contribute to soy flavor. The short chain fatty acids identified were reported to be at a concentration too low to have an effect. However, they can be synergistic at subthreshold levels in causing off flavors. These other compounds all have odors and flavors that could contribute to soy off flavor.

TABLE 6 Major Volatile Compounds Contributing to Soy Flavor

Alcohols	Aldehydes	Amines
2-Pentanol	Acetaldehyde	Ammonia
Iso-pentanol	Hexanal	Methylamine
Hexanol	Heptanal	Dimethylamine
Heptanol	2-Heptenal	Piperidine
1-Octen-3-ol	2,4-Decadienal	Cadaverine
Ketones	**Phenolic Acids**	
2-Pentanone	Syringic	
2-Hexanone	Vanillic	
Ethylvinyl Ketone	Ferulic	
	Gentisic	
	Salicylic	
	Para-Coumaric	
	Para-Hydroxygenzoic	
	Isochlorogenic	
	Chlorogenic	

The aldehydes listed have green notes that are associated with the green notes in soy. 2,4-Decadienal gives a fatty oil like odor and flavor. 1-Octen-3-ol has a green mushroom like flavor. The amines are fishy and putrid like. Many of the phenolic acids are astringent, burnt/bitter like. Many of these compounds can be synergistic at subthreshold levels and can support a particular flavor even though they may not be suspect in themselves.

In addition, part of the typical flavor of soy protein can be associated with nonvolatile compounds. Bitter flavors have been associated with certain peptides (Mabrouk 1976). Since meat analogs are subjected to heat during processing, this also gives rise to flavor development. Qvist et al (1974), studied soy among other proteins and identified over 150 compounds from head space samples. His investigation included retorted soy isolate, unheated isolate, isolate heated in water, and isolate heated in the presence of pork fat and potato starch. The processing was carried out at 120° C for 40 minutes.

Table 7 lists some of the volatiles where the concentration increased on heating. Since some of the threshold values were not available, it was not clear whether all volatiles increased. Also, new compounds were detected in the heated samples resulting from thermally induced reactions. In most cases the concentrations of volatiles are well above their odor thresholds. Taste thresholds are usually above odor thresholds. Although these compounds are reported at less than 10 ppm, many of them could provide portions of the off flavor associated with soy. It is interesting to note that the last 5 compounds are reported in meat, but due to other chemicals present, they do not provide enough characteristic flavor to be effective.

One of the differences between meat and cholesterol free meat analogs is the presence of animal fat versus vegetable fat. Wasserman et al (1967) investigated the organoleptic identification of beef, veal, lamb and pork as affected by fat. The results of his test indicate that fat may be a factor in the identification of meat flavor; however, the results were not conclusive in all cases. Animal fats contain amino acids, proteins, sugars and salts and react in the meat system during heat processing to produce flavor compounds (Wasserman 1972). On the other hand, vegetable fat can develop typical off flavors during heat processing. However, more important is the fact that vegetable oil alone cannot develop flavors like animal fat due to its chemical and physical properties.

TABLE 7 Volatile Compounds in the Headspace Gas of Soy Protein (Promine D)

	Promine D			
Compound	Unheated	Heated in H_2O	Heated with fat, starch, & H_2O	Odor thresholds, ppb in air (v/v)
Ethanal	2500	4250	4400	210
1-Butanal	17	125	108	9
1-Pentanal	68	780	720	12
1-Hexanal	260	1710	1390	4.5
1-Heptanal	2.2	41	21	3
1-Octanal	3.9	26	8.5	0.7
1-Nonanal	3.0	16	7.7	1
2-Methylpropanal	62	230	210	1
3-Methylbutanal	5.6	36	52	0.2
2,4-Decadienal	0.3	0.2	0.1	0.07
2-Pentylfuran	95	2700	110	2000
Hydrogen sulfide	2.7	1690	1200	4.7
Methanethiol		47	26	2.1
Dimethyl sulfide	2.8	43	29	1.0
2-Propanethiol		1.1	1.1	0.45

C. APPLICATION OF MEAT FLAVORS IN MEAT ANALOGS

It is apparent that the chemistry of meat flavors and analog ingredients are very complex. Application of meat flavors in meat analogs include interaction of flavor with analog ingredients and effects of processing. It is well documented that soy protein has good binding properties (Kinsella 1976). This property is useful in providing functional characteristics such as adhesion, fiber formation, viscosity, etc. Soy protein binds fat and water which is important in meat extenders and meat analogs. The major properties related to the interaction of soy protein and water include absorption, viscosity, gelation, syneresis and surface tension (Kinsella 1976). These fat and water binding properties of soy protein could affect added artificial fat and water soluble flavors. The mechanisms of these interactions are not fully understood and a better understanding of this phenomenon is needed.

Gremli (1974) studied the interaction of certain classes of food approved flavor chemicals with soy isolate (Table 8). The classes included aldehydes, phenolic alcohols and ketones, and heterocyclic ketones. The reaction was carried out at room temperature and headspace gas was removed at regular intervals and identified. The alcohols were reported not to react with the protein. Aldehydes and ketones were absorbed by the protein from 0-100%. Although there appears to be some correlation between retention and molecular weight within a chemical class, more work would be needed to arrive at any conclusions.

TABLE 8

Compound	% Retention
Hexanal	37-44
Heptanal	62-70
Octanal	83-85
Nonanal	90-93
Decanal	94-97
Undecanal	96-100
Dodecanal	94-100
2-Phenylmethylketone	9-21
(4-Methyl-2-Phenyl-Methylketone)	45-55
(3,5-Dimethyl-2-phenyl)-methylketone	48-60

TABLE 8 (Cont'd)

Compound	% Retention
4-Methyl-3-penten-2-one	25-39
6-Methyl-3-penten-2-one	36-47
3-Octen-2-one	16-29
6-Nonen-2-one	42-51
2-Furylmethylketone	0
(5-Methyl-2-furyl)-methylketone	0
(2,5-Dimethyl-3-furyl)-propylketone	58-71
2-Furylbutylketone	55-64
2-Hexanone	5-16
2-Heptanone	9-22
2-Octanone	29-43
2-Nonanone	54-61
2-Decanone	59-68
2-Undecanone	28-39
2-Dodecanone	17-32
2-Hexenal	68-75
2-Heptenal	82-88
2,6-Nonadienal	90-98
2,4-Nonadienal	92-97
2-Decenal	100
2-Dodecenal	100

reproduced with permission

In another part of the experiment, increasing amounts of heptanal and nonanone were added to the soy isolate and the retention determined (Table 9). An equilibrium seemed to exist between bound and free volatiles which was independent of the amount added. Based on this data, one could expect that flavors could be developed that would take into account the factors shown. However, in actual application, meat analogs are much more complex and the meat flavors contain many classes of compounds that were not studied. Nevertheless, the model study by Gremli indicates quantitative data on volatile retention, and may be useful in further investigations.

TABLE 9 Retention of Increasing Amounts of Volatile in 5% Protein Solution

Mg Heptanal/100 ml Protein Solution	Percentage Retention	Mg Volatile Bound/g Protein
1	70-75	ca. 0.14
2	71-75	ca. 0.28
5	68-74	ca. 0.7
10	62-70	ca. 1.4
50	66-73	ca. 7.0
Mg nonanone/100 ml		
2	58-64	ca. 0.2
5	58-68	ca. 0.6
10	54-61	ca. 1.2
15	57-66	ca. 1.8
20	55-60	ca. 2.4

reproduced with permission

In another study, Franzen et al (1974) reviewed experiments conducted on individual compounds such as salt, carbohydrates and proteins and their effect on the volatility of flavor compounds. In general his conclusions are similar to Gremli, that is, soy proteins decreased the concentration of head space volatiles in aqueous systems. The proteins showed no consistent or predictable effects with regard to the magnitude of flavor binding and no one flavor was preferentially bound by soy protein isolate and concentrate.

Probably the most important factor affecting the flavor of meat analogs is processing--both heat and freezing affect intensity and quality of flavor. This phenomena can sometimes be overcome by increasing added flavor; however, it has its limitations.

In natural meat flavor systems heat processing develops desirable flavor from precursors depending on the type, degree and time of heating.

As in the preparation of meat, over or under consumer processing also affects the flavor of meat analogs. Specific instructions are provided to give optimum flavor. Underprocessing sometimes does not allow the intended flavor to develop, while overprocessing normally tends to distort flavor and reduce flavor intensity by flashing off volatile flavor components and produces "toasted" flavors.

Another variable that affects added flavor is analog formulation changes. Small changes in the formulation can surprisingly change the added flavor profile. Flavor formulations may have to be modified continuously until analog formulas have been finalized.

It has been demonstrated that the development of meat flavors and their application in meat analogs is extremely complex and will not be fully understood due to the number of possible physical and chemical interactions taking place. Nevertheless, much progress has been made in improving these analogs in terms of flavor and texture. As we gain a better understanding of both meat flavors and their application in analogs, further improvement will take place. There is little doubt that the development of meat flavors for this application is a tremendous challenge for the flavor chemist.

For those of you who are not familiar with meat analogs, the following list (Table 10) will give you an idea of the scope of these products.

This list represents the most popular types of products available; however, there are approximately 100 different frozen, refrigerated and canned analog products available in the U.S. today.

TABLE 10 Types of Meat Analogs Available

Bacon	Chicken	Scallops
Breakfast Sausages	Tuna	Salisbury Steak
Ham Slices	Hamburger	Turkey
Salami	Steaklet	Hot Dogs
Bologna	Chili	Corned Beef

REFERENCES

Chang, S. S., Recent Advances in Meat Flavor Research, 1976.

Demeyer, D. et al, J. Food Science, 39, 293, 1974.

Dwivedi, B. K., Meat Flavor, CRC Critical Reviews in Food Technology, 1975.

Franzen, K. L., Kinsella, J. E., J. Agr. Food Chem., Vol. 22, No. 4, 1974.

Gremli, H. A., Interaction of Flavor Compounds with Soy Protein, J. Am. Oil Chemists' Soc., Vol. 51, 1974.

Horan, F. E., Soy Protein Products and Their Production, J. Am. Oil Chemists' Soc., Vol. 51, 1974.

Kellor, R. L., Defatted Soy Flour and Grits, J. Am. Oil Chemists' Soc., Vol. 51, 1974.

Kinsella, J. E., Functional Properties of Proteins in Foods: A Survey, CRC Crit. Rev. in Food Sci. and Nutrition, 7 (3), 1976.

Mabrouk, A. F., Non-Volatile Nitrogen and Sulfur Compounds in Red Meats and Their Relation to Flavor and Taste, Advances in Chemistry Series, 1976.

Maga, J.A., J. Agr. Food Chem., Vol. 21, No. 5, 1973.

Mattick, L. R., Hand, D. B., J. Agr. Food Chem., 17, 15, 1969.

May, C. G., U. S. Patent No. 2,934,435, 1960.

Qvist, I. H., von Sydow, E. C. F., J. Agr. Food Chem., Vol 22, No. 6, 1974.

Shibamoto, T., Russell, G. F., J. Agr. Food Chem., Vol. 25, No. 1, 1977.

Wasserman, A. E., J. Agr. Food Chem., Vol. 20, No. 4, 1972.

Wasserman, A. E., Talley, F., J. of Food Sci., Vol. 33, No. 2, 1967.

FLAVOR DEVELOPEMENT OF WHITE AND PROVINCIAL GREEK BREAD.TRADITIONAL AND RURAL BREADMAKING PROCESS.

E.C.Voudouris
M.E.Komaitis
Department of Food Chemistry,
University of Athens,
Athens, Greece

B.Pattakou
Cereal Institute,
Thessaloniki,Greece

I.INTRODUCTION

The extremely complex nature of bread flavor is illustrated by the fact that more than 150 compounds have been identified(1).Bread flavor is a function of both ingredients and processing.The importance of the fermentation and the baking process in the formation of the bread flavor was clearly shown by Baker et al.(2).Drapron(3) and El-Dash(4) showed that the enzymes contribute to the overall bread flavor.Maillard-type reactions in the crust contribute directly to the aroma of bread.The conditions of breadmaking and their influence on the bread flavor were extensively stu-

ISBN 0-12-169060-1

died by Nago(5), who found out that the way and the time of dough mixing, the duration of fermentation, the quantity of dough, and the addition of improvers ... contribute to the developement and preservation of bread aroma.

Bread prepared in small bakeries or home has better taste and aroma than that produced in modern plants. This can be attributed to the short baking period employed in plants, whistle the duration of fermentation in homes is longer.

In this work a special kind of bread, called "troz", was prepared by aparticular technique. In this paper the breadmaking process used and the preliminary results of the identification of the bread components are recorded.

II. MATERIALS AND METHODS

For the preparation of the bread "troz" flour from wheat (T. Aestivum) was used. The qualitative characteristics are shown in Table 1.

TABLE 1. Qualitative Characteristics of two Types of Flour used in Experimental Baking Tests.

Flour Type	Wet Gluten %	Protein Content % (N X 5,7)	Ash Content %	Maltose Index mg /10g. fl.
70% extr. (bread wheat)	28.5	11.28	0.50	1.54
85% extr. (durum wheat)	35	13.24	0.95	1.85

The breadmaking for both types of flour was carried out according to the method used by the laboratory of Chemistry and Technology of the Institute of Cereals: 300g of flour, 5g of dough and 6g of sodium chloride were mixed for 5 minutes in a farinograph with addition of water until 500 units Brabender were reached. The produced mass was placed in oven 30 C and relative moisture 80% for $1^{1/2}$ hours. Baking took place at 220 C for 50 minutes.

A.Preparation of "Troz" Leaven

To a recent extract of petals of wild hop, wheat and maize flour and a small quantity of wheat bran was added.The tight mass produced was cut to small pieces and dried.This leaven is active for one year.

B.Breadmaking of "Troz" Bread

A proper quantity of leaven,empirically determined, was dissolved in warm water and to this proper quantity of flour was added.The produced mass was allowed to ferment overnight at 25 C. After a vigorous fermentation the baking process took place.

C.Separation of Flavor Constituents

The vapour developed during the baking process passed through a condencer and was collected in a cooled beaker containing water. After this the condenced vapours were extracted with diethyl ether.

As soon as possible after baking the crust was separated finely divided and then extracted with diethyl ether.

Both extracts were examined by thin layer and column chromatography.

1. <u>Fractionation of extract on silica gel.</u> The technique used (6) for the fractionation of the ex-

tracts has as follows: Silica gel (Merck type 60, 70-230mesh ASTM) was dried for 3 hours at 120 C. A homogenous suspension was prepared from 2g of silica gel and 15ml methylene chloride. The suspension was transferred to a glass column having 150mm length and 10mm diameter. The column was eluted with 10ml of methylene chloride and then 200μl of the condenced extract were introduced. Fractionation took place with solvents having different polarities(Table 2).

TABLE 2. Solvents used for the Fractionation on Silica Gel.

Solvents	Ratio	Volumes ml
Methylene chloride	1	30
Methylene chloride + Hexane	8/2	20
Methylene chloride + Hexane	6/4	20
Methylene chloride + Hexane	4/6	20
Hexane	1	30
Hexane + Diethyl ether	9/1	20
Hexane + Diethyl ether	8/2	20
Hexane + Diethyl ether	6/4	20
Hexane + Diethyl ether	4/6	20
Hexane + Diethyl ether	2/8	20
Hexane + Diethyl ether	1/9	20
Diethyl ether	1	50

The flow rate was adjusted to 0.5ml/min. The solvent of the various fractions was evaporated to 10 ml.

2. Identification. At first, the fractions were examined by gas liquid chromatography on a Carbowax

20M. The identification of the constituents was made by comparison of their retention times with those of the reference compounds. Linked gas liquid chromatography - mass spectrometer was also used.

Type of column: Carbowax 20 M 5% on Chromosorb W AWDMCS 8/100 mesh

Column temperature: 230 C

Detector FID temperature 350 C

Inlet Temperature 170 C

Flow rate 30 ml/min. He

Sample volume 1 μl

III. RESULTS AND DISCUSSION

The breads produced by the different breadmaking procedures were subjected to sensory evaluation and the results are shown in Table 3.

TABLE 3. Taste and Aroma Evaluation of Breads produced by Different Breadmaking Procedures.

Bread	Taste: Basic tastes perceptible by four judges				Aroma: Preference by four judges			
	1	2	3	4	1	2	3	4
No 1 white	imp. so.or sw.	imp. so.or sw.	imp. so.or sw.	imp. so.or sw.	md	sd	md	md
"Troz"	slig. so.	slig. so.	imp. so.or sw.	imp. so.or sw.	vd	d	vd	vd
No 2yellow	slig. sw.	m. sw.	slig. sw.	slig. sw.	md	md	md	md

Note : imp. imperceptible

so. sour
sw. sweet
slig. slighty
md moderately desirable
vd very desirable
m. moderately
d desirable

Dough and bread characteristics are shown in Table 4.

TABLE 4. Dough and Bread Characteristics produced by Different Breadmaking Procedures.

Breadmaking procedure	Dough PH		Time(hr)		Yield %	
	star.	fin.	Fer.	Pr.	BW.	LV.
No 1.Fer. with yeast.Bread Wheat Flour	5.3	5.1	2	1.5	133	389
"Troz".Fer. with yeast "troz"	5.7	4.2	20	4	137	400
No 2.Fer. with yeast.Durum Wheat flour	5.2	4.9	2	1.5	129	335

Note :
star. starting
fin. final
Fer. Fermentation
Pr. Proofing
BW. Bread weight per 100g flour
LV. Loaf volume (cc/100g flour

From Table 3 it is obvious that the aroma of the "troz" bread is more desirable than that of the other kinds of bread.

From Table 4 it can be seen that the yield is better for the "troz" bread.

The preliminary results on the identification of the constituents that contribute to the bread aroma are shown in Table 5.

TABLE 5. Compounds identified in Bread Crust

Alcohols	Ketones	Aldehydes	Acids
ethanol	2-propanone	propanal	acetic
1-propanol	2-butanone	butanal	propionic
1-butanol	2-pentanone	pentanal	butyric
1-pentanol	2-hexanone	hexanal	valeric
1-hexanol	2-heptanone	heptanal	heptanoic
2,3butanediol	2,3butanedi-one	octanal	octanoic
1,4butanediol	3-methylbu-tan2-one	nonanal	nonanoic
2-pentanol	2,3pentandi-one	benzaldehyde	decanoic
3-pentanol		but-2-enal	
		non-2-enal	

Esters	Lactones	Hydrocarbons
ethyl formate	γ-butyrolactone	toluene
methyl acetate	γ-hexalactone	limonene
ethyl acetate	γ-nonalactone	o-xylene

The complex mass spectra recorded indicate that many more compounds have to be identified.

Comparative studies on the constituents of the three kinds of bread are in progress and will appe-ar in a later paper.

IV. SUMMARY

A report on the results of the investigation of a special kind of bread,called "troz" is presented.

In addition comparative studies on the sensory evaluation,dough and bread characteristics are re-ferred as well as, the preliminary results on the identification of the constituents that contribute to the overall bread aroma.

REFERENCES

1. Maga,J.A., Critical Reviews in Food Technology, 5,1: 55-142 (1974).
2. Baker,J.C. and Mize,M.D.,Cereal Chemistry, 16: 295-297 (1939).

3.Drapron,R., Estratto dalla Rivisia Italiana Essence, Profumi, Piante officinaly, Aromi, Saponi, Cosmetiki, Aerosol, Maggio (1977).

4. El-Dash,A.A., The Bakers Digest, 26-31 (1971).

5. Nago, M.C., These, L`Universitè de Paris,(1976).

6. Maarse H., Van Os F., Flavor Ind.(london),4,(II), 477-484.

SOME RECENT ADVANCES IN THE KNOWLEDGE OF CHEESE FLAVOR

Jacques Adda[1]
Sylviane Roger
Jean-Pierre Dumont

Laboratoire des Arômes
Institut National de la Recherche Agronomique
Jouy-en-Josas, France

I. INTRODUCTION

Cheese is a product which the consumer buys for its flavor more than anything else. Unfortunately large-scale production has often led to rather bland products and one of the major preoccupations of the dairy industry is to find some way to improve the flavor characteristics of their products within this production system. This is why precise knowledge of ripening mechanisms becomes essential but the very concept of cheese flavor has not yet been defined.

Many interesting results have been published, even recently, on Cheddar cheese flavor. Roughly speaking we can consider blue cheese flavor as elucidated since methylketones have been recognized as the character impact compounds, but we have only a few results on other types of cheeses such as mold and surface ripened cheeses which are produced on a large scale in Europe (over 200,000 t Camembert are produced every year in France alone) and are gaining increasing popularity all over the world.

For any type of cheese the ripening process can be described as a series of reactions related to lactic acid metabolism, lipolysis, oxidation and proteolysis which can be followed by reactions between the products formed in each pathway. Some of these reactions are probably purely chemical, but most of the time microorganisms play a significant role through their enzymatic system although the importance of enzymes in now being questionned, at least in Cheddar.

[1]Supported by D.G.R.S.T. Grant 75 7 574

ISBN 0-12-169060-1

II - SOME CHEESES TAKEN AS MODELS

In this paper we will summarize recent progress in the knowledge of different cheeses flavor. Several cheeses will be taken as model, each of them being characterized by different microflorae. When possible we will try to relate the presence of characteristic flavor compounds to a specific microflora.

A. Mold ripened cheeses

Camembert will be taken as an example of cheese in which Penicillium caseicolum plays a major role. Although free fatty acids are present in a very large quantity they should not be considered as character impact compounds in good quality cheeses. Nevertheless they contribute to the overall flavor and serve as a substrate for the formation of methyl ketones and 2-alkanols. Under abnormal conditions, for example in cheeses made with refrigerated milk with a high count of lipolytic psychrotrophic bacteria, methyl ketones and 2-alkanols attain a very high level and one can even find quite unusual unsaturated methyl ketones such as 2-undecenone and tridicenone beside 2-nonanone and 2-nonenone which normally exist as major carbonyls in a well ripened cheese (1). The real importance of 2-nonanone is not questionable : If we bring together the value of 3.4 μm 2-nonanone per 10 g fat given by Schwartz and Parks (2) and the taste threshold of methylketones published by others (3) we must admit that in normal quality cheese 2-nonanone plays by itself a significant role with an aroma value close to 2. With the taste threshold value (0.9 ppm) obtained by our group using curd as a diluting medium we find a still higher aroma value of nearly 14.

More significant is oct-1-en-3-ol which plays a highly specific role (4, 5, 6) and accounts for the pleasant mushroom note always present in that type of cheese. Different research workers (7, 8) have shown that several Penicillium sp. are able to produce oct-1-en-3-ol by oxidation of unsaturated fatty acids an a typical off-flavor has been observed in cheese with strains of P. caseicolum which produce too large amount of oct-1-en-3-ol.

In well ripened cheese a floral note can easily be detected, which comes from the presence of large amounts of 2-phenylethanol and β-phenylethylacetate (5), both having a pleasant rose aroma. These compounds can be considered as characteristic of soft cheeses even if we are not yet able to pick out the microorganisms which produce them.

The same remark also applies to 1,3-dimethoxybenzene which has been identified in every sample of mold ripened soft cheese. This compound, the origin and role of which, are not yet clearly understood, has a pleasant hazelnut shell aroma and its threshold values are now being investigated by our group.

It would be unfair not to mention methylcinnamate which has been identified in Camembert, described (4) together with oct-1-en-3-ol as an impact character compound. However we have never been able to find this compound in any sample throughout our own work.

N-isobutylacetamide has recently been identified both in mold and surface ripened cheese. This compound which could originate from the dipeptide val-gly is interesting because of its bitter taste and we are presently measuring the bitterness of the pure synthetic. Although cheese flavor normally involves some bitterness it very quickly becomes a major defect. Up till now bitterness has been explained by the presence of bitter peptides as many proteases are able to produce bitter peptides from pure casein. However bitter peptides have never been isolated from cheese. The identidication of small, ninhydrine negative molecules such as N-isobutylacetamide could offer an alternative.

Some work has also been done on the sulfur compounds which are responsible for the garlic note of ripened cheese. Beside the usual sulfur volatile compounds, namely hydrogen sulfide, methyl sulfide, DMDS and dimethyl trisulfide several other molecules have been identified (9)

2,4-dithiapentane
3,4-trithiahexane
2,4,5-trithiahexane
and 3-(methylthio)-2,4-dithiapentane.

Some of these compounds have also been identified in other foods ; 2,4-dithiapentane in Gouda cheese (10), 2,4,5-trithiahexane in *Brassica* sp. (11). The identity of the isolate compounds was confirmed by comparing mass spectral and chromatographic data with those of pure synthetic compounds (Tables I, II). These compounds were also used, when pure enough, to study their flavor properties and their odor threshold with a dairy medium : Stock solutions of pure compounds were made using double glass distilled water and ethanol ; aliquots were further diluted in a liquid cheese obtained by mixing thoroughly fresh pasteurized unripened heavy cream, low heat powder of milk proteins obtained by membrane ultrafiltration and water (2:1:1, w/w). Dilutions were directly done in a glass stoppered one liter flask fitted with an inflated plastic bag

TABLE I. Retention time of sulfur compounds relative to DMDS[a].

Compound	Relative retention	Compound	Relative retention
- hydrogen sulfide	.08	- 2-methylthiophene	1.05
- methanethiol	.23	- 3-methylthiophene	1.09
- ethanethiol	.44	- methylthiopropanoate	1.12
- methyl sulfide	.48	- 2,3-dithiapentane	1.26
- carbon disulfide	.50	- dipropyl sulfide	1.34
- 2-propanethiol	.55	- 2,5-dimethylthiophene	1.38
- ethylene sulfide	.65	- methylthiobutyrate	1.46
- ethylmethyl sulfide	.67	- 2,4-dithiapentane	1.60
- 1-propene-3-thiol	.68	- 3,4-dithiahexane	1.67
- propylene sulfide	.76	- methylthio-2-methylbutyrate	1.73
- 2-butanethiol	.77	- methylthio-3-methylbutyrate	1.74
- 2-methyl-1-propanethiol	.78	- methional	2.16
- methylthioformate	.79	- 2,3,4-trithiapentane	2.32
- diethyl sulfide	.82	- dibutyl sulfide	3.01
- thiophene	.82	- 2-methylthiophene-3-one	3.14
- methylpropyl sulfide	.84	- 3,5-dimethyltrithiolane(cis)	4.98
- butanethiol	.87	- 3,5-dimethyltrithiolane (t)	5.24
- methylthioacetate	.89	- 2,4,5-trithiahexane	5.53
- DMDS	1.00		

[a]glass column 6.8 m x 0.002 m i.d. - Chrom.G.AW.DMCS 80/100 mesh 5 p.cent Igepal Co 630 - 10 to 110°C at 4°C/min after 15 min isothermal at 10°C - FPD (12).

---

to compensate for inhalated air (13). The flasks were allowed to equilibrate overnight at room temperature. Just before the session, the stopper was replaced by a glass mask.

During each session, 3 different concentrations and a blank were compared for odour with a control blank sample in the form of paired-tests. The four samples to be tested were presented in four replicate to 9 panelists in a randomized order. The judges were asked which sample contained the odorous material, if any. The percentage of correct responses was plotted against odour compound concentration. A 70 % correct response was used as the threshold criterion. Results are given in Table III.

Table II. Mass Spectra of some Sulfur Compounds

- 2,3-dithiapentane
  108(100), 80(95), 29(53), 64(37), 45(36), 27(32), 47(18), 79(16), 110(16), 46(16)
- 2,4-dithiapentane
  61(100), 45(50), 108(34), 35(22), 46(20), 47(17), 63(6).
- 2,3,4-trithiapentane
  126(100), 79(59), 45(59), 47(38), 64(30), 46(20), 111(18), 80(17), 128(13), 61(12)
- 2,4,5-trithiahexane
  61(100), 45(39), 35(23), 46(17), 27(14), 47(12), 63(7), 140(6)
- methylthioformate
  76(100), 47(85), 45(31), 48(25), 29(15), 46(11), 78(6)
- methylthioacetate
  43(100), 90(48), 45(18), 47(15), 42(8), 46(6), 48(6), 75(4), 44(3)
- methylthiopropanoate
  57(100), 29(89), 104(40), 27(25), 47(20), 45(14), 75(11), 48(9), 61(7), 46(5)
- methylthiobutyrate
  43(100), 71(52), 27(48), 41(38), 103(17), 39(12), 47(12), 118(11), 45(9), 75(7)
- methylthio-2-methylbutyrate (±)
  57(100), 85(92), 41(76), 29(61), 56(24), 75(21), 27(18), 47(16), 39(13), 55(10), 132(7)
- methylthio-2-methylbutyrate (+)
  57(100), 29(57), 85(52), 41(50), 27(15), 56(13), 75(11), 47(9), 39(9), 55(6), 132(4)
- methylthio-3-methylbutyrate
  57(100), 41(80), 85(76), 29(59), 43(52), 75(22), 39(21), 27(20), 47(15), 42(11), 132(4)
- 2-methyl-thiophene-3-one
  60(100), 116(57), 45(20), 59(19), 27(11), 88(9), 74(2), 101(1).

TABLE III. Olfactory results

	Odor quality	Odor threshold[a]		
		experimental	theoritical	litterature
2,3-dithiapentane	garlic	6.0		
2,4-dithiapentane	garlic	60.0	0.3	.3 water 3 oil (10)
2,3,4-trithiapentane	over ripened cheese	0.1	0.07	.01 water(11)
methylthioacetate	cooking cauli flower	5.0	7.0	
methylthiopropanoate	cheesy	100.0	2.0	
methylthiobutyrate	chives	200.0	0.6	
methylthio-2-methylbutyrate (±)	wild strawberry	100.0		
methylthio-2-methylbutyrate (-)	rubbery	100.0		
methylthio-3-methylbutyrate	fruity	100.0		

[a]Units are in ppb

B. Bacterial Surface Ripened Cheeses

Now we will look at surface ripened cheeses with Pont l'Evêque as an example. They are characterized by the growth of a smear coat in which orange pigmented coryneform bacteria and micrococci are well represented and the picture is quite different : components arising from protein degradation play a major role and although we still find 2-phenyl ethanol and β-phenyl acetate (14) the flavor is greatly influenced by the presence of large amounts of phenol, cresol, acetophenone and by sulfur compounds.

The importance of hydrogen sulfide and methanethiol in surface ripened cheeses was recognized a long time ago (15) but since then several methylthioesters have been identified in Pont l'Evêque (16).

Methylthioacetate, propionate, butyrate, and methylbutyrate have been isolated and their identity confirmed by comparing retention time and mass fragmentation with those of pure synthetics (Table I, II).

The flavor properties and the odor thresholds of the synthetics have been measured using the same techniques as above and the results appear in Table III. Beside our own experimental results we have also indicated the theoritical values calculated (17) from molecular factors obtained by CG (18) for some of our synthetics.

Consequently we are now trying in cooperation with coworkers from University of Clermont, to isolate microorganisms able to produce methylthioacetate. A few preliminary results can be mentionned : Several strains of *Brevibacterium linens* have been screened. They were all producing high amounts of methanethiol which is in agreement with recent observations (19) but most of them failed to produce methylthioacetate. The production of a small amount was observed with one strain isolated from Brie. High amounts of methylthioacetate have been obtained when growing several strains of micrococci on a medium in which methanethiol was added.

This work is in progress and a full report will be published elsewhere (20).

C. Cheeses with Internal Propionibacteria Fermentation

Methylthioacetate has not only been identified in Pont l'Evêque but also in different types of Gruyère (21, 22).

These are interesting as models : As a matter of fact they possess some similarity both with surface ripened cheeses as the development of a halophilic surface flora is permitted, and with Emmental, which is also characterized by a propionic fermentation.

The overall flavor of both Emmental and Gruyère is well documented (6, 21, 22, 23) : 1-alkanols and esters both seems to play a major role together with propionic acid. More unusual compounds such as pyrazines, the origin of which is not clearly understood have been repeatedly identified (21, 24).

A better understanding of the formation of the flavor compounds was obtained by a separate study of the core and the external part of the cheese, by comparing volatiles from single strains culture of propionibacteria with volatiles from cheeses made with the addition of the same strains, and by extracting volatiles straight from suspensions of smear organisms. It appears that some of the flavor compounds can be specific of the propionibacteria while others are produced by the surface flora.

Propionibacteria proved able to produce large amounts of methyl sulfide and a less common compound 2-methylthiophene-3-one which has been isolated both in cultures and cheeses. The same applies to propanoates and diacetyl. Propionibacteria can also produce phenol, acetophenone and cresol when grown on milk but the presence of these compounds in Gruyère cheeses also results from the proteolytic activity of the smear bacteria. This can be illustrated by the fact that acetophenone is much more abundant in the rind and the outer parts of the cheese than in the core. The same variation has been observed for methanethiol which is found at a very high concentration in the rind.

Some other compounds such as dimethyltrisulfide, 3-octanone and ethyl hexanol have been identified in suspensions of the smear bacteria.

III - CONCLUSION

Despite some progress we still know very little about cheese aroma. Nevertheless we believe that further advancement should be possible by a joint effort between flavor chemist and dairy technologist. Unfortunately, very often, these effort are not fully rewarded as enough attention is not given to profile assessment methods for evaluating the flavor contribution of aroma a constituents.

ACKNOWLEDGMENTS

Grateful acknowlegment is made to Dr A. Cuer, Dr G.Daufin and to Prof. A. Kergomard (Lab. Chimie Organique Biologique de l'Université de Clermont) who accepted to synthesize methylthioesters and others sulfur compounds. We thank Prof. Gelin (INSA Lyon) for synthesizing N-isobutylacetamide and 2-methylthiophene-3-one.

REFERENCES

1. Dumont J.P., Delespaul G., Miguot B., Adda J. Le Lait, 57, 531-532, 619-630 (1977).

2. Schwartz D.P., Parks O.W. J. Dairy Sci., 46, 1136 (1963)

3. Siek T.J., Albin Inga A., Sather Lois A., Lindsay R.C. J. Food Sci. 34, 165-267 (1969)

4. Moinas M., Groux M., Horman I. Le Lait 53, 529-530, 601 (1973)

5. Dumont J-P., Roger Sylviane, Cerf Paule, Adda J. Le Lait 54, 538, 501-516 (1974)

6. Groux M., Moinas M. Le Lait 54, 531-532 (1974)

7. Kaminski E., Stawicki S., Wasowicz E. Appl. Microbiol. 27, 6, 1001-1004 (1974)

8. Richard-Molard D., Cahagnier B., Poisson J., Drapron R. Ann. Technol. Agric. 25 (1) 29, 44 (1976)

9. Dumont J-P., Roger Sylviane, Adda J. Le Lait 56, 559-560, 595-599 (1976)

10. Sloot D., Harkes P.D. J. Agr. Food Chem. 23, 2, 356 (1975)

11. Buttery R.G., Guadagni D.G., Ling Louisa C., Seifert R.M., Lipton W. J. Agr. Food Chem. 24, 4, 829 (1976)

12. Jansen H.E., Strating J., Westra W.M.
J. Inst. Brew., 77, 154 (1971)

13. Laffort P. in Theorics of Odors and Odor Measurement
Tanyolac N.N. Ed., Maidenhead (G.B.), Technivision
(1968) pp.247-270.

14. Dumont J-P., Roger Sylviane, Adda J.
Le Lait 54, 531-532, 31-43 (1974)

15. Grill H. Jr., Patton S., Cone J.F.
J. Dairy Sci. 49, 403 (1966)

16. Dumont J-P., Degas Christiane, Adda J. (1976)
Le Lait 56, 553-54, 177-180 (1976)

17. Patte F. Thesis, Université Claude Bernard, Lyon (1978)

18. Laffort P., Patte F.
J. Chromat. 126, 625-639 (1976)

19. Sharpe Elisabeth M., Law B.A., Phillips B.A.
J. Gen. Microbiol. 94, 430 (1976)

20. Cuer Annie, Daufin G., Kergomard A., Dumont J-P., Adda J.
To be published.

21. Dumont J-P., Pradel G., Roger Sylviane, Adda J.
Le Lait 56, 551-552 (1976)

22. Dumont J-P., Adda J.
J. Agric. Food Chem. 26, 2 (1978)

23. Langler J.E.
Thesis, Corvallis, Oregon State University (1966)

24. Sloot D., Hofman H.J.
J. Agr. Food Chem. 23, 2, 358 (1975).

FLAVOR ASPECTS OF CHOCOLATE

Dominique Reymond
Walter Rostagno

Nestlé Products Technical Assistance Co. Ltd.
Lausanne (Switzerland)

The delicate flavor of a chocolate bar is due to the summation of skillful techniques. Fundamental knowledge has been gained until now on several unit operations which play an important role in flavor development and in the final texture of the chocolate bars.

Flavor precursors are developed in the cocoa beans during the fermentation and the drying steps. Several analytical indices allow to follow changes in composition induced by these operations; several ratios such as catechins/soluble tannins, glucose+fructose/sucrose and soluble nitrogen/total nitrogen were shown to present useful indices (1). Varietal differences and changes in protein during pod ripening and fermentation were characterized by detailed amino-acid analysis (2). Histological changes during cocoa fermentation show a transformation of phases so that, at the end of the operation, hydrophilic cytoplasmic constituents are included in a continuous lipidic phase (3). Classes of flavor precursors developed during this process were shown to include carbohydrates, flavonoids, catechins, phenolic acids and amino-acids (4, 5).

The flavor of cocoa is developed during a roasting step where beans are submitted to a mild heat treatment. Main reactions during the roasting step involve Maillard reactions, Amadori rearrangements and Strecker degradations (6). Pyrazines are particularly generated during the thermal process and their content was shown to vary according to the importance of the roasting temperature (7). Roasting

ISBN 0-12-169060-1

of thin-layers of cocoa mass allow to characterize more precisely the importance of time/temperature treatment in developing optimal cocoa flavor (6). Free amino-acids undergo also a Strecker degradation leading to carbonylic compounds, which themselves may react to give aldol condensation products. The presence of 5-Methyl-2-phenyl-hexenal, an important flavor contributor, can be such-wise explained (8). Some volatile sulphur derivatives may be attributed to the thermal decomposition of methyl-S-methioninesulphonium salt present in raw cocoa (9). Important taste contributors to the bitterness of cocoa are also developed by thermal cyclisation of peptides (10) which play a synergic role with theobromine.

Various chromatographic techniques were used for characterizing the complex volatile mixture which contains important flavor contributors to cocoa aroma; in a recent review, we mention the use of differential gas chromatographic techniques (11) which allow to select constituents involved in specific flavors of cocoa beans from various origins. More elaborate mathematical methodology allows to select important flavor contributors by taking into consideration sensorial evaluation data. For routine quality control, simpler methods can be used; one of them relies on measuring the ultraviolet absorption of steam distillates and allows to characterize the degree of desodorization of cocoa butters (12), the quality of milk chocolate samples and of various ingredients (13).

Roasted cocoa nibs are then milled; the resulting cocoa liquor can be characterized by using the same quality indices as those used for roasted cocoa beans. The cocoa liquor, crystal sugar and milk solids are then mixed and refined together; various technological processes are used for this operation. Continuous lines are now in use (14) so that mixing, comminuting and flavor development are optimized. Future perspective in chocolate technology also involves multistage comminution of chocolate mass (15).

To obtain a smooth flavor of chocolate bar, a last refining step called "conching" is applied during which some volatile constituents are lost and various aggregates are dispersed so that the viscosity of chocolate mass is reduced; the optimal conching time can be determined by the time at which lowest yield value is obtained (16). During the first 24 hours of conching, strong changes in physical and sensory properties are observed (17), but the particle size distribution of solid particles is determined by the anterior milling and refining operations. This particle size distribution plays a considerable role in assessing chocolate bar acceptance to the consumer (18). Granulometric analytical methods were reviewed (19) and correlation experiments showed the importance of particle size distribution in determining preferences of a taste panel (20).

The "conched" chocolate mass is then submitted to a tempering operation during which a suitable crystalline form of polymorphic cocoa butter is favoured. This operation can be analytically followed by applying a differential microcalorimetric method (21). This precrystallization step will ensure a correct hardness to moulded chocolate bars (22).

All these physico-chemical methods are, however, not sufficient for determining the overall quality of chocolate bars which can be only assessed by sensory evaluation. Trained taste panel judgements may be expressed under the form of flavor profiles (23). These judgements can then be submitted to multidimensional classification using, for instance, the principal component analysis (24). Other statistical methods can also be applied (25). These mathematical methodologies allow to characterize the sensorial quality of chocolate bars.

REFERENCES

1. Bracco, U., Grailhe, N., Rostagno, W., Egli, R.H., _J. Sci. Fd. Agr._ 20, 713 (1969).
2. Keeney, P.G., Zak, D.L., _Irst International Congress on Cocoa and Chocolate Research_ (Editor Institut für Lebensmitteltechnologie und Verpackung, München, West Germany), page 135 (1974).
3. Biehl, B., Passern, U., Passern, D., _J. Sci. Fd. Agr._ 28, 41 (1977).
4. Rohan, T.A., _Gordian_ No. 1, 5, 53, 111 (1970).
5. Mohr, W., Röhrle, M., Severin, T., _Fette, Seifen, Anstrichm._ 73, 515 (1971).
6. Mohr, W., _Irst International Congress on Cocoa and Chocolate Research_ (Editor Institut für Lebensmitteltechnologie und Verpackung, München, West Germany), page 114 (1974).
7. Keeney, P.G., _J. Am. Oil Chem. Soc._ 49, 567 (1972).
8. Van Praag, M., Stein, H.S., Tibbetts, M.S., _J. Agri. Fd. Chem._ 16, 1005 (1968).
9. Lopez, A.S., Quesnel, V.C., _J. Sci. Fd. Agr._ 27, 85 (1976).
10. Pickenhagen, W., Dietrich, P., Keil, B., Polonsky, J., Nouaille, F., Lederer, E., _Helv. Chim. Acta_ 58, 1078 (1975).
11. Reymond, D., _Chemtech_ 7, 664 (1977).
12. Rostagno, W., Reymond, D., Viani, R., _Intern. Choc. Rev._ 25, 346 (1970).
13. Kleinert, J., _Kakao und Zucker_ 28, 218, 224, 228 (1976).
14. Kleinert, J., _Internat. Rev. Sugar and Confectionery_ 28, 10, 23 (1975).
15. Niediek, E.A., _Süsswaren_ 21, (9), 316, 319, 322, 324, 326, 328, 330, 332, 334 (1977).
16. Chevalley, J., _J. texture studies_ 6, 177 (1975).
17. Sommer, K., _Irst International Congress on Cocoa and Chocolate Research_ (Editor Institut für Lebensmitteltechnologie und Verpackung, München, West Germany), page 287 (1974).
18. Beesley, J., _Sensory properties of foods_ (Editors G.G. Birch, J.G. Brennan, K.J. Parker, Applied Science publishers, London), page 291 (1977).

19. Friedrich, H., Heidenreich, E., Schuldt, U., Lebensmittel-Industrie 22, 351 (1975).
20. Rostagno, W., Chevalley, J., Manufacturing Confectioner 49, (5), 81 (1969).
21. Chevalley, J., Rostagno, W., Egli, R.H., Internat. Choc. Rev. 25, 3 (1970).
22. Kleinert, J., CCB Rev. for Chocolate, Confectionery and Bakery 1, (2), 3 (1976).
23. Daget, N., Irst International Congress on Cocoa and Chocolate Research (Editor Institut für Lebensmitteltechnologie und Verpackung, München, West Germany), page 319 (1974).
24. Vuataz, L., Sotek, J., Rahim, H.M., Proc. IV Internat. Congress Food Sci. and Technol. 1, 68 (1974).
25. Randebrock, R.E., Deutsche Lebensmittel Rundschau 73, 286 (1977).

COCOA SUBSTITUTION

Willi Grab
Walter Brugger
Derek Taylor

Givaudan Dübendorf Ltd
Dübendorf
Switzerland

I would like to present some thoughts related to the question why and how cocoa could be substituted. After a short survey of the history, I will review the composition of cocoa and chocolate and I shall define the properties of the different constituents with respect to their function and their contribution to the perception of cocoa. Possible substitutes will be discussed in the same way with special emphasis to the flavour and its importance.

INTRODUCTION

Since the early 6th century the Mayas had their cocoa plantations in Central America. At that time, cocoa beans had two functions: they were used to prepare a delicious beverage and also served as a form of payment. In the 16th century, chocolate came to Europe. Then in 1875 Daniel Peters made the first milk chocolate in Switzerland and 1880

ISBN 0-12-169060-1

Rodolphe Lindt subsequently made a soft melting chocolate by introducing the conching process and by adding cocoa butter to the chocolate mass.

Since that time the basic treatment of cocoa and the composition of chocolate has not changed. Of course new technologies have been applied, and many new applications have been found in modern food technology. But we should always remember that the chocolate of today is an historical product developed a century ago and the question whether and how to substitute cocoa may sound rather provocative.

But is it not a mere accident that chocolate is based on cocoa powder, milk and sugar? Of course not! These products, combined with an appropriate technology yield an excellent food.

So why look for a substitution? Is it not possible to introduce new raw materials and new technologies into the concept "cocoa" and "chocolate" respectively? Can we not take the basic idea of chocolate and design a new food with similar properties but based on other raw materials?

WHAT IS "COCOA"?

Let me go back to cocoa. How are cocoa products made? Why do we like chocolate? Cocoa is a natural product from the tree Theobroma Cocoa and is grown in the hot regions of the world. The beans cannot be eaten as such directly from the tree. They have firstly to be fermented and dried. By this process a very important change in the chemical composition occurs: most of the astringent polyhydroxiphenols are converted into tasteless polymers, the phlobaphenes. At the same time important flavour components and flavour precursors are formed. Roasting the dried beans leads subsequently to the desired delicious flavour. The addition of sugar, milkproducts and cocoa-butter to the cocoa mass gives a chocolate mass, which after conching or similar processing and crystallization of the fat is converted into a delicious nutritional food.

When chocolate is eaten, all human senses become activated:

- the eyes are caught by silky lustre and natural brown colour;
- the ears perceive the sound of cracking the hard and at the same time soft texture;
- the tongue is pleased by the smoothly melting texture;
- the sense of temperature is smoothened by a slight cooling effect, triggered by the melting fat crystals and milk sugar;
- the taste is bitter, sweet, astringent and even certain savoury notes can be detected;
- finally the smell is the most attractive property. We perceive the typical cocoa flavour: sweet, flowery, green, honey, roasted, caramel, milk notes.

Since chocolate activates all the sensations in a well balanced manner, it is liked by most people. In addition, chocolate can be regarded as a very nutritious food: the high energy content of about 2200 kilojoules per 100 g, which means 1/5th to 1/8th of the daily energy demand, resulting from the sugar and fat content, makes chocolate one of the richest food products (which may, on the other hand, be an important drawback if you like chocolate but if you do not need it).

In a vast number of other food products, cocoa is used: the most important fields are beverages and couverture. For ice cream, puddings, yoghurts and biscuits cocoa is important to impart flavour and colour only.

COMPOSITION

Cocoa is composed of thousands of components. Some of them are known:

Approximate Composition of Cocoa Mass

Fat	55,7 %	Ash	2,7 %
Carbohydrates	18,0 %	Water	2,0 %
Proteins (raw)	11,9 %	Acids	1,6 %
Polyhydroxiphenols	6,2 %	Purines	1,4 %
Others (vitamins, flavour components etc.)			0,5 %

The main components: fat, carbohydrates and proteins contribute to the texture and nutritional value. The polyhydroxiphenols bring the colour. These are already complex mixtures of hundreds of components.

Very important from our, the flavour industry's point of view, are the traces of flavour substances. Today about 400 volatile chemicals are known to occur in cocoa. Most, but not all of them are responsible for the overall flavour. They can be formed in several stages beginning from the tree to the finished product (growth, fermentation, drying, roasting, Dutch process, conching and combination with other raw materials).

The most important flavour components belong to the following classes:
aldehydes, heterocycles, acids and terpenes. What seems exciting to us is, that in the cocoa flavour we have a very broad range of different and at the same time highly important flavour components, originating from the enzymatic and non enzymatic processes. Only few other food products exhibit such a broad and complicated flavour profile.

What are the positive properties of the most important components?

cocoa butter:
- has a high energy content
- gives texture
- has smooth melting characteristics
- helps to smoothen the flavour

cocoa powder:
- gives colour
- gives structure (mass)
- gives taste and flavour

purines:
- react with the central nervous system to trigger a stimulating effect
- impart to the bitter taste together with other components

sugar (in chocolate):
- has a high energy content
- contributes to the structure and
- imparts sweetness

What are the drawbacks of these components?

cocoa-butter:
- high price

- (high energy content), some-times you may take the high energy content as a drawback
- (low) melting point may pose problems in hot countries or in certain applications
- for ice cream coatings, other fats behave better than cocoa-butter

cocoa powder:

- variable, some-times high price
- poor solubility
- low protein and mineral content

We have all the elements to weigh the pros and contras of substituting chocolate. Can we substitute some of the components to the effect that less desirable properties are exchanged for desirable ones. Cocoa is a natural product and the crop fluctuates from year to year. Being a commodity item it has become a speculation object. This results in price fluctuations and I think, nobody has forgotten the last year with its tremendously rising prices.

APPLICATION

Today our modern food technology and our know-how permit to create new food products related to chocolate and with equally attractive properties but starting with other raw materials. The following food types can stand for a vast number of possible other applications:

Chocolate: We shall see later how chocolate-like food can be made without cocoa. Nevertheless, I think, chocolate is not the best product to be matched. It has reached a very high quality standard image and being an historical product it is well protected by some food laws, at least in some countries. But an imitation chocolate can serve as a basis for other new food, it can be produced from different raw materials enriched with nutritious elements such as proteins, vitamins and trace elements. And the most important: it may be flavoured in many different ways.

Beverages: The outstanding properties imparted by cocoa are colour, flavour and the slight stimulating effect. All of these benefits can be substituted by other raw materials.

Confectionery and dessert products: Since chocolate mass, chocolate or couverture is used only for some of its properties, a substitution would well be indicated.

Nutritious food: This may be the most important class of products, where we can combine the chocolate technology with other non traditional raw materials.

SUBSTITUTION

To answer the question what and how to substitute, we have to identify all the desired properties and possible raw materials in a morphological frame. From this we can extrapolate the necessary combination of raw materials to get the desired substitution product or we even can create a new food with additional taylor-made properties. Of course, this frame is by no means complete, other elements may be added.

Regarding the replacement of cocoa-butter by other, less expensive fats I need not say very much, since this is a broad field being covered by a number of well known companies. Fats of different natural origin are compared with cocoa-butter regarding their specific properties. In most cases the characteristics of the whole fat are unsatisfactory, but by fractional crystallization, blending and other procedures known in the fat industry, an acceptable product can be made. Today the fat industry offers a large number of products with the desired properties on the market. Most are based on fractions of palmoil, sheafat and borneo tallow (Tenkawang).

	STRUCTURAL PROPERTIES				SENSORIC PROPERTIES					NUTRITIONAL VALUE				PRICE	
	MELT	SMOOTHNESS	STRUCTURE	MASS	FLAVOUR	SWEET	BITTER	COLOUR	STIMULATION	ENERGY	PROTEINS	VITAMINES	TRACE ELEMENTS	UNSTABLE	STABLE
CHOCOLATE	X	X	X	X	X	X	X	X	X	X	X	(x)	(x)	X	
COCOA POWDER				X	X		X	X	X		(x)			X	
COCOA BUTTER	X	X	X	X						X				X	
MILK SOLIDS				X	(x)	(x)				X	X	X	X		X
SUGAR				X		X				X					X
ST.JOHN'S BREAD				X	(x)			X		(x)					(x)
STARCH				X						X					X
YEAST				(x)	(x)		(x)	(x)			X				X
VEGETABLE FAT	X	X	X	X						X					X
VEGET. PROTEINS				X						(x)	X				X
SYNTHET.FLAVOUR					X	(x)	X		(x)						X
CHOCLESS	X	X	X	X	X	X	(x)	X		X					X
CHOCOPRO	X	X	X	X	X	X	X	X	(x)	X	X	X	X		X

The substitution of cocoa powder is more difficult than the replacement of cocoa butter, because it brings the most important part of cocoa: flavour and colour. A few products are known to be relatively good cocoa powder replacer, e.g. roasted carob bean powder. The carob tree, Ceratonia Siliqua is cultivated here in Greece. In a recent patent the use of brewers yeast as a cocoa replacer was claimed. Other roasted products can be used to impart colour but most of these products still lack the flavour.

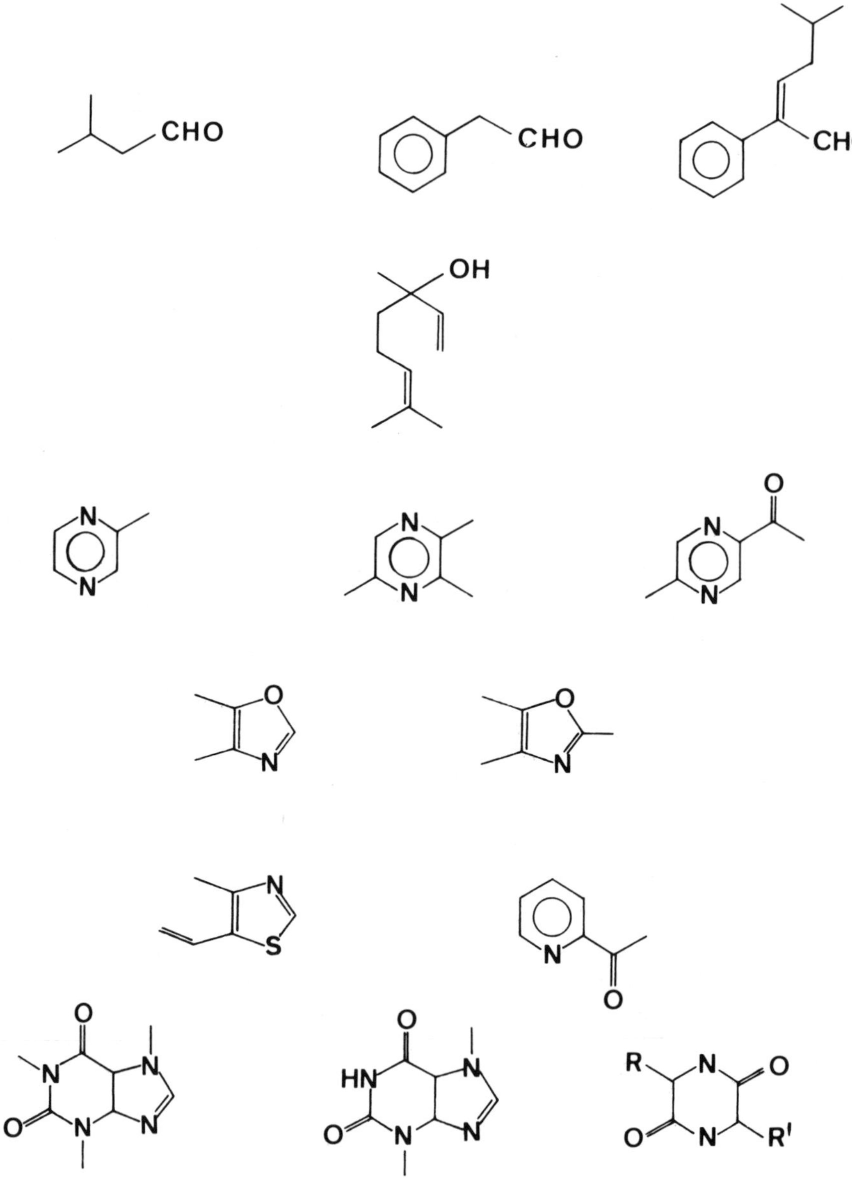

Some Important Flavour Chemicals in Cocoa

FLAVOUR

As we have already seen, the delicious cocoa flavour is a complicated mixture of hundreds of components resulting from biological and technological processes. Enzymatic reactions in the fermentation step and non enzymatic browning reactions in the following roasting process lead to the desired flavour.

The most important flavour components are the aldehydes like isovaleraldehyde, phenylacetaldehyde, their condensation products and linalool for the sweet flowery green cocoa note; low molecular weight pyrazines, oxazoles, acetylpyridin, methylvinylthiazol for the roasted cocoa notes. The purines theobromin and coffein together with diketopiperazines impart the bitter taste and the stimulating effect.

If we want to imitate the cocoa flavour, we have three possibilities:

1) we can produce the flavour by blending pure synthetic chemicals. The lead is given by an analysis of the natural cocoa flavour.

2) we can imitate the roasting process, starting with synthetic or natural precursors like sugars and aminoacids.

3) we can extract the flavours from inexpensive waste materials like cocoa shells.

CONCLUSION

At the end I should like to show you our approach to a chocolate substitute we called it "Chocless". The target was actually to find an application medium in which our flavours could be tested. We improved our approach constantly; out of this work we arrived at the idea:

"more pleasure for less money"

Our recipe is based on that of ordinary milk-chocolate, but we substituted the expensive cocoa-butter and the expensive cocoa powder by two other natural products, a commercially available vegetable fat and roasted St. John's bread flour. A chocolate flavour is added to impart the basic chocolate note. With various flavours like orange, hazelnut etc. and other ingredients like low priced nuts or other texture imparting products we change the top note, the texture and hence the overall perception. To obtain a smooth, softly melting texture and the characteristic "bite" we had to apply a conching process for 2 - 3 hours.

The basic principles of this food can be varied within such a vast limit, that an indefinite number of different end products can be thought of, some of them only faintly reminiscent of traditional chocolate. We believe that we have opened the window for a new food with its own merits. Whether it is considered a substitution of chocolate, is not important at the end. It has many attractive properties, which will help to make it known. Ethical reasons require that it should not be offered as chocolate and most national food laws provide that the consumer must not be deceived.

BIOMOLECULAR STUDIES OF A CLASS OF SWEET COMPOUNDS

Lloyd N. Ferguson
Rose W. Bragg
Yvonne Chow
Susan Howell
Manque Winters

Department of Chemistry
California State University Los Angeles
Los Angeles, California

In spite of the fact that many scientists of various disciplines are studying the sense of taste, we still cannot predict the taste of a given compound or design the structure of a sweet compound with near certainty. We still rely upon a "taste and see" approach in which we synthesize many types or analogs in search of an ideal member.

The fact that the tasting process is quick, reversible, and involves low energies, indicates that the substrate-receptor interaction does not involve covalent bond formation. A prevelant view is that it is a stereospecific, low energy process, probably involving H-bonds, van der Waals, and dispersion forces. Two types of processes of this type are adsorption and charge-transfer complexation. Accordingly, measurements were made to explore the latter possibility.

We used sweet and tasteless isomeric m-nitroanilines or m-nitrophenols as the donor component. These were paired with well known acceptor molecules[1] such as polynitro aromatics, quinones, or molecules containing several CN, COOH, NO, NO_2 or C=O groups. CT-complex formation was measured using colorimetry, uv, ir, or nmr spectrophotometry, or index of

[1]Ferguson, L.N., Organic Molecular Structure, Willard Grant Press, Boston, 1975, p 406.

ISBN 0-12-169060-1

refraction data. From the variety of measurements made, we are satisfied that CT complexation is not the critical process for a sweet taste sensation.

What we would like to have is an experimental set of data which could be used to give a single parameter for predicting the sweet or nonsweet taste of a given molecule. The generally accepted view that is emerging is that the initial event in taste descrimination occurs at the surface of taste receptors probably involving a stereospecific adsorption. With this in mind, a wide variety of electronic and adsorption properties of the 2- and 4-substituted 5-nitroanilines have been measured. Some of the data are given on Tables 1-3. It can be seen there that the sweet isomers are weaker bases, have much smaller differences between ground state (pKa) and excited state (pKa*) basicities, whatever that means, are harder to oxidize, and have larger Rf values on polar as well as nonpolar absorbents, than the nonsweet isomers. Most of the electronic effects can be rationalized in terms of physical-organic concepts.

Several attempts have been made to find a parameter which correlates some property of the hydrophobic portion of these molecules with taste. Deutsch and Hansch[2] used hydrophobic and Hammett substituent constants, Kier[3] used polarizabilities, and we have used basicities and Hammett constants of the substituents. Good correlation equations can be developed but always there are exceptions which implicate steric factors. For this reason, we made extensive adsorption studies in the hope of finding a system which alone, or combined with electronic properties, would distinguish sweet from nonsweet compounds. For example, TLC measurements were made using plates spread with β-cyclodextrin impregnated with selected salts ($MgCl_2$, $MnCl_2$, $ZnCl_2$, KCl, or $AgNO_3$), d- or l- amino acids, aspartame, neohesperidin dihydrochalcone, or with naringen. Water, phosphate buffer, organic liquids, or salt solutions were used as developing solvents. Invariably there are differences between sweet and nonsweet isomers, but there is no consistent pattern with sweetness.

---

[2]Deutsch, L.W. and Hansch, C., Nature <u>211</u>, 75 (1966).
[3]Kier, L.B., J. Pharm. Sciences <u>61</u>, 1394 (1972).

TABLE 1. Basicities of Sweet (2-Y) and Nonsweet (4-Y) 5-Nitroanilines

O_2N–C$_6$H$_2$(NH$_2$)–Y (positions 6, 4)

2-Y	4-Y	RS	pKa	pKa*	pKa-pKa*
F		40	1.09	0.77	0.32
	F		2.36	-11.19	13.55
Cl		375	0.64	0.14	0.50
	Cl		1.93	-9.37	11.30
Br		714	0.52	0.12	0.40
	Br		1.80	-8.65	10.45
CH_3		298	2.30	1.77	0.53
	CH_3		2.86	-8.04	11.90
OCH_3		167	2.49	2.36	0.13
	OCH_3		3.36	-6.29	9.65

TABLE 2. Redox Potentials of Sweet (2-Y) and Nonsweet (4-Y) 5-Nitroanilines

2-Y	4-Y	RS	$E_{1/2}$[a] (NO_2) (mv)	$E_{1/2}$[b] (NH_2) (mv)
F		40		0.040
	F			-.010
Cl		375	0.35	0.030
	Cl			-.013
Br		714	.33	0.020
	Br		.34	-.025
CH_3		298	.41	-.015
	CH_3			-.058
OCH_3		167	.43	
	OCH_3		.42	

[a] The more negative the value, the harder to reduce.[4]
[b] The more negative the value, the easier to oxidize.[5]
[4] Lawrence, A.R. and Ferguson, L.N., Nature 183,1469 (1959).
[5] Morris, J.B., Howard University, unpublished results.

TABLE 3. N-H Chemical Shift and TLC Rf Values for Some 2- and 4-Substituted 5-Nitroanilines

2-Y	4-Y	RS	δ_{NH_2}[c] (ppm)	Rf (talc)	Rf[d] (β-CD)
F		40	5.88	0.57	0.40
	F		5.64	.29	.18
Cl		375	6.07	.66	.51
	Cl		5.89	.34	.22
Br		714	5.98	.72	.60
	Br		5.95	.42	.23
CH_3		298	5.55	.52	.37
	CH_3		5.50	.54	.31
OCH_3		167	5.40		
	OCH_3		5.23		

What is particularly hard to account for among these compounds, in terms of polarization, H-bonding, and the A-H, B hypothesis is the fact that compounds 1 and 2 are sweet whereas 4 and 5 are not.

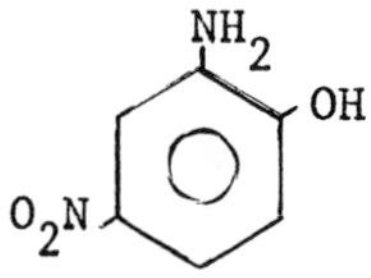

NH_2, OH, O_2N — 1

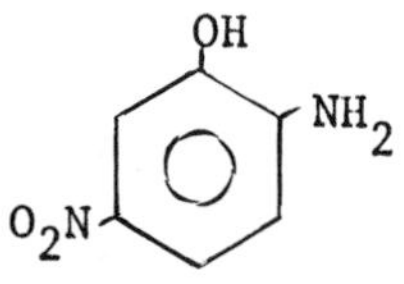

OH, NH_2, O_2N — 2

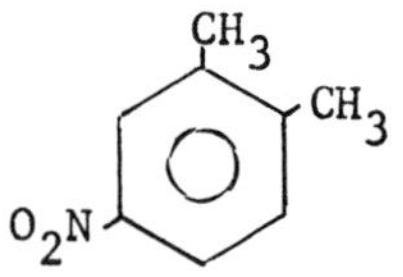

CH_3, CH_3, O_2N — 3

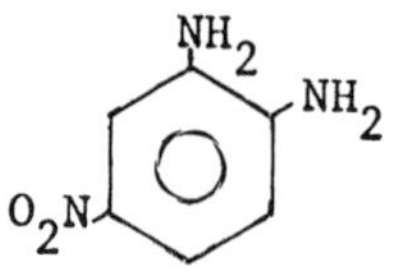

NH_2, NH_2, O_2N — 4

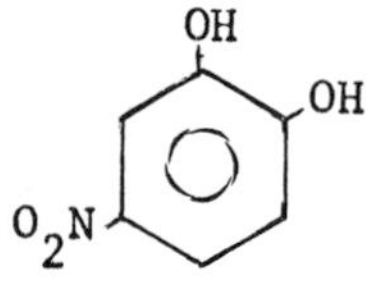

OH, OH, O_2N — 5

[c]nmr chemical shift.
[d]β-cyclodextrin.

The structural differences between 1 and 2 on the one hand and 4 and 5 on the other, are very subtle. The loss of the sweet taste cannot be attributed to the presence of two identical groups because 3 is sweet. It is possible that the 4-H in these 5-nitroanilines and -phenols is not the A-H of the A-H, B system as proposed in the literature. For instance, the sweet compounds 6 and 7 have substituents in that position, although it is easy to pick out a different A-H, B system in their structures. Also, NMR spectra of a series of these 5-nitroanilines indicate that the electron density and therefore the acidity at the 4-position does not parallel their relative sweetness. Thus, the chemical shifts for the H-4 protons listed in Table 4 do not correlate directly with the RS values.

NH_2, O_2N, SO_2NH_2

6

NH_2, O_2N, COOH

7

TABLE 4. Proton Chemical Shifts of 5-Nitroanilines[6]

O_2N, 6, NH_2, X, 4, 3

		Chemical Shifts (ppm)			
X	NH_2	H-3	H-4	H-6	RS
F	5.88	7.19	7.42	7.68	40
OCH_3	5.40	6.97	7.52	7.56	167
CH_3	5.55	7.17	7.34	7.50	298
Cl	6.07	7.51	7.37	7.70	375
Br	5.98	7660	7.25	7.66	714

[6]We are grateful to Dr. Kelvin Shen for assistance in interpreting the spectra.

At this time, we are looking for a difference in H bonding characteristics of compounds 1, 2 and 3 vs compounds 4 and 5 in a system which can serve as proton donor and proton acceptor and compare this with one which only serves as a proton acceptor.

ACKNOWLEDGMENTS

The authors are grateful for partial support from the National Science Foundation and the National Institutes of Health through its Minority Biomedical Research Program.

INTENSE SWEETNESS OF NATURAL ORIGIN

G. E. Inglett

Northern Regional Research Center, Federal Research, Science and Education Administration, U.S. Department of Agriculture, Peoria, Illinois

Many people around the world use intensely sweet materials of natural origin. Plant parts containing intensely sweet principles are used for sweetening foods or they are used for medicinal purposes as a part of some societies' daily lives. In one culture, an intensely sweet leaf is used as a part of a special ceremony. Intense sweetness of natural origin is tasted in these cultures today as it has been done throughout the ages (25).

PHYLLODULCIN

A sweet tea, Amacha, is served at Hanamatsuri, the flower festival celebrating the birth of Buddha. Amacha is the dried leaves of *Hydranga macrophylla* Seringe var. *Thunbergii* Makino. The sweet principle, phyllodulcin, was isolated and the structure was determined by Asahina and Asano (3-5). The absolute configuration of phyllodulcin was shown to be the 3R configuration at the asymmetric center at C(3) by identification of a malic acid from ozonized phyllodulcin (2). Phyllodulcin is the first representative of a natural isocoumarin. More information is needed about the intensity and sweetness quality of phyllodulcin. Some 3,4-dihydroisocoumarins were reported to have a sweet taste (74). Recently, an analog of phyllodulcin, 2-(3-hydroxy-4-methoxyphenyl)-1,3-benzodioxan, was found to be intensely sweet (17).

ISBN 0-12-169060-1

STEVIOSIDE

The sweet herb of Paraguay (Yerba dulce), called variously Caa-ehe, Azuca-caa, Kaa-he-e, and Ca-a-yupe by the Guarani, has long been the source of an intense sweetener. Natives use the leaves of this small shrub to sweeten their bitter drinks. The plant was first given the botanical name *Eupatorium rebaudianum*, but this was later changed to *Stevia rebaudiana* Bertoni; recently, the name *Stevia rebaudiana* (Bert.) Hemsl has appeared. For the sweet, crystalline glycoside that has been extracted from the leaves of *S. rebaudiana*, the name stevioside was adopted by the Union Internationale de Chimie in 1921. Historical accounts of knowledge on stevioside and proposals for cultivation of *S. rebaudiana* for commercial use of stevioside as a sweetener have been reviewed by Bell (8), Fletcher (20), and Nieman (47). Wood *et al.* (73) reported a method for extracting stevioside in 7% yield from air-dried leaves of the Paraguayan plant.

Bridel and Lavieille (10) reported stevioside as a white, crystalline, hygroscopic powder approximately 300 times sweeter than cane sugar. Very small amounts on the tongue gave a delectable sweetness, like the leaves of the plant; however, large amounts tasted sweet at first, then distinctly bitter. On the other hand, no bitter taste was attributed to stevioside by Nieman. By taste panel, Pilgrim and Schutz (48) determined a relative sweetness of 280 for stevioside as against 306 for saccharin (sucrose = 1.00).

Bridel and Lavieille (11) showed that stevioside is rapidly hydrolyzed by enzymatic material extracted from the vineyard snail, *Helix pomatia*, giving rise to 3 moles of D-glucose and 1 mole of a tasteless, acidic aglycone which they named steviol. Acidic hydrolysis gave the same percentage of D-glucose but a different aglycone, named isosteviol. Their work on the constitution of stevioside was repeated, confirmed, and extended by Wood *et al.* (73). The identity and positions of the sugars linked to steviol were fixed by Wood *et al.* (73) and by Vis and Fletcher (67). The absolute configuration of the diterpenoid aglycone was finally resolved by Mosettig *et al.* (45).

The disaccharide of stevioside is sophorose (2-*O*-β-*D*-glucopyranosyl-β-*D*-glucopyranose). It is linked to the tertiary α-hydroxyl at C-13 of steviol, whereas the monosaccharide, β-*D*-glucopyranose, is condensed with the sterically hindered α-carboxyl at C-4. The two sugars are, therefore, appended to the same side of the rigid aglycone at opposite ends. Alkaline splitting of the unusual *D*-glucose-to-carboxyl linkage produced levoglucosan (1,6-anhydro-β-*D*-glucopyranose) and the sophoroside of steviol, which was devoid of sweetness, Wood *et al.* (73), Vis and Fletcher (67).

Pomaret and Lavieille (49) reported that stevioside readily passes through human elimination channels in its original form. It did not appear to be toxic to guinea pigs, rabbits, or chickens. Furthermore, there are no recorded reports of ill effects in Paraguayan users of the leaves of *S. rebaudiana*. Nevertheless, the long-term effects of ingestion of stevioside would have to be investigated carefully before it could be considered for human use as a sweetener in the United States. The diterpenoid aglycone, steviol, has shown specific physiological activity (66) and weak antiandrogenic effects (18). It remains to be proved that stevioside does not split to form *any* steviol in the human digestive tract.

Also from the leaves of *Stevia rebaudiana*, two new sweet glucosides, rebaudiosides A and B, were isolated besides the known glucosides, stevioside and steviolbioside. On the basis of IR, MS, IH, and ^{13}C NMR as well as chemical evidences, the structure of rebaudioside B was assigned as 13-O-[β-glucosyl(1-2)-β-glucosyl(1-3)]-β-glucosyl-steviol and rebaudioside A was formulated as its β-glucosyl ester (36). In an attempt to produce stevioside as a substitute for synthetic sweeteners, *Stevia rebaudiana* is cultivated extensively in Japan.

GLYCYRRHIZIN

Licorice, well-known for centuries and widely used, is obtained from the roots of *Glycyrrhiza glabra*, a small shrub grown and hand-harvested in Europe and

Central Asia. The roots contain from 6 to 14% glycyrrhizin. Licorice is the only botanical possessing significant amounts of glycyrrhizin.

Glycyrrhizic acid exists in licorice root as the calcium-potassium salt in association with other constituents, such as starch, gums, sugars, proteins, asparagine, flavonoids, and resins (46). Although this glycoside is difficult to free of nitrogen, minerals, color, and licorice flavor, it has been isolated in pure form. Glycyrrhizic acid is a glycoside of the triterpene, glycyrrhetic acid, which is condensed with O-β-D-glucuronosyl-(1'→2)-β-D-glucuronic acid. Colorless, crystalline glycyrrhizic acid was first isolated by Tschirch and Cederberg (57). Although the empirical formula and optical inactivity that they reported were incorrect, the other properties given and their isolation procedures have been repeatedly reproduced (58,41,42,71). A "very sweet" taste was ascribed to the free, tribasic acid. It was practically insoluble in cold water and soluble in hot water. The free acid and its ammonium and potassium salts formed pasty gels upon cooling of the warm aqueous solutions; however, they could be recrystallized from glacial acetic acid or alcohol. An "intensely sweet" taste was ascribed to the water-soluble salts. Tschirch and Cederberg (57) hydrolyzed potassium glycyrrhizinate in dilute sulfuric acid to obtain the crystalline aglycone, glycyrrhetic acid. It was insoluble in water and tasteless. Glucuronic acid was identified in the mother liquor by Tschrich and Gauchmann (58).

Voss *et al.* (68) first obtained the correct empirical formula for glycyrrhizin, and Voss and Pfirschke (69) showed that the glucuronic acid found is linked to the hydroxyl group of glycyrrhetic acid as a disaccharide. Lythgoe and Trippett (41) proved the (1'→2)-interglycosidic linkage, also the pyranoside rings of the diglucuronic acid moiety, by periodate oxidation and methylation analysis. From optical rotation data, they could agree with Voss and Pfirschke's supposition that the interglycosidic linkage is beta. Both the interglycosidic linkage and the linkage to the aglycone were shown to be beta by Marsh and Levvy (42). They hydrolyzed pure ammonium glycrrhizinate with a β-glucuronidase isolated from mouse liver. This enzyme was shown to activate only β-glucuronides in pyranoside ring-form. Vovan and Dumazert (72) have confirmed that the only sugar present in pure, crystalline glycyrrhizic acid (71) is glucuronic acid. Voss and Butter (70) showed that the C-20 carboxyl group is not sterically hindered. All three carboxyl groups are readily methylated (12).

The absolute configuration of the aglycone, glycyrrhetic acid, is determined as the result of investigations too numerous to cite completely. Important contributions were made by Ruzicka et al. (50-54), Voss et al. (68,70), and Beaton and Spring (6). Although two isomers (18-α and 18-β) have been isolated, Beaton and Spring have indicated that only the 18-β-glycyrrhetic acid is the natural isomer that occurs in glycyrrhizin.

The aglycone of glycyrrhizin is closely related to β-amyrin, a triterpene; however, it gives medicinal effects similar to those of deoxycorticosterone acetate. Both glycyrrhizin and glycyrrhetic acid, and their respective methyl esters, show a low hemolytic activity in comparison with some other triterpenoid saponins (55).

Countercurrent extraction process, in which the most concentrated liquor passes over fresh root while fresh water scrubs the nearly spent root, yields a variety of licorice products including block licorice, licorice powder, liquid extracts, and glycyrrhizin. Ammoniated glycyrrhizin is manufactured by a special process in which only the glycyrrhizin is removed from the total extract.

The extracts are universally employed in the flavoring and sweetening of pipe, cigarette, and chewing tobaccos (14), and they are regularly used in confectionery manufacture. Some segments of the flavor industry have long utilized these extracts (15) in root beer, chocolate, vanilla, liqueur, and other flavors. Ammonium glycyrrhizin (AG), the fully ammoniated salt of glycyrrhizic acid, is commercially available as a high-purity, spray-dried, brown powder. Further treatment and repeated crystallizations yield the more costly, colorless salt, monoammonium glycyrrhizinate (MAG). Both derivatives have the same degree of sweetness, but they differ markedly from each other in solubility properties and sensitivity to pH. AG is the sweetest substance on the FDA list of natural GRAS flavors. AG is 50 times sweeter than sucrose. In the presence of sucrose, AG is around 100 times sweeter than sucrose alone.

LO HAN FRUIT (MOMORDICA GROSVENORI) SWINGLE

Lo Han Kuo (Lo Han fruit), from *Momordica grosvenori* Swingle, is a dried fruit from Southern China. The fruits are gourd-like, 6-11 cm long by 3-4 cm broad, dark brown, broadly ellipsoid, ovoid, or subglobose, with broadly rounded ends and a very thin rind (0.5-0.8 mm thick). It has 3 double locules (compartments or cavities), each with 2 rows of seeds (about 10-12 in a row). The brownish-gray pulp dries to a light fibrous mass. The seeds are light brownish-gray, flattened (15-18 mm long, 10-12 mm broad, and 3-4 mm at edges) with a depressed area in the center of each side. Characterization of the tissue, as well as the taxonomic status of the plant, was described by Swingle (56) who reported that the species was introduced to the United States through the Division of Plant Exploration and Introduction, Bureau of Plant Industry. Swingle also reported that 1000 tons of the green fruits were delivered every year to the drying sheds at Kweilin (Kwangsi Province). The fruits lost much weight in drying and were then carefully packed in boxes and shipped to Canton, but large amounts were also exported to Chinese communities outside China. The dried fruit is a valued folk medicine used for colds, sore throats, and minor stomach and intestinal troubles. Lee (39) found that the sweet principle could be extracted by water from either the fibrous pulps or from the thin rinds of Lo Han Kuo; 50% ethanol was also found to be a good extractant. Rinds afforded a more easily purified extract. Sweetness of Lo Han sweetener was accompanied by a lingering taste described as licorice-like, somewhat similar to that of stevioside, glycyrrhizin, and the dihydrochalcones. Structural studies indicate the sweetener to be a triterpenoid glycoside with 5 or 6 glucose units (40). The purified sweetener has a more pleasant sweet taste than the impure material. The purest sample is about 400 times sweeter than sucrose.

OSLADIN

The sweet taste of rhizomes of the widely distributed fern, *Polypodium vulgare* L., has attracted the interest of many chemists and pharmacists. Van der Vijver and Uffelie (59) and others have shown that the sweet substance is not glycyrrhizin, as was once proposed. Many constituents of the rhizomes have been isolated, but the substance that resembles saccharin in sweetness was isolated and identified only recently (34,32). The name osladin is based on the Czech name for polypody, osladic. Osladin comprised only 0.03% of the dry weight of the rhizomes. Its chemical structure was revealed as a bis-glycoside of a new type of steriodal saponin.

The glycoside that results by replacement of the monosaccharide radical with hydrogen was isolated separately and named polypodosaponin. Its absolute configuration was determined by Jizba *et al*. (33). They made no comment on its taste.

The disaccharide of osladin was shown to be neohesperidose, 2-*O*-α-L-rhamnopyranosyl-β-D-glucopyranose. The glycosidic linkage was shown to be beta by cleavage of neohesperidin, with a specific β-glucosidase from *Aspergillus wentii*. Therefore, neohesperidose is in the same glycosidic configuration in osladin as is found in the intensely sweet neohesperidin dihydrochalcone. Neohesperidose is only slightly sweet, if at all (35,22,23). The configuration of the glycosidic linkage of the monosaccharide, L-rhamnose, has not been determined. It is probably α-L- because this corresponds to the β-D-linkage usually found in natural glycosides. The molecular structure of osladin resembles that of stevioside; it shows a (1'→2)-linked disaccharide at one end of the aglycone and a monosaccharide at the other.

The C-26-*O*-methyl polypodosaponin shows hemolytic activity and a strong inhibition of fungal growth (33). Even without regard to biological activity, the very low concentration of osladin found in polypody rhizomes dims the prospects for developing osladin as a commercial sweetener.

MIRACLE FRUIT (*SYNSEPALUM DULCIFICUM*)

An important approach to sweet taste perception is the study of the strange properties of the miracle fruit (*Synsepalum dulcificum*). Although the miracle fruit's capacity to cause sour foods to taste sweet has been known in the literature since 1852 (16), scientific investigations of the fruit were not made until Inglett and his associates found some experimental evidence that the active principle was macromolecular (28). This berry posseses a taste-modifying substance that causes sour foods such as lemons, limes, grapefruit, rhubarb, and strawberries to taste delightfully sweet. Even dilute organic and mineral acids will induce a sweet taste. The berries are chewed by West Africans for their sweetening effect of some sour foods. The quality of the sweetness induced by this taste modifier is very similar to sucrose.

Preliminary studies on miracle fruit by Inglett *et al.* (28) were intended to isolate, characterize, and synthesize the taste-modifying substance. The fruit was processed to give a stable concentrate. Since the active principle appeared polymeric, the material was incompatible with the sponsoring organization's mission, so research was discontinued.

Subsequently, the taste-modifying principle was independently isolated by two different research groups in 1968 (37,13). Kurihara and Beidler (37) separated the active principle from the fruit's pulp with a carbonate buffer (pH 10.5). Destroying the active principle by trypsin and pronase suggested its proteinaceous character. The taste-modifying protein was also separated from the fruit's pulp with highly basic compounds, salmine and spermine (13), and further purified by gel filtration. The taste-modifying protein was called "miraculin" by the Netherland researchers.

Amino acid composition of this taste-modifying protein is listed as amino acid residue per 100 total residues: lysine, 7.9; histidine, 1.8; arginine, 4.7; aspartic acid, 11.3; theonine, 6.1; serine, 6.1; glutamic acid, 9.2; proline, 6.0; glycine, 9.8; alanine, 6.3; half cystine, 2.3; valine, 8.0; methionine, 1.0; isoleucine, 4.7; leucine, 6.5; tyrosine, 3.6; and phenylalanine, 5.0. The protein has 6.7% sugar components, identified as L-arabinose and D-xylose. The basic glycoprotein has a molecular weight between 42,000 and 44,000.

The purified glycoprotein has no inherent taste. Sweetening of acid taste was observed at 5 X 10^{-8}M concentration of the glycoprotein solution and reached a maximum at 4 X 10^{-7}M. The sweetening effect at 10^{-7}M glycoprotein concentration slowly declines over a period up to 2 hours. The intrinsic activity of this glycoprotein, with the greatest effective affinity known for any sweetener, is slightly less than that of sucrose.

Taste-modifying activity is lost when amino acids and possible imidazoles are modified (7). Modification of sulfhydryl and carboxylic acid groups, as well as tryptophan and tyrosine residues, does not lose more than 25% of the normal activity. Not only does the 30 seconds or more latency of sweetening action need additional investigation, but also how to increase intrinsic activity.

Two theories have been advanced to explain the taste-modifying activity. Dzendolet (19) suggested that anions of some acids, such as citric acid, are sweet but are inhibited by the sour taste. The taste-modifying glycoprotein, by blocking the sour receptor sites, allows the sweet taste of the anion to be perceived. Kurihara and Beidler (38) proposed that the glycoprotein binds to the receptor membrane near the sweet receptor site, so that it fits the sugar groups attached to the glycoprotein when its structure is modified by acid to produce a sweet taste.

With increasing pressure for an excellent nonnutritive sweetener in recent years, a corporation launched a venture on dietetic foods using miracle fruit concentrate as its source of latent sweetness. The FDA denied a petition for affirmation that miracle fruit and its extracts and concentrates are generally recognized as safe (1). They stated also that the substances cannot be approved as food additives because there is insufficient evidence of acceptable toxicological studies, insufficient history of consumption of the substances in the U.S. to demonstrate that no known hazards exist, and insufficient data to establish the conditions of safe use as food additives.

SERENDIPITY BERRIES (*DIOSCOREOPHYLLUM CUMMINSII* DIELS)

While studying various natural sweeteners previously mentioned, the author discovered the intense sweetness of some red berries from West Africa. The fruit was called the serendipity berry, and its botanical name, *Dioscoreophyllum cumminsii* was established many months later (29,31,27).

Serendipity berries are indigenous to tropical West Africa. The *Dioscoreophyllum cumminsii* Diels plant grows from Guinea to the Cameroons and is also found in Central Africa. It grows in the rain forest during the rainy season from approximately July to October. The serendipity berries are borne by hairy climbing vines sometimes 15 ft long and 1/8 to 3/16 in. in diameter. The berries are red in color, approximately 1/2 in. long, and grow in grapelike clusters with approximately 50 to 100 berries in each bunch. The tough outer skin of the berry encloses a white, semisolid, mucilaginous material surrounding a friable thorny seed. In spite of its intense sweetness, the fruit is not commonly cultivated or used by Nigerians (30).

Researchers at the Monell Senses Center and the Unilever Research Laboratorium, working independently, confirmed the protein nature of the serendipity sweetener (43,60,63,44). The serendipity berry sweetener has been called monellin (43) and serendip (26). Amino acid composition of monellin was determined by Van der Wel and Loeve (63) and Morris *et al.* (44). The most outstanding observation is the complete absence of histidine. The absence of mono- or dimethyl derivatives of lysine or arginine was also concluded by both groups. Monellin is composed of two dissimilar polypeptide chains with known amino acid sequences that are noncovalently associated (9,24).

The sweetness of monellin is approximately 2500 times sweeter than sucrose on a weight basis. Its isoelectric point is between 9.03 and 9.26 as determined by preparative isoelectric focusing (Table I). The molecular weight of the sweetener is 11,000. Sweetness is lost on heating the protein at 50°C (pH 3.2), 65°C (pH 5.0), and 55°C (pH 7.2).

TABLE I. Some Physical Properties of Monellin

Criteria	Monellin[a]	Monellin[b]
Isoelectric point	9.03	9.26
Molecular weight	11,500	10,500
$A_{1\,cm}^{1\%}$ (pH 5.6, 278 nm)	16.2	---
Sweetness intensity (times sweeter than sucrose)		
On a molar basis	8.4×10^4	---
On a weight basis	2500	---
Temperature (°C) above which sweetness disappears		
at pH 3.2	50	---
5.0	65	---
7.2	55	---

[a]Van der Wel (61).
[b]Morris *et al.* (44).

KATEMFE (*THAUMATOCOCCUS DANIELLI*)

Besides studies on miracle fruit and the serendipity berry, a large variety of plant materials were examined systematically by Inglett and May (30) for intensity and quality of sweetness. Another African fruit containing an intense sweetener was katemfe, or the miraculous fruit of the Sudan. Botanically the plant is *Thaumatococcus danielli* of the family Marantaceae. Inside the fruit, three large black seeds are surrounded by a transparent jelly and a light yellow aril at the base of each seed. The mucilaginous material around the seeds is intensely sweet and causes other foods to taste sweet. The seeds were observed to be present in trading canoes in West Africa as early as 1839, and were reported to be used by the native tribes to sweeten bread, fruits, palm wine, and tea. Preliminary studies have indicated a substance similar to the serendipity berry sweetener (30). Katemfe yields two sweet-tasting proteins (61,62) which they called thaumatin I and II.

Physical constants for the sweet proteins of katemfe are given in Table II. Like serendip, these protein sweeteners are heat sensitive and undergo irreversible heat denaturation. Because denaturation coincides with the loss of sweetness, the groups underlying the conformational change must also be responsible for generating their sweet taste. At least part of the intact tertiary protein structure must be required for the sweet taste.

A process for extraction of the thaumatins from the fruit was reported (21) and commercial interest in this sweetener is developing. The purified sweetener is 1600 times sweeter than sucrose on a weight basis.

Thaumatin I contains 193 amino acids (61) polyacrylamide gel electrophoresis in the presence of sodium dodecyl sulfate indicated that the protein is a single polypeptide chain with alanine as the N-terminal amino acid (62). Thaumatin I was crystallized and physical characteristics and diffraction data of the crystals were obtained (65).

TABLE II. Physical Properties of the Sweet-Tasting Proteins from Katemfe[a]

Criteria	Thaumatin I	Thaumatin II
Isoelectric point	12	12
Molecular weight	21,000 ± 600	20,400 ± 600
$A^{1\%}_{1cm}$ (pH 5-6, 278 nm)	7·69	7·53
Sweetness intensity (times sweeter than sucrose):		
on a molar basis	1×10^5	1×10^5
on a weight basis	1600	1600
Temperature (°C) above which sweetness disappears at:		
pH 3·2	55	55
5·0	75	75
7·2	65	65

[a]Source: Van der Wel (1973).

Sweetness of acetylated thaumatins decreased with the increasing number of acetylated amino groups. The sweet taste disappeared completely when four amino groups were acetylated. Methylated thaumatin with seven modified lysine residues had a sweetness intensity practically equal to that of the original thaumatin (64).

REFERENCES

1. Anon. Miracle fruit petition denied. Federal Register, May 24, 1977.
2. Arakawa, H., and Nakazaki, M. Chem. Ind. (London) 671 (1959).
3. Asahina, Y., and Asano, J. Ber. Dtsch. Chem. Ges. 62:171-177 (German) (1929).
4. Asahina, Y., and Asano, J. Ber. Dtsch. Chem. Ges. 63:429-437 (German) (1930).
5. Asahina, Y., and Asano, J. Ber. Dtsch. Chem. Ges. 64:1252 (German) (1931).
6. Beaton, J. M., and Spring, F. S. J. Chem. Soc. 3126-3129 (1955).
7. Beidler, L. M. Biophysics of sweetness, in "Symposium: Sweeteners" (G. E. Inglett, ed.), Avi Publishing Co., Inc., Westport, Conn., pp. 10-22, 1974.
8. Bell, F. Chem. Ind. (London) 897-898 (July 17, 1954).
9. Bohak, Z., and Li, S-L. Biochim. Biophys. Acta 427:153-170 (1976).
10. Bridel, M., and Lavieille, R. Compt. Rend. 192:1123-1125; J. Pharm. Chim. 14(3):99-113; 14(4):154-161 (French) (1931A).
11. Bridel, M., and Lavieille, R. Compt. Rend. 193:72-74 (1931); Bull Soc. Chim. Biol. 13:636-655 (French) (1931B).
12. Brieskorn, C. H., and Sax, H. Arch. Pharm. (Weinheim, Ger.) 303:905-912 (German) (1970).
13. Brouwer, J. N., van der Wel, H., Francke, A., and Henning, G. J. Nature (London) 220:373-374 (1968).
14. Cook, M. K. Flavour Ind. 1(12):831-832 (1970).
15. Cook, M. K. Flavour Ind. 2(3):155-156 (1971).

16. Daniell, W. F. Pharm. J. 11:445-448 (1852).
17. Dick, W. E., Jr., and Hodge, J. E. J. Agric. Food Chem. (in press) (1978).
18. Dorfman, R. I., and Nes, W. R. Endocrinology 67:282-285 (1960).
19. Dzendolet, E. Percept. Psychophys. 6:187-188 (1969).
20. Fletcher, H. G., Jr. Chem. Dig. 14(7):18 (July-August, 1955).
21. Higginbotham, J. D. U.S. Pat. 4,011,206. March 8, 1977.
22. Horowitz, R. M., and Gentili, B. U.S. Pat. 3,087,821 (April 30, 1963).
23. Horowitz, R. M., and Gentili, B. J. Agric. Food Chem. 17:696-700 (1969).
24. Hudson, G., and Biemann, K. Biochim. Biophys. Res. Commun. 71:212-220 (1976).
25. Inglett, G. E. J. Toxicol. Environ. Health. 2:207-214 (1976).
26. Inglett, G. E. Protein Sweeteners, in "The Chemistry and Biochemistry of Plant Proteins," (J. B. Harborne, C. F. Van Sumere, and J. G. Vaughn, eds.), Academic Press, Inc., Limited, London (1975).
27. Inglett, G. E. Sweeteners: New Challenges and Concepts, in "Symposium: Sweeteners," Avi Publishing, Westport, Conn. (1974).
28. Inglett, G. E., Dowling, B., Albrecht, J. J., and Hoglan, F. A. J. Agric. Food Chem. 13:284-287 (1965).
29. Inglett, G. E., and Findlay, J. C. Serendipity berry--source of a new macromolecular sweetener (Abstr. Pap. 75A presented to the Division of Agriculture and Food Chemistry, 154th Am. Chem. Soc. Meeting, Chicago, Ill.) (1967).
30. Inglett, G. E., and May, J. F. Econ. Bot. 22:326-331 (1968).
31. Inglett, G. E., and May, J. F. J. Food Res. 34:408-411 (1969).
32. Jizba, J., Dolejs, L., Herout, V., and Sorm, F. Tetrahedron Lett. No. 18, 1329-1332 (1971A).
33. Jizba, J., _et al_. Chem. Ber. 104:837-846 (1971B).

34. Jizba, J., and Herout, V. Collect. Czech. Chem. Commun. 32:2867-2874 (English) (1967).
35. Koeppen, B. H. Tetrahedron 24:4963-4966 (1968).
36. Kohda, H., et al. Phytochemistry 15:981-983 (1976).
37. Kurihara, K., and Beidler, L. M. Science 161:1241-1243 (1968).
38. Kurihara, K., and Beidler, L. M. Nature (London) 222:1176-1179 (1969).
39. Lee, C. H. Experientia 31(5):533-534 (1975).
40. Lee, C. H. Personal communication. General Foods Corporation, Tarrytown, N.Y. (1977).
41. Lythgoe, B., and Trippett, S. J. Chem. Soc. 1983-1990 (1950).
42. Marsh, C. A., and Levvy, G. A. Biochem. J. 63:9-14 (1956).
43. Morris, J. A., and Cagan, R. H. Biochim. Biophys. Acta 261:114-122 (1972).
44. Morris, J. A., Martenson, R., Deibler, G., and Cagan, R. H. J. Biol. Chem. 248:534-539 (1973).
45. Mosettig, E., et al. J. Am. Chem. Soc. 85:2305-2309 (1963).
46. Nieman, C. Licorice. Advances in Food Research, Vol. 7. Academic Press, New York (1957).
47. Nieman, C. Zucker. Suesswaren-Wirtsch. 11:124-126, 236-238 (German) (1958).
48. Pilgrim, F. J., and Schutz, H. G. Cited by A. R. Lawrence and L. N. Ferguson, Exploratory physiochemical studies on the sense of taste. Nature (London) 183:1469-1471 (1959).
49. Pomaret, M., and Lavieille, R. Bull Soc. Chim. Biol. 13:1248-1252 (French) (1931).
50. Ruzicka, L., Furter, M., and Leuenberger, H. Helv. Chim. Acta 20:312-325 (German) (1937A).
51. Ruzicka, L., Jeger, O., and Ingold, W. Helv. Chim. Acta 26:2278-2282 (German) (1943).
52. Ruzicka, L., and Leuenberger, H. Helv. Chim. Acta 19:1402-1406 (German) (1936).
53. Ruzicka, L., Leuenberger, H., and Schellenberg, H. Helv. Chim. Acta 20:1271-1282 (German) (1937B).
54. Ruzicka, L., and Marxer, A. Helv. Chim. Acta 22:195-201 (German) (1939).

55. Schloesser, E., and Wulff, G. Z. Naturforsch. Teil B 24:1284-1290 (German) (1969).
56. Swingle, W. T. J. Arnold Arbor. Harv. Univ. 22:198 (1941).
57. Tschirch, A., and Cederberg, H. Arch. der Pharm. 245:97-111; Chem. Zentr. 1907(I):1799-1800 (German) (1907).
58. Tschirch, A., and Gauchmann, S. Arch. der Pharm. 246:545-558; Chem. Zentr. 1908(II):1604-1605 (German) (1908).
59. van der Vijver, L. M., and Uffelie, O. F. Pharm. Weekbl. 101:1137-1139; Chem. Abstr. 66:52936 (English) (1966).
60. van der Wel, H. FEBS Lett. 21:88-90 (1972).
61. van der Wel, H. Katemfe, serendipity berry, miracle fruit, in "Symposium: Sweeteners" Inglett (ed.), Avi Publishing Co., Westport, Conn. (1974).
62. van der Wel, H., and Loeve, K. Eur. J. Biochem. 31:221-225 (1972).
63. van der Wel, H., and Loeve, K. FEBS Lett. 29:181-184 (1973).
64. van der Wel, H., and Bel, W. J. Chem. Senses and Flavor 2:211-218 (1976).
65. van der Wel, H., Soest, Van T. C., and Royers, E. C. FEBS Lett. 56(2):316-317 (1975).
66. Vignais, P. V., Duee, E. D., Vignais, P. M., and Huet, J. Biochim. Biophys. Acta 118:465-483 (1966).
67. Vis, E., and Fletcher, H. G., Jr. J. Am. Chem. Soc. 78:4709-4710 (1956).
68. Voss, W., Klein, P., and Sauer, H. Ber. 70B:1212-1218 (German) (1937A).
69. Voss, W., and Pfirschke, J. Ber. 70B:132-137 (German) (1937B).
70. Voss, W., and Butter, G. Ber. 70B:1212-1218 (German) (1937C).
71. Vovan, L., and Dumazert, C. Bull. Soc. Pharm. Marseille 19:41-44 (French) (1970A).
72. Vovan, L., and Dumazert, C. Bull. Soc. Pharm. Marseille 19:45-50 (French) (1970B).
73. Wood, H. B., Allerton, R., Diehl, H. W., and Fletcher, H. G., Jr. J. Org. Chem. 20:875-883 (1955).
74. Yamato, M., et. al. J. Pharm. Soc. Jpn. (Yakagaku Zasshi) 92, I. 367-370, II. 535-538; III. 850-853 (Japanese) (1972).

FLAVOR INTERACTIONS WITH FOOD INGREDIENTS FROM HEADSPACE ANALYSIS MEASUREMENTS

Fouad Z. Saleeb and John G. Pickup

General Foods Corporation

Quantitative headspace analysis of volatile organic compounds in equilibrium with food ingredients can be used as a valuable tool for investigating the nature of flavor interactions with these systems. The variation of headspace concentration with temperature of a given volatile component is governed by the nature and energy of interaction of that particular component with the substrate. Such pressure (concentration) - temperature measurements can be readily used to determine whether these volatiles are present simply as solutes absorbed in liquid and solid substrates, or are held by adsorption forces.

The above technique was used to investigate the nature of interaction of pyridine, α-pinene, ethyl acetate and methyl ethyl ketone with corn oil, pectin, Capsul, hydroxypropyl cellulose and polyvinyl pyrrolidone.

The data showed that these volatile compounds are physically adsorbed on the solid substrates.

ISBN 0-12-169060-1

The calculated isosteric heats of adsorption are always higher than the corresponding heat of liquefaction of these volatile compounds.

I. INTRODUCTION

Natural and fabricated foods are complex and usually non-uniform systems. The perceived flavor of these foods is the result, in practically every case, of the combined sensation caused by numerous minor and major classes of flavor compounds. When a given food is kept in a closed container, the volatile components generally approach an equilibrium distribution between the food substrates and the headspace above it. However, it is often found that the reported headspace composition (1-5) bears little or no relationship to the actual volatile composition in the bulk of the food system. The reason for this apparent discrepancy is that flavor components interact in various ways and to different degrees with the major and minor food ingredients (polysaccharides, proteins, fats, etc.). For a food system in an equilibrium state, the concentration of a given volatile component in the headspace is not only a function of its concentration in the food system, at a given temperature, but it is also determined by the nature and energies of the interactions of the component with the other food ingredients (6,7). Such flavor interactions would affect the perception of foods when sniffed at different temperatures or upon ingestion. During food consumption the flavor perceived is significantly dependent on the extent and rate of release of flavors in a non-equilibrium system (8). Maier (9) has reported on the nature of binding of volatile aroma components to various foodstuffs. Very recently Ahmed et al (10) found that the flavor threshold of d-limonene is significantly changed when determined in aqueous solutions of the non-volatile citrus juice constituents (acids, sugars, pectin). Interactions of the flavor with these constituents is responsible for the modification of the threshold level. Quantitative data on the energies of adsorption of several adsorbate molecules on to lactose were provided by Nickerson (11). Gas chromatographic-retention time data were used to calculate heats of adsorption in the temperature range 100-120°C. In the field of surface chemistry, it is well established that isosteric heats of adsorption can be readily and accurately estimated from equilibrium gas adsorption measurements at more than one temperature.

This study is an introductory part of studies designed to provide information on the nature of interactions of typical volatile flavor components with dry food substrates. From equilibrium head-space analysis at different temperatures, the energies of interaction (sorption) of the volatile compounds were calculated. The method requires no knowledge of surface area or total volatile content in the food system. However, a knowledge of the saturated vapor pressure of each volatile organic component at any temperature, is helpful to avoid confusion between heats of adsorption and latent heats of liquefaction when multilayer adsorption is taking place in the system.

II. EXPERIMENTAL

A. Materials

The following substrates were used as typical food ingredients. Corn oil from CPC International, Engelwood Cliffs, N.J.; Klucel GF, hydroxypropyl cellulose, from Hercules Inc., Wilmington, Del.; pectin, a rapid set pectin from General Foods Corp., White Plains, N.Y.; Capsul, a modified food starch from National Starch & Chemical Corp., Plainfield, N.J.; and polyvinyl pyrrolidone (PVP), molecular weight 10,000, from Aldrich Chemical Co., Milwaukee, Wis. The volatile organic compounds, α-pinene, pyridine, ethyl acetate, and methyl ethyl ketone were used as received.

B. Methods

The volatile components were added directly, as liquids, to corn oil at levels of 0.2 to 0.5%. After thorough mixing in a closed bottle, 2 ml aliquots of the flavored oil were transferred to 250 ml bottles equipped with teflon speta for GC analysis. The contents of the bottles were allowed to equilibrate for 24 hours at different temperatures ranging from 0 to 50 $^{\circ}$C. At the end of this period each bottle was analyzed for the equilibrium vapor pressure concentration of the volatile component in the headspace.

The solid substrates were allowed to equilibrate with the volatile components by adsorption from the gaseous phase. This was achieved by placing a known weight of the solid in close proximity to a given weight of the volatile liquid compound (0.2 to 0.5% by weight of solid) in a small covered desiccator. After the complete transfer of the volatile compound (usually overnight), four grams of the flavored solid were transferred to a 250 ml stoppered bottle and analyzed in the same way as the corn oil bottle. An alternative procedure was to use a microsyringe to add the volatile compound (as liquid) directly to the stoppered bottles containing the dry solid. Both methods of volatile addition produced the same results.

Headspace analyses were carried out using a Hewlett Packard Model 5730A gas chromatograph equipped with a flame ionization detector and a HP3380A integrator. A 12 ft. by 1/8 in. glass column packed with 10% SP-1000 on 100/120 Chromosorb WAW was used. The GC was also used to measure the saturated vapor pressure (p_o) of the volatile compounds at 20°C. These measurements were used to determine the relative vapor pressure of the volatile components after equilibration with the solid substrates. In addition knowledge of the saturated vapor pressures was helpful in order to take precautions to prevent volatiles from condensing in the sampling syringes, especially when analyzing samples equilibrated at temperatures above ambient.

III. RESULTS AND DISCUSSION

A. Heats of Sorption, Theory and Calculations

When a solid or liquid substrate (e.g., a food) is allowed to come in contact with a vapor, the concentration of the gas molecules is always found to be greater in the immediate vicinity of the surface than in the free gas phase, regardless of the nature of the gas or surface. The atoms at the surface of any solid or liquid are subject to unbalanced forces of attraction normal to the surface plane, the balance of forces being partially restored by the adsorption of gas molecules.

The diffusion coefficients of the volatile molecules into most solid substrates are often very small at ambient temperature, and adsorption occurs mainly at the outer surface layer of these solids. However, if the vapor molecules diffuse (penetrate) readily into the bulk structure of the substrate (solid or liquid), then we are dealing with an absorption process. Absorption is definitely the case with volatile compounds when retained by liquid substrates such as vegetable oils. The term sorption is usually applied to cases in which both the above processes may be occurring simultaneously.

Measurements of equilibrium vapor pressure (p) of the volatile components as a function of the food system temperature (T in $^{\circ}$K) provide valuable information on the nature of interaction of these volatiles with the food substrate. Plots of log p vs. 1/T for a given fixed volatile content will yield usually a straight line, provided the heat of adsorption does not vary significantly over the temperature range studied. The heats of adsorption, at a given amount adsorbed, may be calculated from the slopes of the isosteres, using the Clausius-Clapeyron equation in the form

$$\frac{d\ln p}{d(1/T)} = -\frac{q_{st}}{R} \qquad (1)$$

Where R is the gas constant, and q_{st} is the isosteric heat of adsorption, a differential quantity which varies with the degree of surface coverage and hence with the amount of volatiles in the system. If adsorption is not taking place in the food system, the slope of the linear plot of ln p vs. 1/T gives the latent heat of vaporization of that particular volatile component, provided its activity is reasonably constant over the given temperature range. The isosteric heats of adsorption are generally higher than the latent heats of condensation but may approach the latter when multilayer adsorption is taking place.

B. Determination of Heats of Sorption

Figures 1-3 give some typical data for the variation of the equilibrium vapor pressure, as a function of temperature, of pyridine, α-pinene and ethyl acetate when sorbed onto Capsul, PVP, Klucel, pectin and corn oil. In these graphs the equilibrium vapor pressure (p), at any absolute temperature T, is presented as GC counts/1 cc of the headspace. In practically all cases studied (cf Figs.

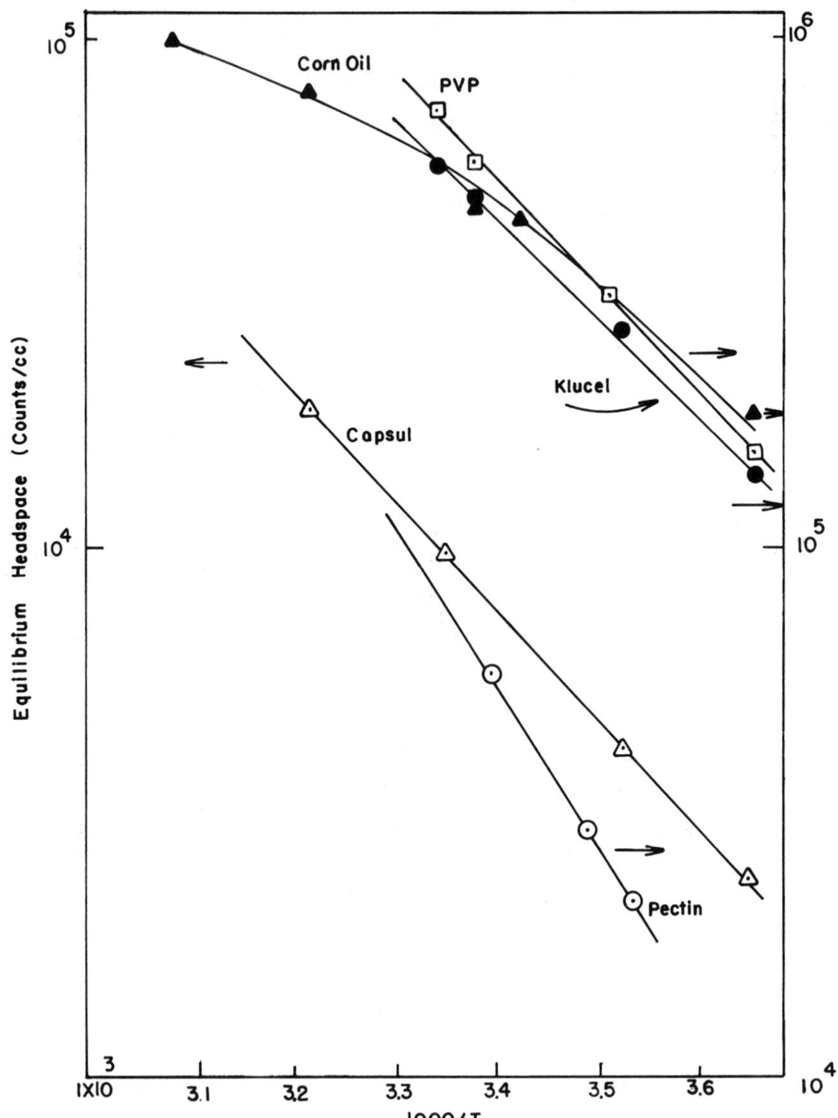

Fig. 1. Effect of temperature on equilibrium headspace concentration of ethyl acetate.

1-3), the experimental points for each system fall on a straight line. The slopes of these linear plots were used to calculate the heats of sorption using equation 1.

Tables I to IV give the calculated heats of sorption of ethyl acetate, pyridine, α-pinene and methyl ethyl ketone on the five substrates used in this study. Included also in the tables are the latent heats of liquefaction of the particular volatile compounds (12). The relative vapor pressures (p/p_o) at 20°C are also provided for all the systems studied.

1. Volatile Interactions with Liquid Substrates

The figures and the tables show that when oil is the substrate, the calculated heat of sorption is, in some cases, equivalent to the published latent heat of condensation of that volatile component (12). The vapor pressure of the volatile compound over the oil, at any temperature (t), is expressed by Henry's Law as follows:

$$p = p_o \gamma N \qquad (2)$$

where γ is the activity coefficient of the solute in oil, p_o is the saturated vapor pressure of the pure adsorbate at temperature T and N is the mole fraction of the solute in the oil (i.e., number of molecules of solute over the total number of moles of oil and solute). Since N is practically constant in the present system (i.e., the amount of solute in the headspace is a very small fraction of its total concentration), p is directly proportional to p_o, at any given temperature, provided that γ

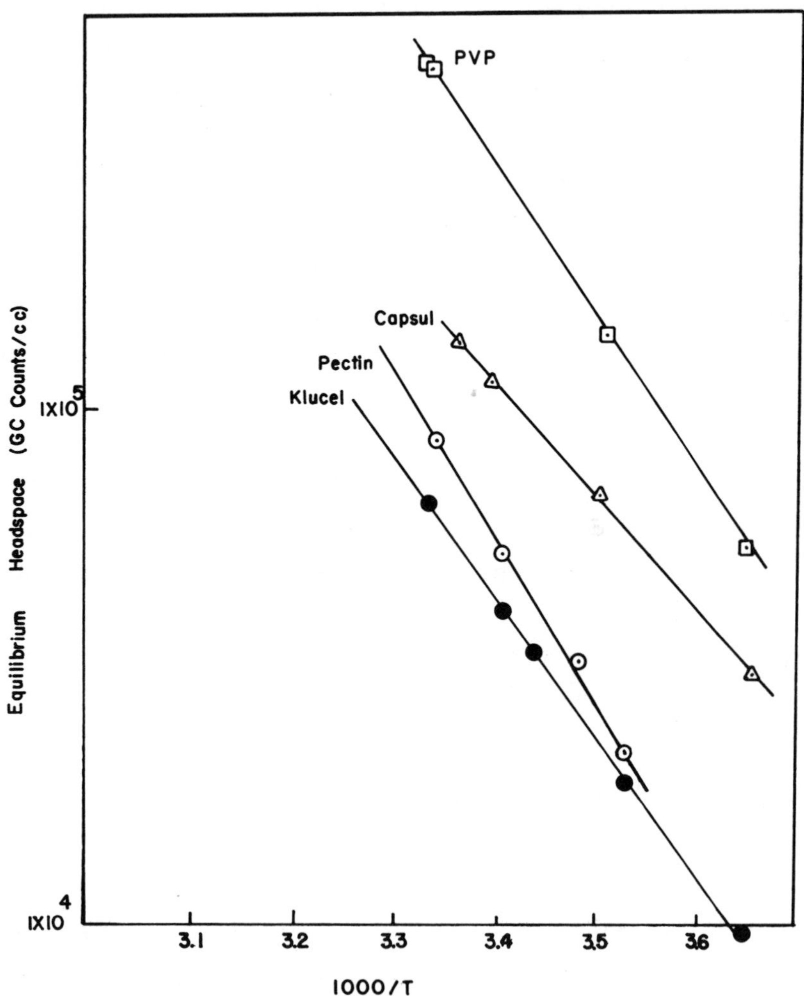

Fig. 2. Effect of temperature on equilibrium headspace concentration of pyridine.

remains constant over the temperature range studied. If that assumption is true then equation 1 should provide the latent heat of condensation of the volatile solute. However, the constancy of γ over the present temperature range, 0 to 50°C, is doubtful, especially since practically all binary liquid systems are non-ideal.

An increase in temperature almost always has the effect of making any system approach more closely ideal behavior, whereas deminution of temperature increases the deviations whether positive or negative. Changes in the slopes of the curves when plotting vapor pressure against temperature is very evident with ethyl acetate (Fig. 1) and to a lesser extent with pyridine. In both cases the calculated heats of sorption, at any point on the curves, are always equal to or less than the corresponding latent heat of vaporization. Table III and Fig. 3 show that when α-pinene is the volatile component, good agreement between the theoretical and experimental value of heat of vaporization is evident.

2. <u>Volatile Interactions with Dry Solid Substrates</u>

The calculated heats of sorption of the four organic volatiles on the four solids (Capsul, PVP, Klucel and pectin) indicate that all these volatiles are physically adsorbed on the particular solids, i.e., the adsorption is reversible.

The magnitude of the calculated isosteric heats of adsorption agrees with reported data on adsorption of gases on a variety of solids (13). The heats of physical adsorption are generally larger but of the same order of magnitude as the latent heat of liquefaction of the adsorbate, and

Table I

HEATS OF SORPTION OF ETHYL ACETATE ($-\Delta Hv$ = 8.30 Kcal/mole)

Sorbent	Experimental Heats (Kcal/mole)	Relative Vapor Pressure (p/p_o) at 20°C
Corn Oil	⩽ 8.30	-
Capsul	10.04	0.0050
PVP	9.73	0.0364
Klucel	9.04	0.0303
Pectin	14.77	0.0036

Table II

HEATS OF SORPTION OF PYRIDINE ($-\Delta Hv$ = 9.65 Kcal/mole)

Corn Oil	⩽ 9.65	-
Capsul	9.72	0.0192
PVP	13.18	0.0498
Klucel	12.20	0.0074
Pectin	14.41	0.0096

Table III

HEATS OF SORPTION OF α-PINENE ($-\Delta Hv$ = 9.81 Kcal/mole)

Corn Oil	9.89	-
Capsul	15.20	0.0095
PVP	15.25	0.0051
Klucel	10.22	0.1450
Pectin	14.04	0.0041

Table IV

HEATS OF SORPTION OF METHYL ETHYL KETONE ($-\Delta Hv$=8.15 Kcal/mole)

PVP	8.70	0.0276
Klucel	8.91	0.0213

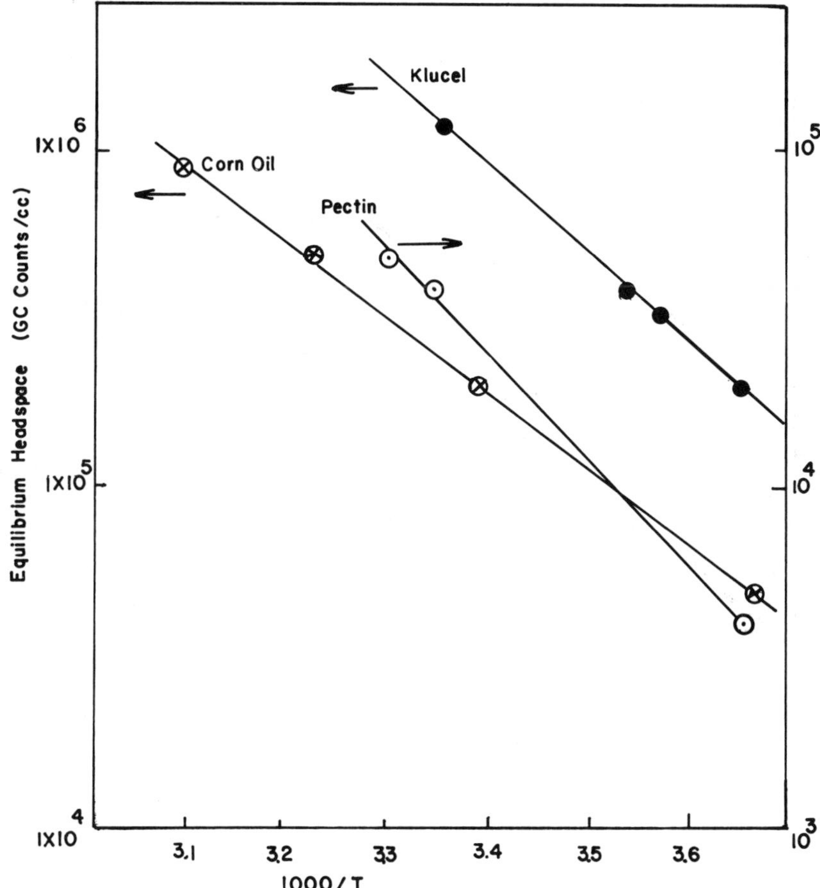

Fig. 3. Effect of temperature on equilibrium headspace concentration of α-pinene.

are rarely more than two or three times as large.

The heats of physical adsorption generally decreases steadily with increasing amount adsorbed and tend to approach the heat of liquefaction of the adsorbate as p approaches p_o (i.e., at multi-layer adsorption). Figure 4 is a representative plot of the variation of q_{st} with surface coverage. Curves a and b are typical plots for adsorption on uniform and non-uniform (hetereogeneous) surfaces respectively. In both cases the heat of adsorption approaches the heat of liquefaction when surface coverage exceeds a monolayer. Monolayer formation usually occurs by the time approximately one-tenth of the saturated pressure of the adsorbate is reached ($p/p \simeq 0.1$), regardless of its absolute value. Inspection of the data presented in Tables I and IV reveals, in most cases, the inverse relationship between the estimated value of q_{st} and p/p_o for a given adsorbate. These data are consistant with the general trend shown in Fig.4. The dependence of q_{st} on surface coverage is clearly demonstrated by the fairly non-polar adsorbate α-pinene (Table III). When the relative vapor pressure of α-pinene reaches 0.145 (at 20°C) its heats of adsorption on Klucel is only about 0.4 Kcal/mole higher than its latent heat of liquefaction. Tables I, II and IV show that when the fairly polar adsorbates ethyl acetate, pyridine and methyl ethyl ketone are used, monolayer adsorption appears to be reached at much lower vapor pressures as compared with α-pinene. Headspace-temperature measurements should, therefore, be determined at volatile loadings that give p/p_o values fairly lower than 0.1 in order to avoid any confusion in interpreting the heats of sorption (adsorption or absorption). The above limitation on the relative vapor pressure of volatile compounds is rarely reached in food systems. The volatile content of foods is far lower than the levels used in this work.

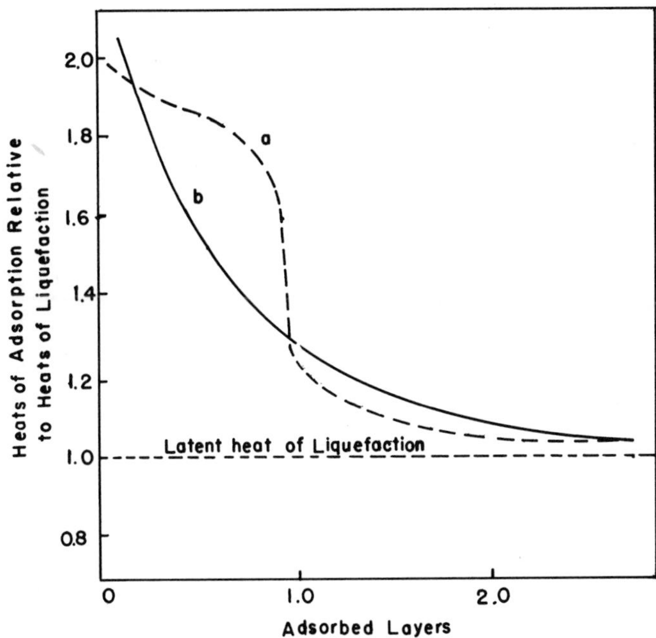

Fig. 4. Schematic dependence of heats of adsorption on surface coverage.

It should be pointed out that the magnitude of the heat of adsorption of a given vapor on different solids is also determined by the type of adsorption forces which occur between the adsorbent and the adsorbate molecules. Non-specific forces (van der Waals or dispersion) and/or specific forces (primarily hydrogen bonding and dipole interactions) contribute to the heats of adsorption. The surface of the food systems used in this work is very hetereogenous in nature. A small number of very active sites may be available for adsorption and contribute most to the estimated heats of adsorption.

3. Complex Food Systems

In the present study simple model volatile and food substrates were used. Real food systems are made up of more than one sorbent and sorbate. Competition for surface sites as well as sorbate-sorbate interactions are expected to occur. However, it was demonstrated experimentally, for non-food systems, that the magnitude of heat of adsorption is the main factor in determining which volatile molecules will be preferentially adsorbed (14). In addition Hill (15) has shown that equations of the Clausius-Clapeyron type (cf Equation 1) are valid in describing adsorption from a mixture of gases. These findings are encouraging in that the present techniques and data analysis can be extended to real food systems.

The present data on model food systems show that a few quantitative measurements of the equilibrium headspace at more than one temperature can provide very useful information on the heats of interaction of volatiles with foods. These measurements can contribute to our understanding of factors affecting retention and release of flavor components of foods. A quick glance at Figs. 1 to 3 indicates the changes that can be expected in the headspace composition of foods at different temperatures. These changes in headspace composition are a result of the differences in heats of sorption of different volatiles on the various food ingredients.

IV. REFERENCES

1. Vitzthum, O.G., and Werkhoff, P., in "Analysis of Foods and Beverages, Headspace Techniques" (G. Charalambous, Ed.), p. 115. Academic Press, New York, 1978.

2. Radford, T., Kawashima, K., Friedel, P.K., Pope, L.E., and Gianturco, M., J. Agric. Food Chem. 22: 1066 (1974).

3. Voilley, A., Simatos, D., and Loncin, M., Lebensm. - Wiss. u.-Technol. 10: 45 (1977).

4. Buttery, R.G., Bomben, J.L., Guadagni, D.G. and Ling, L.C., J. Agric. Food Chem. 19: 1045 (1971

5. Hougen, F.W., Quillian, M.S., and Curran, W.A., J. Agric. Food Chem. 19: 182 (1971).

6. Gremli, H.A., J. Am. Oil Chemists' Soc. 51: 95 (1974).

7. Issenberg, P., Greenstein, G., and Boskovic', M., J. Food Sci. 33: 621 (1968).

8. McNulty, P.B., and Karel, M., J. Food Techn. 8: 309, 319, (1973).

9. Maier, H.C., in "Proc. Intern. Symp. Aroma Research" (H. Maarse and P.J. Groenen, Ed.), p. 143. Pudoc, Wageningen, (1975).

10. Ahmed, E.M., Dennison, R.A., Dougherty, R.H. and Shaw, P.E., J. Agric. Food Chem. 26: 192 (1978).

11. McMullin, S.L., Bernhard, R.A., and Nickerson, T.A., J. Agric. Food Chem. 23: 452 (1975).

12. "Handbook of Chemistry and Physics", (R.C. Weast, Ed.) p. C-734. CRC Press, 57th Edition, (1976).

13. Adamson, A.W. in "Physical Chemistry of Surfaces", p. 565. Interscience Publishers, 2nd Ed., (1967).

14. Arnold, J.R., J. Am. Chem. Soc. 71: 104 (1949).
15. Hill, T.L., J. Chem. Phys. 18: 246 (1950).

OCCURRENCE OF AMADORI AND HEYNS REARRANGEMENT PRODUCTS IN PROCESSED FOODS AND THEIR ROLE IN FLAVOR FORMATION

Godefridus A.M. van den Ouweland
Hein G. Peer
Sing Boen Tjan

Unilever Research Duiven
Zevenaar, The Netherlands

I. INTRODUCTION

During the processing of foods a great variety of chemical reactions occur, among which those between carbonyls and amino compounds are the most important. If the carbonyl compound in this reaction is a reducing sugar and the amino compound an amino acid, a chain of reactions starts which is known as the Maillard reaction. The initial stages in this reaction for both glucose and fructose, being te most abundant reducing sugars in foods, are the following:

$$\underset{\text{glucose}}{CHO{-}(CHOH)_4{-}CH_2OH} + RNH_2 \rightleftharpoons HC(OH)(NH{-}R){-}(CHOH)_4{-}CH_2OH \xrightarrow{-H_2O} CH_2{-}NH{-}R{-}C{=}O{-}(CHOH)_3{-}CH_2OH$$

$$\underset{\text{fructose}}{CH_2OH{-}C{=}O{-}(CHOH)_3{-}CH_2OH} + RNH_2 \rightleftharpoons CH_2OH{-}C(OH)(NH{-}R){-}(CHOH)_3{-}CH_2OH \xrightarrow{-H_2O} CHO{-}HC(NH{-}R){-}(CHOH)_3{-}CH_2OH \;\;\text{and}\;\; CHO{-}C(H)(R{-}NH){-}(CHOH)_3{-}CH_2OH$$

ISBN 0-12-169060-1

TABLE I. ARP's Isolated from Natural Products: (A)pricots, (B)eet Molasses, (C)ured Tobacco Leaves, (L)iver, (LI)quorice

RP derived from	Isolated from				
GLUCOSE and					
- glycine	A[1]	L[2,3]	B[4]		
- alanine	A	L	B	LI[5]	C[6]
- 4-aminobutyric acid	A		B		
- valine					C
- aspartic acid	A	L	B	LI[7]	
- asparagine				LI	C
- proline				LI	C
- tyrosine					C
- phenylalanine					C
FRUCTOSE and					
- glycine		L			
- aspartic acid				LI	

Condensation takes place followed by rearrangement. The rearrangements of glucose and fructose in the literature are often referred to as the Amadori and Heyns rearrangement products (RP's). In recent years a number of RP's have been isolated from processed foods and these are summarized in Table I. The importance of these intermediates in the production of flavors and brown color during processing of foods has been recognized for quite some time now, and the effect of the water content of the food during processing on RP formation is remarkable. In some model experiments of glucose with amino acids it was found that optimal conversion into the RP takes place at a water content of 25-30% as illustrated in Fig. 1.

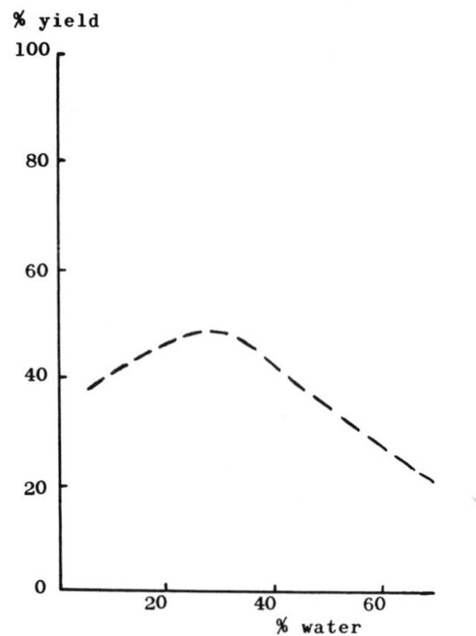

FIGURE 1. Yield of rearrangement products from sugar and amino acid after 1 hour at 100°C

Almost all foods contain glucose, fructose and amino acids and this region of optimal water content is often passed through during processing. In order to extend our knowledge about flavor of foods we investigated many foods for the occurrence of RP's. Our work on tomato powder, black tea and roasted meat is presented below.

II. ISOLATION OF REARRANGEMENT PRODUCTS

A. Tomato Powder

The flavor of tomato powder largely differs from that of fresh tomatoes; the greater part of its flavor is developed during processing of the powder. During spray-drying the region of optimal water content as mentioned in Chapter I is passed through. In addition a loss of glucose and fructose occurs and the amino acid content decreases (the two types of compound react) which is illustrated in Table II. We therefore investigated tomato powder for the occurrence of RP's. The water soluble constituents of tomato powder were dissolved by stirring the powder in water for a short period at 50°C. The non-dissolved material was separated by centrifugation. The aqueous solution so obtained was fractionated on an ion-exchange column into neutral, cationic and anionic constituents. The cationic fraction containing RP's was eluted with an aqueous solution of trichloroacetic acid (TCA) and this was continued until the amino acids aspartic acid and glutamic acid eluted.

TABLE II. Amino Acids in Fresh and Processed Tomato Juices (mg/100 g solids)

Amino acid	Fresh juice	Processed juice
Aspartic acid	488	476
Threonine	135	120
Serine	155	126
Asparagine	273	266
Glutamic acid	2 436	2 326
Glutamine	631	736
Proline	41	9
Glycine	40	10
Alanine	59	28
Valine	49	35
Methionine	65	15
Isoleucine	71	34
Leucine	73	26
Tyrosine	83	23
4-Aminobutyric acid	488	446
Lysine	105	58
Histidine	121	80
Arginine	80	38

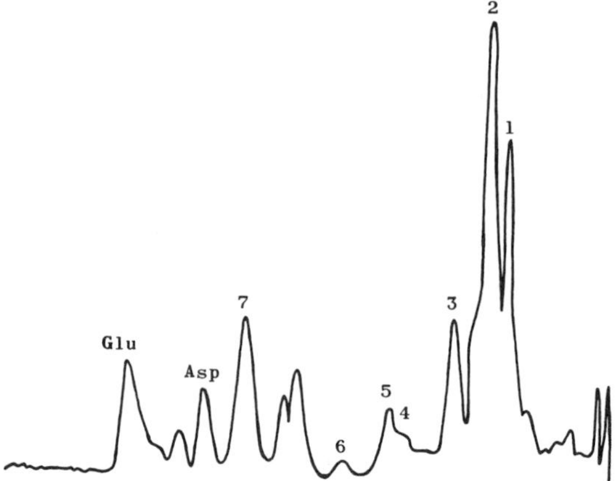

FIGURE 2. High pressure liquid chromatogram of the cationic constituents of tomato powder. RP!s derived from: (1) glucose-aspartic acid, (2) glucose-glutamic acid, (3) fructose-glycine, (4) glucose-glycine, (5) glucose-alanine, (6) fructose-4-aminobutyric acid, (7) glucose-4-aminobutyric acid. CONDITIONS: column, 250 mm stainless steel; adsorbent, Aminex A9; eluent, 0.1 M trichloroacetic acid; detection, refractive index

After removal of TCA and concentration, the cationic fraction was separated into the various constituents by high pressure liquid chromatography (HPLC) on Aminex A9, H^+ form. This technique proved to be very effective for this separation (Fig. 2).

B. Black Tea

Black tea is another example of a food which derives its flavor from processing. Black tea manufacturing involves several operations: plucking - withering - fermentation - firing - packing. During withering membranes become permeable and aroma precursors are formed while during fermentation mainly oxida-

TABLE III. Amino Acids in Fresh Tea Flush and in Black Tea

Amino acid	Amount (mg per 100 g, dry wt.)	
	Fresh tea	Black tea
Alanine	35	25
Valine	50	40
Leucine	84	62
Aspartic acid	176	176
Glutamic acid	394	267
Glutamine	53	30
Threonine	43	26
Serine	102	51
Tyrosine	28	21
Theanine	625	520

tion reactions take place, giving rise to aroma constituents. The firing step normally takes about 20 min, during which the moisture contents of the tea leaf material is reduced from about 60% to about 3%, and the leaf temperature rises rapidly from room temperature to about 65°C and then steadily to about 85°C. During this time the optimal moisture content for RP formation and aroma formation via Maillard is again passed through. When considering the amino acid composition of fresh tea flush and black tea, a number of potential amino acids appears to be available for a rearrangement reaction (Table III)[8].

On analysing the water soluble constituents of black tea in the way as described earlier for tomato powder only the RP derived from glucose and theanine was found:

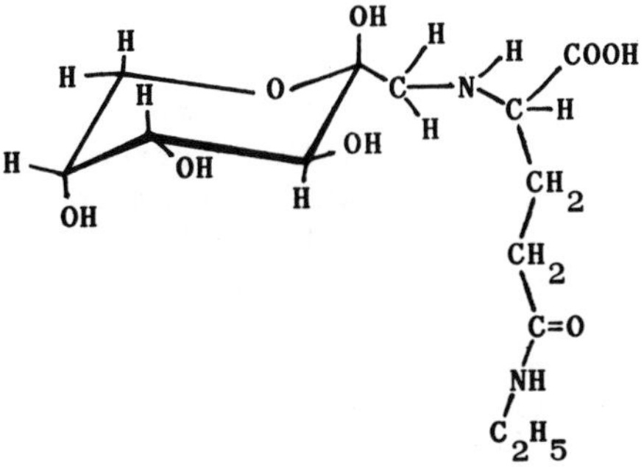

C. Roasted Meat

During the frying of meat approximately 60% of the initial water is lost. In the crust this loss is even higher and again by this operation conditions are created at which reaction of carbohydrates and amino acids is favored. There is also a decrease in amino nitrogen and sugars when a beef diffusate is heated[9], as appears from Table IV.

TABLE IV. Effect of heating at 125°C on the Concentration (μM/ml) of sugars and Amino Nitrogen in a diffusate of a Water Extract of Beef

	Heating time (min)	
	0	60
Amino Nitrogen	6.098	2.416
Glucose	5.503	0.009
Fructose	3.136	0.023
Ribose	1.066	0.010

By using essentially the same isolation procedure as mentioned earlier the RP derived from glucose and glycine could be isolated from roasted meat:

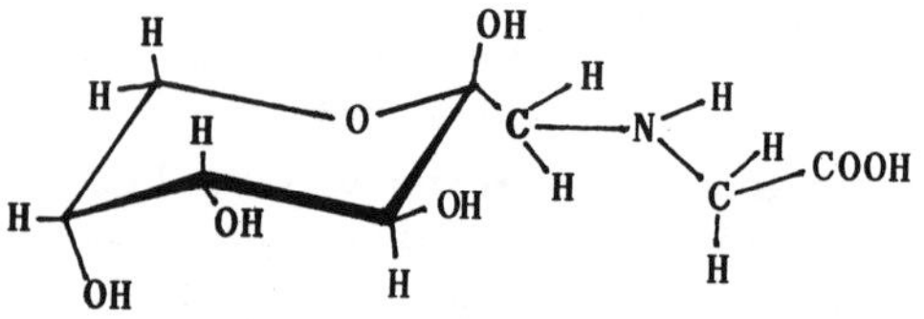

It should be stressed that all the RP's mentioned here, could not be detected in the fresh material which implies that their formation has to take place during the processing.

III. IDENTIFICATION OF RP's BY PROTON MAGNETIC RESONANCE

As shown above a number of RP's have been isolated earlier from various natural products. Their structures were characterized by chemical methods, which often involved acid hydrolysis of the RP followed by identification of the amino acid which is split off in this decomposition reaction.

As an example the structure determination of the RP derived from glucose and glycine will be discussed. Examination of the 220 MHz PMR spectrum of this compound, recorded in D_2O at pH 3, shows that the product exists at least as a mixture of two components (Fig. 3). The assignment of the proton signals

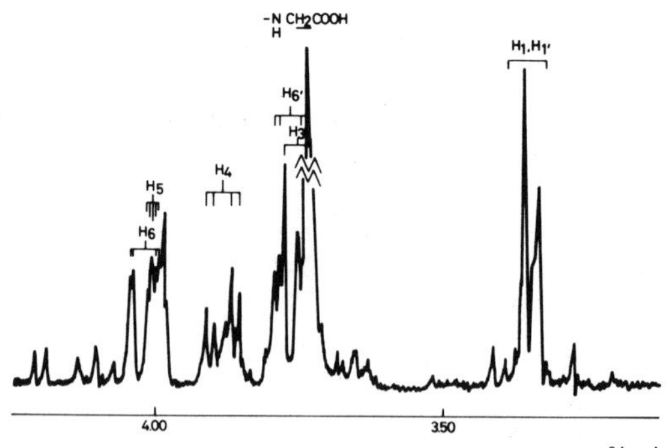

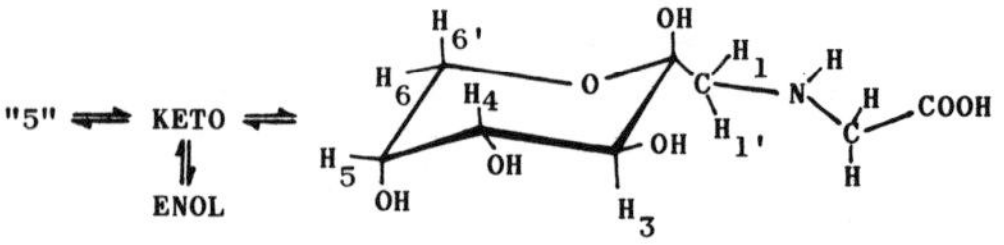

FIGURE 3. 220 MHz PMR spectrum (D_2O, pH ~3) of RP from glucose and glycine

of the major pyranose component is given and it appears that the coupling constants are identical with those of β(D)-fructose. The AB doublet resonating at 3.33 and 3.38 ppm can be assigned to the methylene protons next to the nitrogen atom. The complete exchangeability of these protons with deuterium at pH 9.5 provides conformatory evidence of this assignment and indicates that in D_2O solution an equilibrium exists between the various possible structures, i.e. the pyranose form, furanose form and the open-chain structure. The major component in the mixture has the six-membered structure and exists exclusively in the β(D)-$^2C_5$ conformation and the minor component has the five-membered ring structure as indicated.

Changing the pH from 3 to 9.5 results in upfield shifts of the amino acid protons, while the protons H_3, H_4, H_5, H_6 and $H_{6'}$ of the sugar residue are influenced only slightly as is shown in the 60 MHz spectrum in Fig. 4.

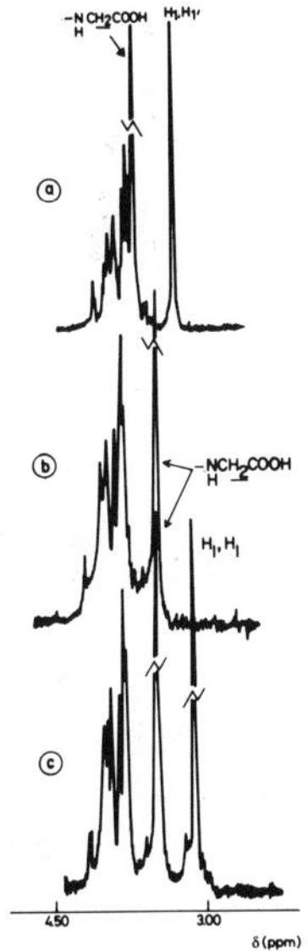

FIGURE 4. 60 MHz PMR spectrum (D_2O) of RP from from glucose and glycine (see Fig.3): (a) pH∼3, (b) pH∼9, (c) freeze-dried aqueous solution of pH 9, dissolved in D_2O and measured at once

Due to the very rapid exchange of protons H_1 and $H_{1'}$ with deuterium at pH 9.5, their upfield shifts have to be recorded as follows. The solution of the RP in water was adjusted with $NaHCO_3$ to pH 9.5 and then freeze-dried. The PMR spectrum was then recorded as quickly as possible in D_2O. Despite partial exchange with deuterium upfield shifts of about 0.2 ppm could be observed. The structures of the RP mentioned earlier and isolated from tomato powder and black tea were elucidated in a similar way.

IV. FORMATION OF FLAVOUR BY THERMAL DEGRADATION OF RP's

To date more has been published on the production of volatile compounds from sugar-amine interaction than on any other single topic concerned with browning. So apparently these degradation reactions are very appealing to many research workers. The early steps in the degradation of RP's proceed by 1,2-enolization and 2,3-enolization:

```
CHO                     HC-OH
|  H                    || H
H2C-N-R       <=>       C-N-R
(CHOH)3                 (CHOH)3        \
CH2OH                   CH2OH           ->  CHO
                                            C=O
                        HC-N-R          ->  CH2      --+
                        || H                (CHOH)3    |
                        C-OH                CH2OH      |
   H                    (CHOH)3                        |---> flavor
H2C-N-R       <=>       CH2OH                          |     compounds
C=O                                                    |
(CHOH)3                                                |
CH2OH                   H2C-N-R             CH3        |
              <=>       |  H                C=O      --+
                        C-OH        ->      C=O
                        ||
                        C-OH
                        (CHOH)2             (CHOH)2
                        CH2OH               CH2OH
```

There are a number of excellent review papers dealing with the chemistry of flavor formation in the Maillard reaction and a few examples of RP degradation will be presented here.

A. Degradation of RP's Derived from Glucose-Fructose and Glycine

Both the RP's of glucose/glycine and fructose/glycine were heated at temperatures above their melting points and the products formed were analysed by GC/MS. Figure 5 shows a stick diagram of the gaschromatogram together with the components formed in both reactions. The majority of the compounds are furan and pyrrole derivatives which indicate that 1,2-enolization is favored. It is interesting to note that in both degradation reactions dihydropyrone derivative <u>35</u> is formed and its formation will be discussed in a following chapter. Figure 6 shows the differences between the two degradation products.

In the fructose/glycine degradation more carbohydrate chain scission products are formed and also nitrogen seems to play a more active role than in the other degradation reaction. The formation of the dihydropyrone derivative <u>35</u> seems to proceed

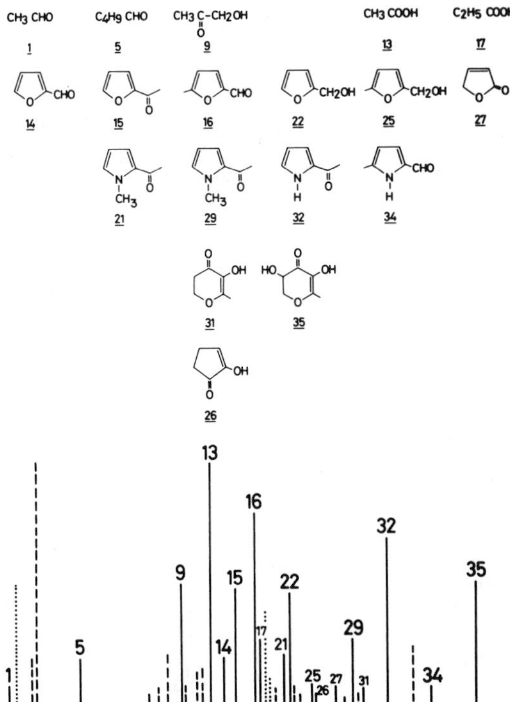

FIGURE 5. Gaschromatogram of the degradation products of RP's of glucose/glycine and fructose/glycine. The numbers indicate the compounds which are formed in both cases. See also Fig. 6

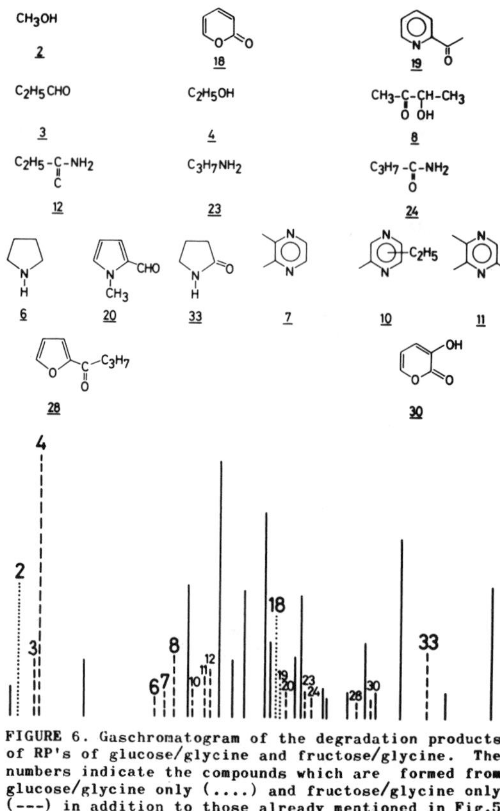

FIGURE 6. Gaschromatogram of the degradation products of RP's of glucose/glycine and fructose/glycine. The numbers indicate the compounds which are formed from glucose/glycine only (....) and fructose/glycine only (---) in addition to those already mentioned in Fig.5

along two different routes in both degradation reactions. In the degradation of the RP from glucose/glycine its formation can be explained by 2,3-enolization of the RP:

$$\begin{array}{l} CHO \\ C{=}O \\ CH_2 \\ (CHOH)_3 \\ CH_2OH \end{array} \;\not\longrightarrow$$

$$\begin{array}{l} CH_3 \\ C{=}O \\ C{=}O \\ (CHOH)_2 \\ CH_2OH \end{array} \longrightarrow \text{HO, O, OH, } CH_3\text{, OH (pyranone)} \longrightarrow \underline{35}$$

$$\begin{array}{l} CH_3 \\ C{=}O \\ HC{=}O \\ + \\ CH_2OH \\ C{=}O \\ CH_2OH \end{array} \longrightarrow \begin{array}{l} CH_3 \\ C{=}O \\ CHOH \\ CHOH \\ C{=}O \\ CH_2OH \end{array} \longrightarrow \begin{array}{l} CH_3 \\ C{=}O \\ CHOH \\ C{=}O \\ CHOH \\ CH_2OH \end{array} \uparrow$$

In the degradation reaction of the RP from fructose/glycine it seels tempting to assume that the dihydropyrone derivative 35 is formed by a condensation reaction of the carbohydrate chain scission products methylglyoxal and dihydroxyacetone.

B. Degradation of the RP Derived from Glucose-Theanine

For the thermal degradation of the RP derived from glucose/theanine the difference between the dry degradation and the degradation in the presence of water (see Figs. 7, 8 and 9)was studied.

When examining the stick diagram of the gaschromatogram of these degradation reactions (Fig. 7) it is seen that degradation in an aqueous medium yields more compounds than the dry degradation and also that different compounds are formed under the different reaction conditions. The compounds formed in both reactions, the structures of which are given, are common brown-

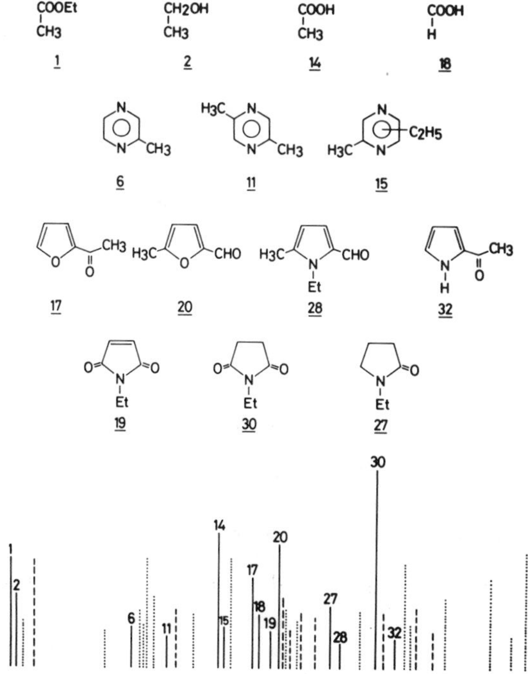

FIGURE 7. Gaschromatogram of the degradation products of the RP of glucose/theanine. The numbers indicate the compounds which are formed both during wet anddry degradation. See also Figs. 8 and 9

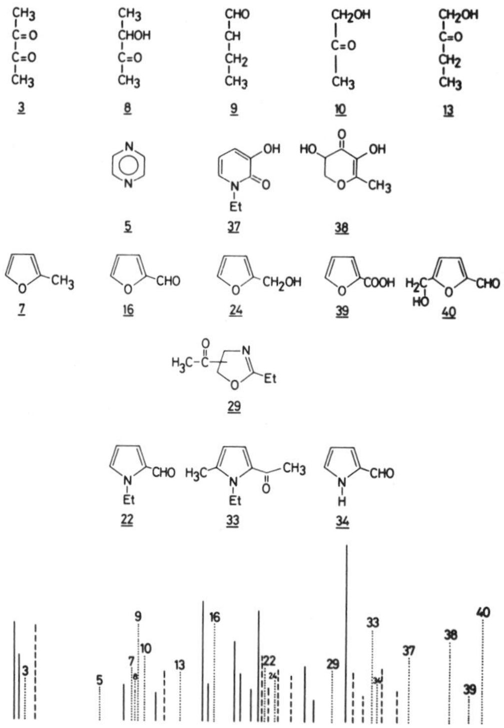

FIGURE 8. Gas chromatogram of the degradation products of the RP of glucose/theanine. The numbers indicate the compounds which are formed during WET degradation in addition to those already mentioned in Fig. 7

ing products together with malonimide and succinimide, for which the uncommon amino acid theanine seems to be responsible, although the latter was also found in a lactose-casein reaction product.

In Fig. 8 the extra compounds formed during wet degradation of the RP of glucose/theanine are given. Apart from the C_3-C_4 carbohydrate-chain scission products the amount of derivatives of furan and pyrrole increases, while the compounds 29 and 37 were not found earlier in browning systems.

The compounds detected only in the dry degradation reaction are given in Fig. 9. There are only heterocyclic compounds from which the formation of compound 31 is hard to explain. From the pyrrole derivatives, isolated from glucose/theanine degradation, the majority is substituted with an ethyl group at the nitrogen atom. It seems therefore likely that during degradation ethylamine is formed from theanine which then interferes with the formation of the flavor compounds. The majority of the compounds found in these degradation reactions also occurs in the aroma of black tea which proves that this type of reaction plays an important role in the flavor formation of black tea.

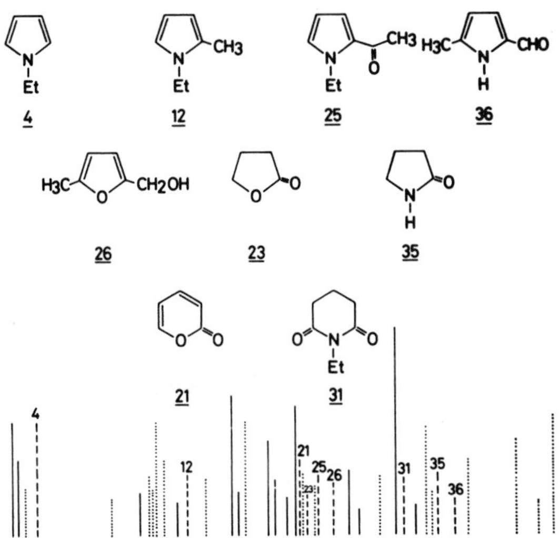

FIGURE 9. Gas chromatogram of the degradation products of the RP of glucose/theanine. The numbers indicate the compounds which are formed during DRY degradation in addition to those already mentioned in Fig. 7

In summary it can be said that HPLC has been demonstrated to be a very effective tool for the isolation of Amadori and Heyns RP's from processed foods, and that with the aid of PMR spectrometry the structures of these products can be elucidated. From the components isolated from the degradation of some rearrangement products it was shown that the type of flavor compound formed is influenced not only by the carbohydrate moiety but also by the presence or absence of water.

REFERENCES

1. Anet, E.F.L.J. and Reynolds, T.M., Austr. J. Chem. 10: 182 (1975).
2. Heyns, K. and Paulsen, H., Ann. 622: 160 (1959).
3. Borsook, H., Abrams, A. and Lowy, P.H., J. Biol. Chem. 215: 111 (1955).
4. Carruthers, A., Dutton, J.v. and Oldfield, J.F.T., Intern. Sugar J. 65: 297 (1965).
5. Nishi, H. and Morishita, I., Nippon Nogei Kagaku Kaishi 45: 507 (1971).
6. Yamamoto, K. and Noguchi, M., Agr. Biol. Chem. 37:2185(1973).
7. Vondenhof, T., Glombitza, K.W. and Steiner, M., Z. Lebensm. Unters. Forsch. 152: 345 (1973).
8. Roberts, G.R. and Sanderson, G.W., J. Sci. Fd. Agr. 17: 182 (1966).
9. Wasserman, A.A. and Spinelli A.M., J. Fd. Sci. 35:328 (1970).

FORMATION OF LACTONES AND TERPENOIDS BY MICROORGANISMS

Roland Tressl
Martin Apetz
Ronald Arrieta
Klaus Günter Grünewald

Technische Universität Berlin

1. FORMATION OF LACTONES BY MICROORGANISMS

The list of foods in which lactones have been found is steadily growing. Maga (1), reviewing the flavor literature, demonstrated that nearly all major food classes contain lactones. They are formed by chemical and enzymatic reactions. Lactones are known as flavor-contributing components in many fermentation products where they are formed by microbial reactions.

Figure 1 shows some components which have been characterized in certain alcoholic beverages and in Sporobolomyces odorus. Compounds I-IV in sherry and wine (2-4); V in whisky (5); VI in saké (6); VII-IX in beer (7), and X-XII in culture media from Sporobolomyces odorus (8). There are different metabolic pathways for the microbial formation of lactones.

1.1 Conversion of 4-Oxobutyrate into γ-Butyrolactone, Alkoxy- and Acyl Lactones

γ-Butyrolactone may be formed during yeast fermentation as shown in Figure 2. Glutamate is trans-

[1]Present Address: Technische Universität Berlin
Seestr. 13, D-1000 Berlin 65

ISBN 0-12-169060-1

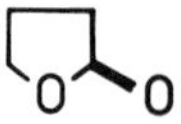

Figure 1 Lactones in fermentation products

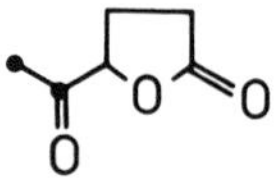

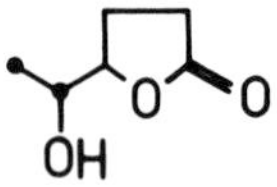

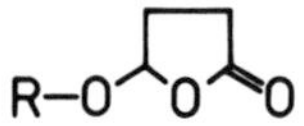

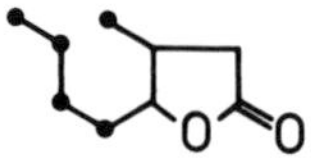

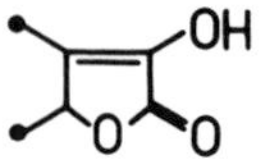

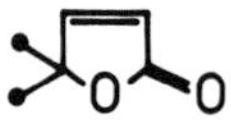

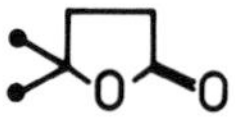

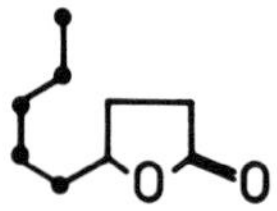

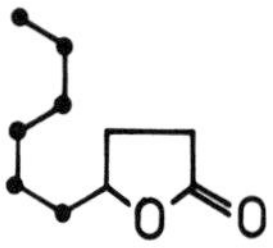

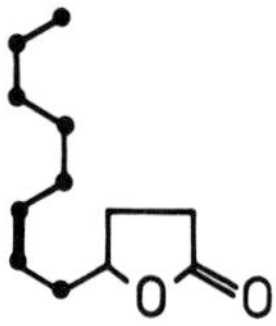

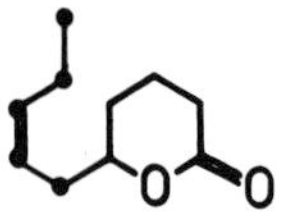

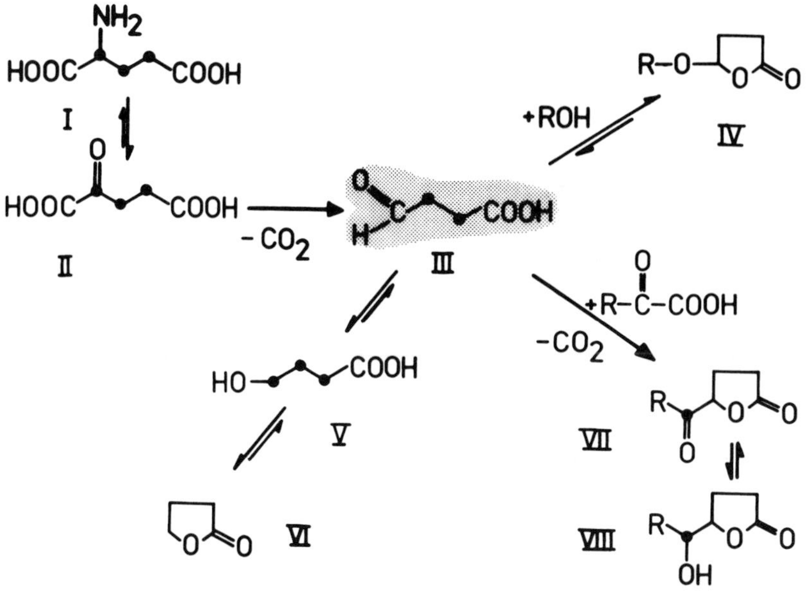

Figure 2 Formation of γ-butyrolactone, alkoxy- and acyl lactones

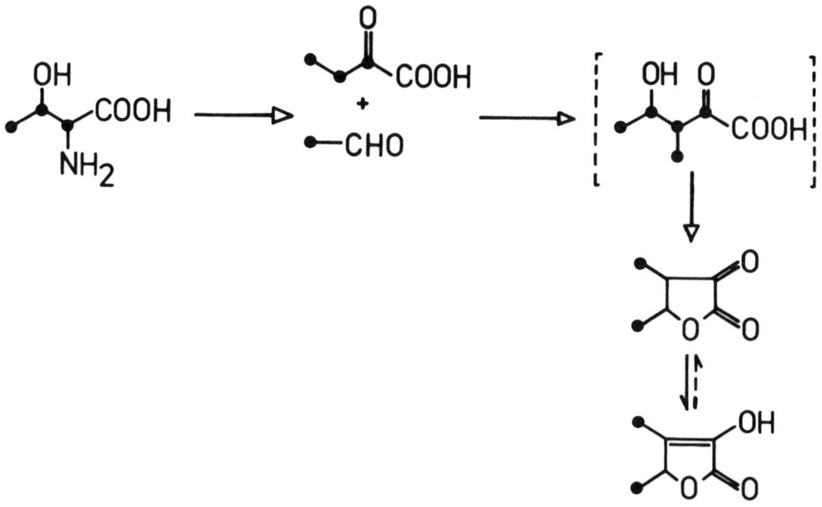

Figure 3 Formation of 3-hydroxy-4,5-dimethyl-2(5H)furanone

formed via 2-oxoglutarate (II) into 4-oxobutyrate (III) which Muller et al. suppose to be a precursor in the formation of alkoxy- and acyl lactones (4). Reduction of 4-oxobutyrate yields γ-butyrolactone which is known in sherry, wine, and beer. By reaction with alcohols alkoxy lactones may be formed, and aldolcondensation with 2-oxoacids leads to acyl lactones. Among the possible 5 to 6 alkoxy lactones (corresponding to ethanol, propanol, isobutanol, 2-methylbutanol, 3-methylbutanol, and 2-phenylethanol) only 4-ethoxy-4-hydroxybutyric acid γ-lactone (5-ethoxydihydro-2(3H)furanone) has been identified (4). 5-Acetyldihydro-2-(3H)furanone (solerone) is the only acyl lactone found in sherry and wine (3). Reduction of solerone leads to optically active 4,5-dihydroxyhexanoic acid γ-lactones which have been characterized in sherries (2).

1.2 Conversion of 2-Oxobutyric Acid into 3-Hydroxy-4,5-dimethyl-2(5H)furanone (HDMF)

Recently Takahashi et al. (6) identified a lactone in aged saké which possessed a burnt flavor. The formation of this compound is shown in Figure 3. Aldolcondensation of 2-oxobutyrate with acetaldehyde leads to the unstable compound which is easily decomposed to acetoin and oxalic acid. The threshold was determined at 50 ppb. The lactone was determined in soy sauce, raochu (6), and wine (9). We investigated the formation of the lactone with Saccharomyces cerevisiae. L-Threonine and α-ketobutyric acid were transformed to some extent into HDMF. In addition we identified the corresponding 5-ethyl component which was characterized by Sulser et al. (10) in protein hydrolysates. Both constituents were also identified as flavor-contributing components in beer.

1.3 Reduction of 4- and 5-Oxoacids by Yeasts

In 1962 Muys et al. (11) showed that 4- and 5-oxoacids are transformed into optically active γ- and δ-lactones by yeasts. 4-Oxoacids are reduced to the corresponding d-4-hydroxyacids which form d-4-alkylbutane lactones. 5-Oxoacids are reduced to d-5-hydroxyacids which undergo cyclisation to d-5-alkylpentane lactones. We investigated

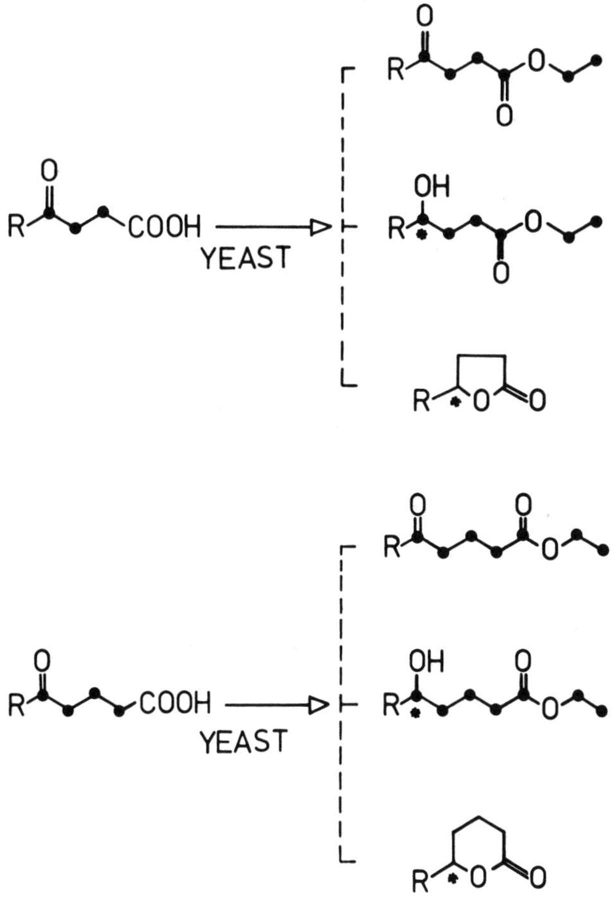

R-CO-CH2-CH2-COOH → YEAST

Figure 4 Transformation of 4- and 5-oxo-acids by Saccharomyces cerevisiae

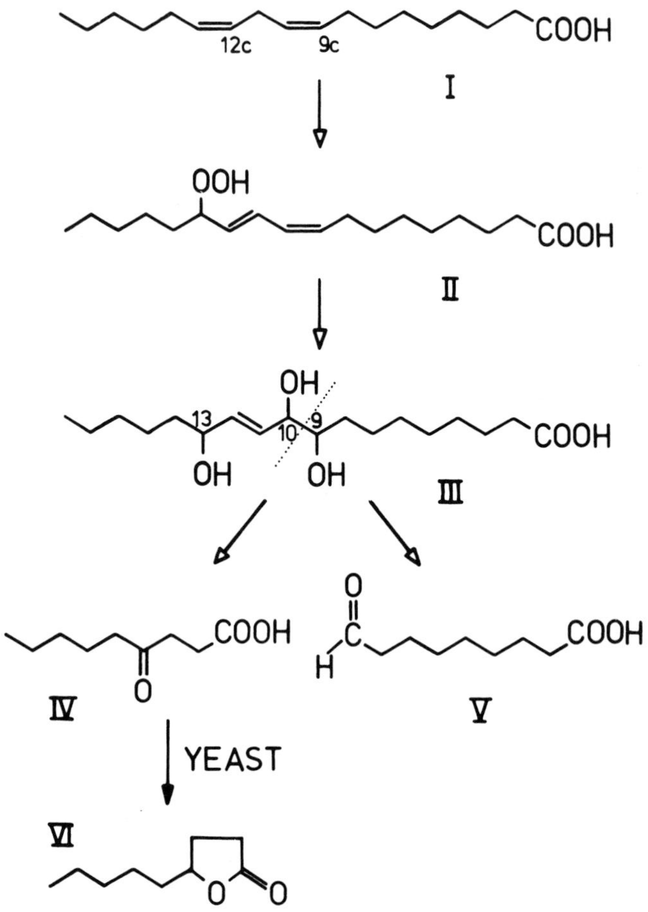

Figure 5 Reaction sheme which may explain the formation of γ-nonalactone in beer

the conversion of certain 4- and 5-oxoacids into lactones with Saccharomyces cerevisiae. As shown in Figure 4 oxoacids are transformed into optically active lactones, oxoacid ethyl esters, and the corresponding d-hydroxyacid ethyl esters.

In beer 4-pentanolide and 4-nonanolide may be formed by this reaction. 4-Oxopentanoic acid (levulinic acid) is a degradation product of glucose, and 4-oxononanoic acid is derived from linoleic acid during malt preparation and wort boiling as shown in Figure 5.

In Figure 6 some lactones which have been identified in beer are summarized. Compound V, a degradation product of humulone, is reduced to some extent during fermentation. Lactones VII, VIII, XII, and XIII are oxidation products of linoleic acid. Compound XV is formed during malting and wort boiling from ß-carotine. The amounts of individual components vary from 10 ppb to 5 ppm depending on the beer-type.

1.4 Formation of Lactones by Sporobolomyces odorus

In 1930 Derx (12) isolated an organism from orange leaves which showed similarity to Sporobolomyces roseus and had a peach-like odor. Tahara et al. (13) identified 4-decanolide and cis-6-dodecen-4-olide which are responsible for the peach-like odor. The authors failed to optimize the production of both lactones.

Figure 7 shows a gas chromatogram of an aroma concentrate from Sporobolomyces odorus grown on glucose. Components 62 and 72 correspond to 4-decanolide and cis-6-dodecen-4-olide. Besides lactones we identified alcohols, acids, hydroxy-, and ketoacids. The results will be published in detail (14).

In Table I some of the characterized lactones are presented. Figure 8 shows the structures of some lactones. Besides saturated lactones from C_6 to C_{11} we observed some components with ω-6- and -ω-3- $^{H}C=C^{H}$ double bonds in the alkyl side chains. To date there are only a few indices on the biosynthesis of lactones in (fruits and) microorganisms.

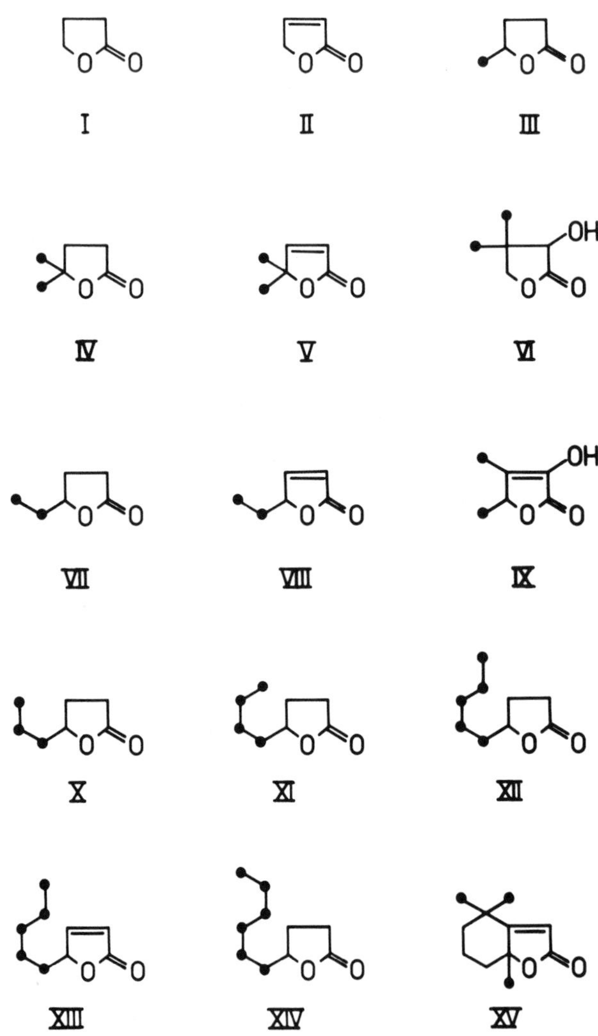

Figure 6 Lactones identified in beer

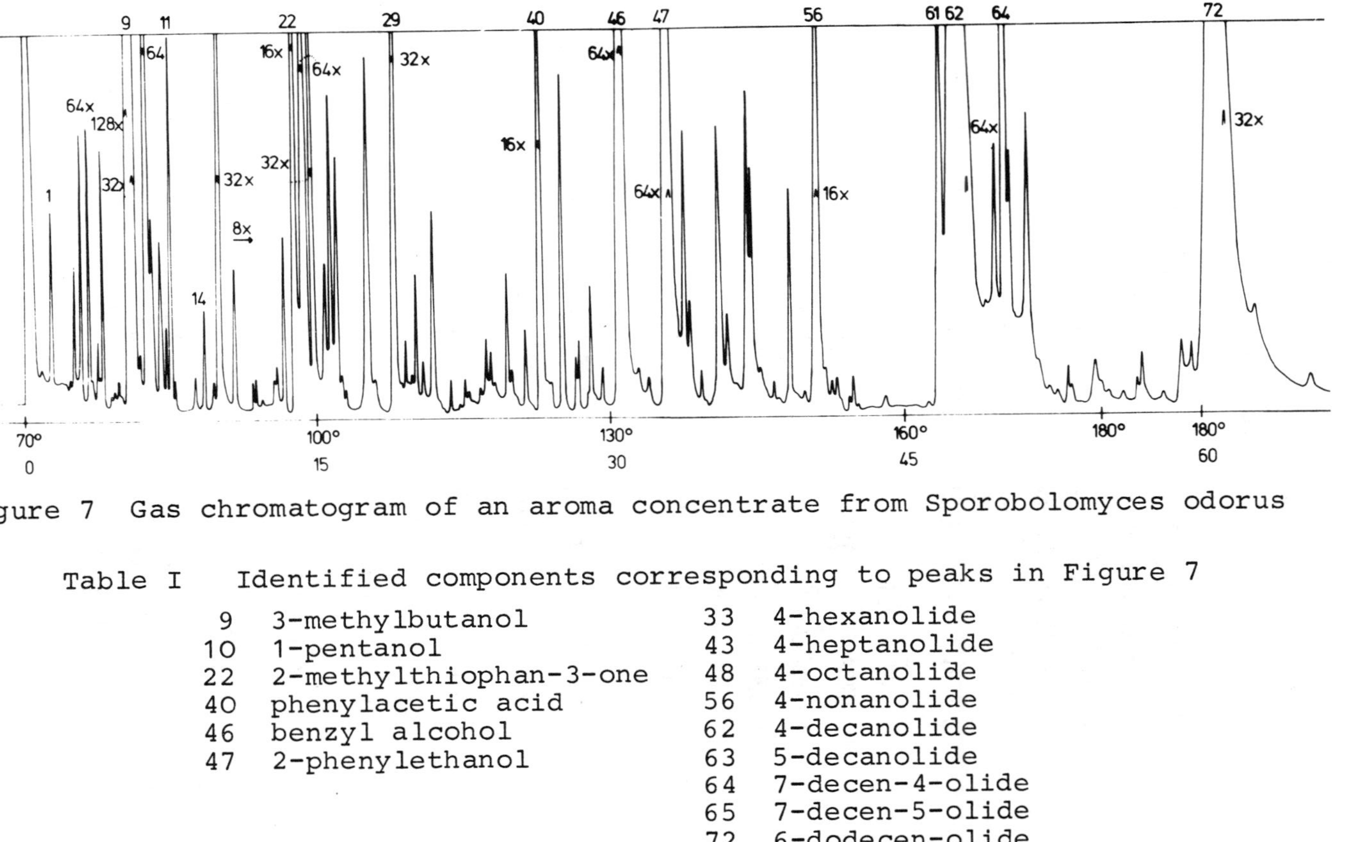

Figure 7 Gas chromatogram of an aroma concentrate from Sporobolomyces odorus

Table I Identified components corresponding to peaks in Figure 7

9	3-methylbutanol	33	4-hexanolide
10	1-pentanol	43	4-heptanolide
22	2-methylthiophan-3-one	48	4-octanolide
40	phenylacetic acid	56	4-nonanolide
46	benzyl alcohol	62	4-decanolide
47	2-phenylethanol	63	5-decanolide
		64	7-decen-4-olide
		65	7-decen-5-olide
		72	6-dodecen-olide

TABLE II CONVERSION OF [1-^{14}C]-DECANOIC ACID BY SPOROBOLOMYCES ODORUS

PRECURSOR (50 μCi)	[1-^{14}C]-DECANOIC ACID
FERMENTATION BROTH	500 ML 10 G CELLS
INCUBATION TIME	6 HR
RADIOACTIVITY IN THE AROMA EXTRACT, %	0,5
DISTRIBUTION OF RADIOACTIVITY AMONG VOLATILE COMPONENTS	(%)
DECANOIC ACID	4
3-HYDROXYDECANOIC ACID	9
4-DECANOLIDE	84
6-DODECEN-4-OLIDE	0,5
STEARIC-, OLEIC ACID	1
LINOLEIC ACID	1,5

TABLE III CONVERSION OF [1-^{14}C]-DODECANOIC ACID BY SPOROBOLOMYCES ODORUS

PRECURSOR (50 μCi)	[1-^{14}C]-DODECANOIC ACID
FERMENTATION BROTH	500 ML 11 G CELLS
INCUBATION TIME	6 HR
RADIOACTIVITY IN THE AROMA EXTRACT, %	0,5
DISTRIBUTION OF RADIOACTIVITY AMONG VOLATILE COMPONENTS	(%)
DODECANOIC ACID	93
3-HYDROXYDECANOIC ACID	1
PALMITIC ACID	4
4-DODECANOLIDE	0,5
6-DODECEN-4-OLIDE	0,5
STEARIC-, OLEIC ACID	1

TABLE IV CONVERSION OF [U-^{14}C]-LINOLEIC ACID BY SPOROBOLOMYCES ODORUS

PRECURSOR (50 μCi)	[U-^{14}C]-LINOLEIC ACID
FERMENTATION BROTH	500 ML 11 G CELLS
INCUBATION TIME	6 HR
RADIOACTIVITY IN THE AROMA EXTRACT, %	0,8

DISTRIBUTION OF RADIOACTIVITY AMONG VOLATILE COMPONENTS

	(%)
1-PENTANOL 1-HEXANOL	55
2,4-NONADIENOL	6
9-OXONONANOIC ACID	16
PALMITIC ACID	6
6-DODECEN-4-OLID	4
LINOLEIC ACID	13

We therefore investigated the formation of 4-decanolide and cis-6-dodecen-4-olide in a series of ^{14}C-labelling experiments. Some of the results are shown in Table II-IV. It can be seen that [1-^{14}C]-decanoic acid is easily transformed into 4-decanolide by Sporobolomyces odorus. The 3-hydroxyacid showed a small amount of radioactivity but 2-/or 3-decenoic acid contained no radioactivity. [1-^{14}C]-dodecanoic acid was not converted into lactones as shown in Table III. The γ-hydroxylation of decanoic acid seems to be specific to Sporobolomyces odorus. Table IV shows the conversion of [U-^{14}C]-linoleic acid into volatiles by Sporobolomyces odorus. Besides the oxidative degradation products of linoleic acid, 6-dodecen-4-olide contained a small amount of radioactivity. The results show that the lactones are formed by two different pathways. Saturated lactones may be formed by γ- or δ-hydroxylation of the corresponding saturated acids. Similar results were obtained by Dimick et al. (15) in labelling experiments of [1-^{14}C] -dodecanoic acid via intramammary infusion to the lactating goat. Unsaturated lactones are obviously derived from unsaturated fatty acids via ß-oxidation as shown by Mizugaki et al. (16). Figure 9 illustrates the formation of 4-decanolide and cis-6-dodecen-4-olide by Sporobolomyces odorus.

2. FORMATION OF TERPENOIDS BY MICROORGANISMS

Monoterpenoids and sesquiterpenoids are primarily plant products and are well known as characteristic components of many essential oils. In 1970 Collins and Halim (17) identified geraniol, citronellol, linalool, neral, citral, geranyl acetate, and citronellyl acetate in Ceratocystis variospora grown on dextrose. Halim and Collins (18) showed that nerol, citronellol, and geraniol are formed by Trametes odorata. Lanza et al. (19) investigated the aroma production by the fungus Ceratocystis moniliformis. Sensory evaluation showed the quality and intensity of the aroma to vary with the composition of the culture medium and with the age of the cultures. Combined gas chromatography-mass spectrometry was utilized to identify volatile constituents in cultures

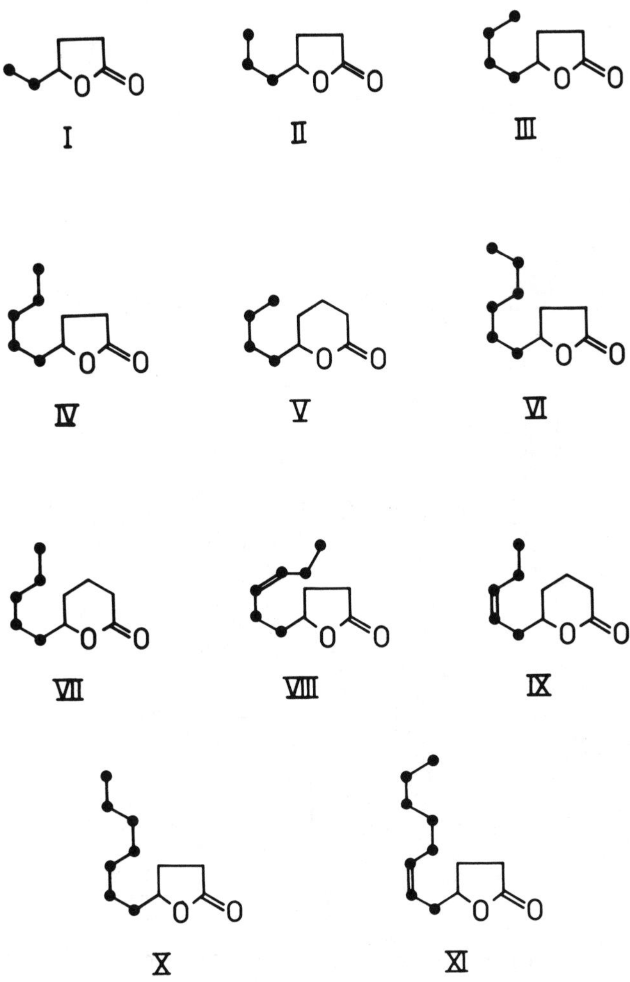

Figure 8 Lactones produced by Sporobolomyces odorus

COOH → O O

I II

COOH

III

COOH

IV V

Figure 9 Possible formation of lactones by Sporobolomyces odorus

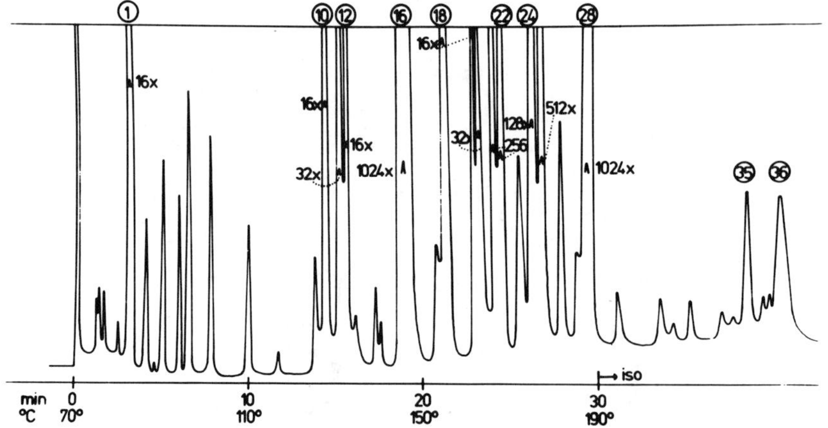

Figure 10 Gas chromatogram of an aroma concentrate from Ceratocystis variospora

Table V Identified components corresponding to Figure 10

2	myrcene	18	citronellic acid
3	limonene	22	α-terpineol
4	ocimene	25	citronellol
5	γ-terpinolene	26	nerol
6	terpinolen	28	geraniol
10	trans-linalooloxid	35	4-decanolide
12	cis-linalooloxid	36	5-decanolide
16	linolool		

characterized as having "fruity-banana", "peach-pear", and "citrus"aromas. Besides aliphatic esters, γ- and δ-decalactones, citronellol, nerol, geraniol, and geranial were identified. Sprecher et al. (20) investigated the production of volatiles by a mutant of Ceratocystis coerulescens. More than 50 constituents were identified (among them: citronellol, α-terpineol, nerol, geraniol, the acetates of these monoterpene alcohols, linalool, nerolidol, farnesol, and 2,3-dihydro-6-trans-farnesol). Geosmin and methylisoborneol possess "earthy" odors and are produced by Actinomycetes (21). These examples show that certain fungi are able to form mono- and sesquiterpenoids from simple carbon sources and may be used as sources for natural flavor compounds. In addition they represent an excellent experimental subject for studying the biosynthesis of monoterpenes.

2.1 Formation of Terpenoids by Ceratocystis variospora

We investigated the production of aroma components by Ceratocystis variospora in shake cultures and in a biostate. The production of volatiles depends on the culture media and the age of the cultures. More than 50 constituents were identified by capillary gas chromatography-mass spectrometry. The results and the methods used will be published in detail (22).

Figure 10 shows a gas chromatogram of an aroma concentrate from Ceratocystis variospora grown on glucose. Some of the (characterized) constituents are summarized in Table V and Figure 11. It can be seen that Ceratocystis variospora produces acyclic and alicyclic hydrocarbones, alcohols, aldehydes, a ketone, acids, acetates, and three sesquiterpenes. Myrcene, limonene, γ-terpinene, terpinolene, citronellic acid, cis- and trans-geranic acid were identified for the first time as metabolites of fungi. Most of the monoterpenoids are produced in the ppm-range (depending on the culture medium).

Croteau (23) studied the biosynthesis of monoterpenes in a series of labelling experiments with peppermint plants. Mevalonic acid pyrophosphate is

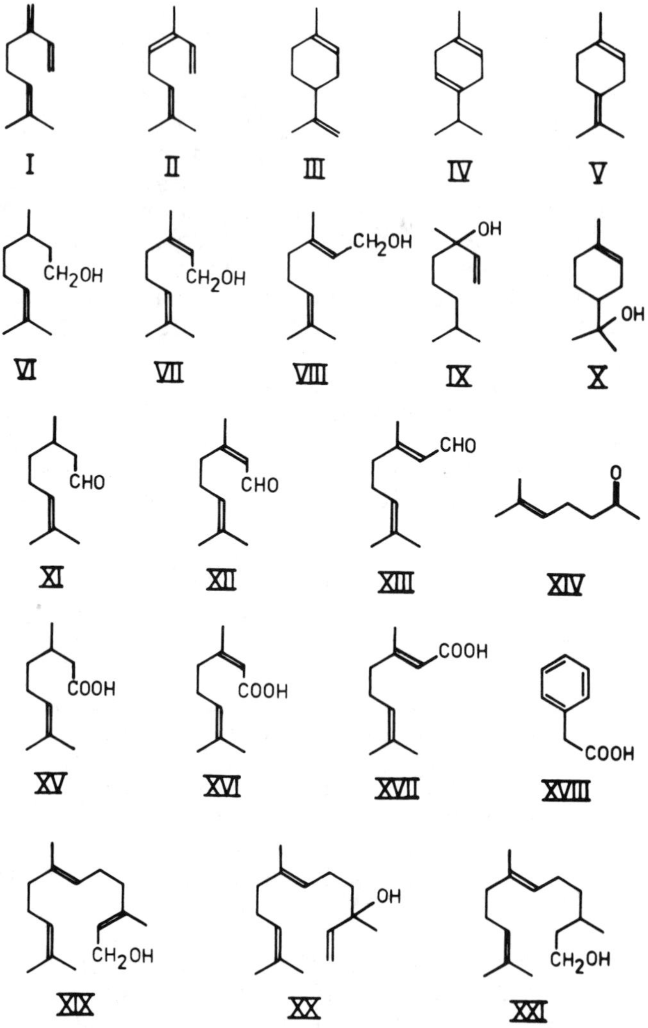

Figure 11 Terpenoids produced by Ceratocystis variospora

first converted to isopentenyl pyrophosphate (IPP), which is isomerized to dimethylallyl pyrophosphate (DMAPP). Condensation of IPP and DMAPP yield geranyl pyrophosphate, the first C_{10} isoprenoid compound (Figure 12). The author suggests that geranyl pyrophosphate is an intermediate in the biosynthesis of acyclic monoterpenes while the cis-isomer, neryl-pyrophosphate is the most likely precursor of alicyclic monoterpenes. Enzymes from peppermint leaf were shown to isomerize geraniol to nerol and geranyl pyrophosphate to nerylpyrophosphate. The trans-cis isomerase appears to be a key enzyme of terpenoid metabolism because it functions at the branchpoint where geranylpyrophosphate is diverted toward cyclic monoterpenes and away from further elongation to farnesyl pyrophosphate. Cell-free preparations from peppermint and spearmint leaves cyclize neryl pyrophosphate to α-terpineol, while geranylpyrophosphate is not cyclized. A key step in the biosynthesis of the p-menthane monoterpenes appears to be the enzymatic dehydration of α-terpineol to terpinolene and limonene. In extracts from carrot root, α-terpineol is dehydrated to terpinolene, which is the major monoterpene of carrot root oil. In peppermint leaf preparation α-terpineol is also converted to terpinolene. However, extracts from spearmint leaf dehydrate α-terpineol mainly to limonene. Terpenolene seems to be a precursor for the C_3-oxygenated monoterpenes like piperitone, pulegone, menthone, and limonene for the corresponding C_2-oxygenated constituents. Our knowledge about the biosynthesis of terpenoids is still very limited. There are only a few experimental proofs (with two or three plant systems). Our results show that fungi like Ceratocystis variospora may be an excellent subject for these studies.

2.2 Transformation of Terpenes by Microorganisms

It is well known that microorgnisms are able to synthesize, transform, and degrade substances which possess very complicated structures. Apart from these products, microorganisms also produce very active enzyme systems which are capable of catalyzing chemical reaction with foreign substances when the latter are added to a culture solution in which the organism is grown. The microbial enzymes assume the

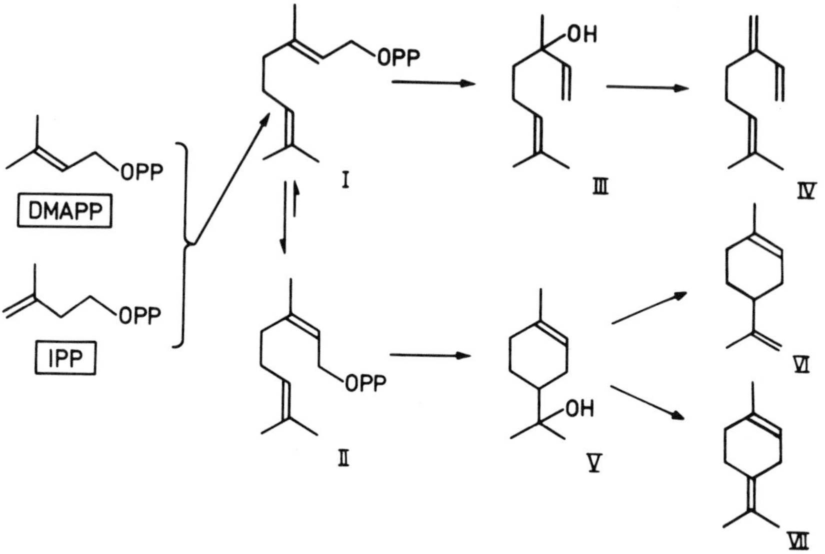

Figure 12 Possible pathway which may explain the formation of monoterpenes

TYPES OF MICROBIAL REACTIONS

OXIDATION

OXYGENATION OF $\geqslant$CH OR $>$C$=$C$<$

DEHYDROGENATION OF $>$CHOH

DEHYDROGENATION OF $>$CH-CH$<$

REDUCTION

HYDROGENATION OF $>$C$=$O

HYDROGENATION OF $>$C$=$C$<$

ISOMERIZATION

ESTERIFICATION / HYDROLYSIS

ACYLATION

TRANSGLYCOSIDATION

METHYLATION

CONDENSATION

CLEAVAGE OF $\geqslant$C-C$\leqslant$

DECARBOXYLATION

DEHYDRATION

AMINATION / DEAMINATION

HALOGENATION

PHOSPHORYLATION

Figure 13 Types of microbial reactions

role of chemical reagents characterized by a high degree of selectivity and stereospecifity. Thus, microorganisms are frequently in a position to carry out conversions in a single step with high yields, whereas many complicated steps would be required using synthetic organic methods. The oxidation of D-sorbit to L-sorbose, which is required for the synthesis of vitamin C, with Acetobacter suboxydans; the selective and stereospecific decarboxylation of meso α,α'-diamino pimelic acid by Bacillus sphaericus permits a convenient synthesis of L-lysine; stereoselective hydroxylations or selective oxidation of steroidal secondary hydroxyl groups to ketones are other important reactions in the synthesis of steroids. Some types of microbial reactions are summarized in Figure 13. Microbial oxidations and reductions seem to be the most interesting reactions because they are selective and stereoselective. Some of these reactions have been studied with terpenoids (24).

Figure 14 shows the stereospecific reduction of (-)-isofenchone quinone to 2-endo-hydroxy-epi-isofenchone by Absidia orchidis (24). (-)-Camphor quinone is transformed into 3-endo- and 3-exo-hydroxy camphor. (-)-Isofenchone is hydroxylated by the same organisms to 6-endo- and 6-exo-hydroxy isofenchone. (+)-fenchone is transformed into 6-exo- and 5-exo-hydroxy fenchone as shown in Figure 14. These examples demonstrate that non-activated carbon atoms, not only of steroidal substrates but also of very different structures, are hydroxylated by the same microbial enzymes in a highly regioselective manner. The hydroxylations which proceed with retention of configuration, require molecular O_2 and NADPH.

Figure 15 illustrates the stereospecific hydroxylation of p-menthane by a Pseudomonas strain (25). The hydrocarbone is transformed into cis-p-menthane-1-ol while the chemical reaction produces 6 isomer menthanols. The mechanism of this reaction was investigated by Tsukamoto et al. (26). They demonstrated that the hydroxylation of p-menthane and other saturated hydrocarbones is a mixed function oxidation requiring O_2 and NADH.

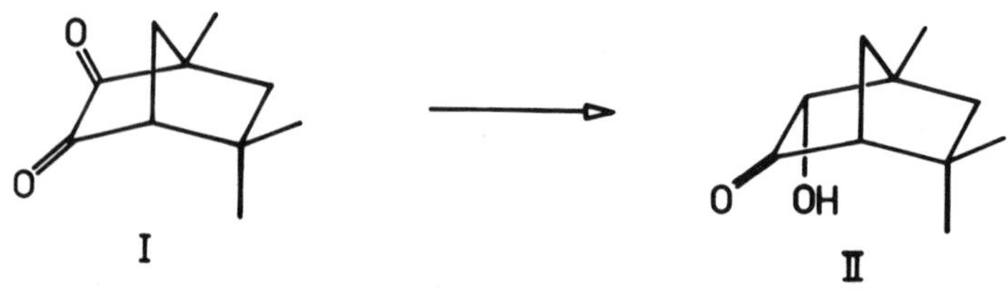

Figure 14 Stereospecific reduction and hydroxylation by Absidia orchidis

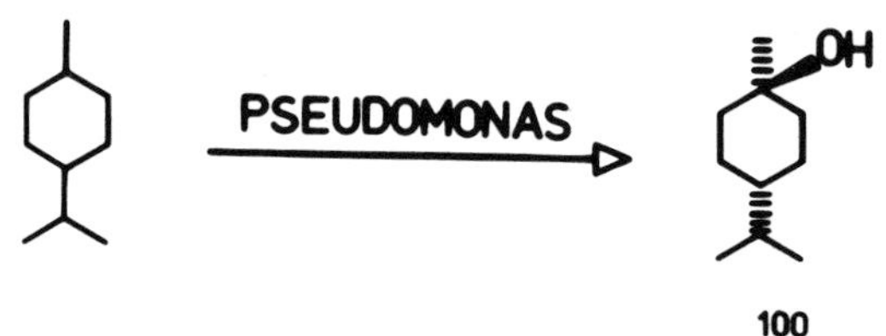

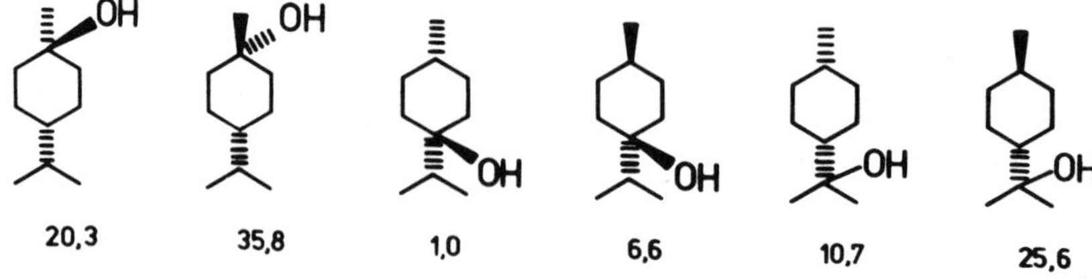

Figure 15 Stereospecific hydroxylation of p-menthane by Pseudomonas

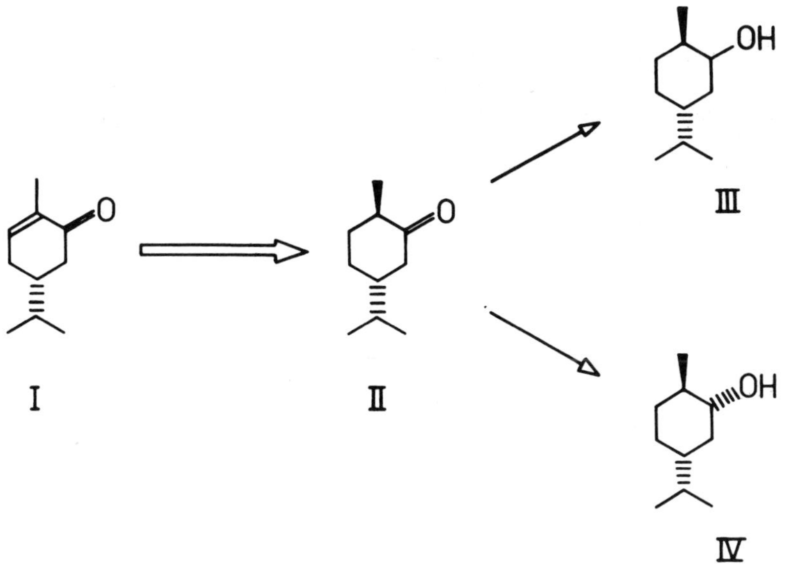

Figure 16 Stereospecific reduction by Pseudomonas ovalis

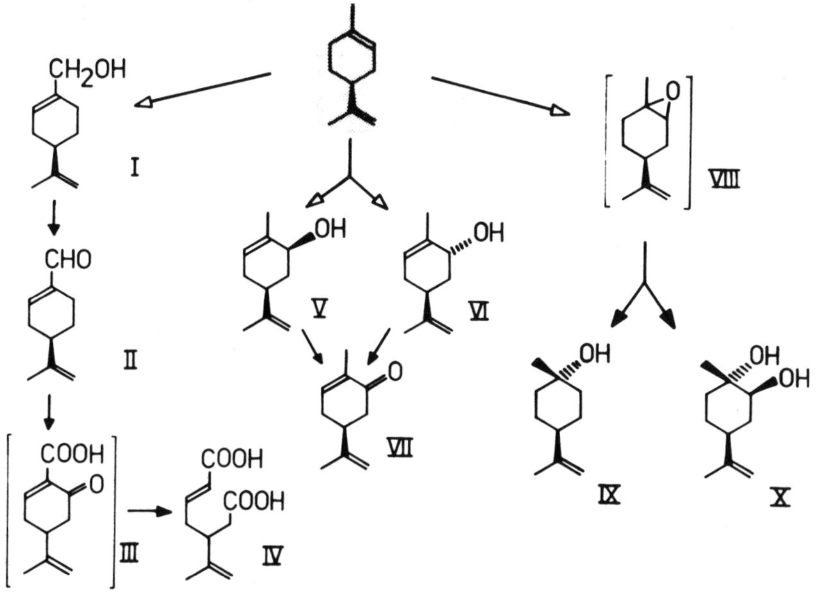

Figure 17 Transformation of limonene by Ceratocystis variospora

The stereoselective reduction of a α,ß-unsaturated ketone is presented in Figure 16. (-)-Carvotanacetone is transformed into (+)-carvomenthone by Pseudomonas ovalis (27). (-)-Carvon is reduced to (+)-dihydrocarvone. The microbial hydrogenation of the >C=C< double bond is stereospecific while the reduction of the carbonyl group did not have high stereospecifity.

Some microorganisms can use monoterpenes as a carbon source. (+)-Camphor was degraded by a strain of Pseudomonas via 5-hydroxycamphor, 2,5-diketocamphane to a keto acid and to a lactone (28). 1-Menthene was used as a carbon source by a strain of Cladiosporum. ß-Isopropyl glutaric acid was characterized as an end product of the degradation (29) Besides this compound trans-p-menthane-1,2-diol was isolated from the growth medium.

We investigated the transformation of limonene by Ceratocystis moniliformis. The organisms grown on a 0,5 % solution of limonene metabolized the monoterpene as shown in Figure 17. There are obviously three different metabolic pathways by which limonene is degraded. Epoxidation leads to limonene-1,2-epoxide which is hydrogenated to ß-terpineol or hydrated to the corresponding transglykol. Hydroxylation of limonene yields cis- and trans-carveol which are oxidized to carvone. In addition limonene is degraded via perilla alcohol and perilla aldehyde to a dicarboxylic acid. A possible endproduct of this branch could be ß-iso-propenyl glutaric acid, comparable to the degradation of 1-menthene by Cladiosporum. This example demonstrates that microorganisms are potent systems in converting monoterpene hydrocarbones into oxygenated terpenes. In addition they represent an excellent experimental subject for studying the biosynthesis of monoterpenes.

ACKNOWLEDGMENTS

This work was supported by Deutsche Forschungsgemeinschaft, Bonn.

REFERENCES

1. Maga, J.A., Critical Reviews in Food Science and Nutrition 8, 1 (1976)
2. Muller, C.J., Maggiora, L., Kepner, R.E., and Webb, A.D., J. Agr. Food Chem. 17:1373 (1969)
3. Augustyn, O.P.H., Van Wyk, C.J., Muller, C.J., Kepner, R.E., and Webb, A.D., J. Agr. Food Chem. 19:1128 (1971)
4. Muller, C.J., Kepner, R.E., and Webb, A.D., J. Agr. Food Chem. 20:193 (1972)
5. Suomalainen, H., and Nykänen, L., Naerings-middelindustrien 1:1 (1970)
6. Takahashi, K., Tadenuma, M., and Sato, S. Agr. Biol. Chem. 40:325 (1976)
7. Tressl, R., and Renner, R., Mschr. Brauerei 8:195 (1975)
8. Tahara, S., and Mizutani, J., Agr. Biol. Chem. 39:281 (1975)
9. Dubois, P., Rigaud, J., and Dekimpe, J., Lebensm.-Wiss. u. -Techn. 9:366 (1976)
10. Sulser, H., De Pizzol, J., and Büchi, W., J. Food Sci. 32:611 (1967)
11. Muys, G.T., Van Der Ven, B., and De Jonge, A.P. Nature 194:995 (1962)
12. Pfaff, H.J. in: "The Yeasts" (J. Lodder ed.), p. 845, North Holland Pub., Co., Amsterdam, 1970
13. Tahara, S., Fujiwara, K., and Mizutani, J., Agr. Biol. Chem. 37:2855 (1973)
14. Tressl, R., and Apetz, M., in preparation

15. Dimick, P.S., Walker, N.J., and Patton, S., Biochem. J. 111:395 (1969)
16. Mizugaki, J., Uchiyama, M., and Okui, S., Biochem. J. 58:273 (1965)
17. Collins, R.P., and Halim, A.F., Lloydia 33:481 (1970)
18. Halim, A.F., and Collins, R.P., Lloydia 34:451 (1971)
19. Lanza, E., Ko, K.H., and Palmer, J.K., J. Agr. Food Chem. 24:1247 (1976)
20. Sprecher, E., Kubeczka, K.-H., and Ratschko, M., Arch. Pharmaz. 308:843 (1975)
21. Buttery, R.G., and Garibaldi, J.A., J. Agr. Food Chem. 24:1246 (1976)
22. Tressl, R., and Apetz, M., in preparation
23. Croteau, R. in: "Geruch- und Geschmackstoffe" (ed. F. Drawert) p. 153, Verlag Hans Carl, Nürnberg, 1975
24. Tamm, C., FEBS Letters 48:7 (1974)
25. Tsukamoto, Y., Nonomura, S., and Sakai, H., Agr. Biol. Chem. 39:617 (1975)
26. Tsukamoto, Y., Nonomura, S., and Sakai, H., Agr. Biol. Chem. 41:435 (1977)
27. Noma, Y., Nonomura, S., and Sakai, H., Agr. Biol. Chem. 38:1637 (1974)
28. Bradshaw, W.H., Conrad, H.E., Corey, E.J., Gunsalus, I.C., and Lednicer, D., J. Am. Chem. Soc. 81:5507 (1959)
29. Mukherjee, B.B., Kraidman, G., and Hill, I.D., Appl. Microbiol. 27:1070 (1974)

THE CHEMISTRY OF THE BLACK-CARAMEL COLOUR SUBSTANCES OF SEVERAL HUMAN FOODS[x]

J.P.J. Casier
A. Ahmadi Zenouz
G.M.J. De Paepe

Department of Food Science and Technology
Katholieke Universiteit Leuven
Leuven, Belgium

The black caramel colour substances of several caramelised foods as coffee, chicory, the breadcrust, molasse and many pure carbohydrates were totally isolated by special solvent dehydration as creamy-white or white pulvers reversibly turning white-black by desolvatation-solvatation with several polar solvents. Sephadex fractionation of the pulvers and their acetates showed the common polymeric nature and origin of all caramel colours, originating by very actively, irreversibly polymerising, low molecular weight, heat degradation products of lower sugars. A study was made of the highest caramel polymer fractions, separated by Sephadex G200 fractionation, from coffee, chicory, and the breadcrust, as well as from the low molecular weight monomer from molasse caramel colour. No taste relationship was found between Sephadex G200 purified caramel fractions from coffee, chicory and breadcrust, and the original products used.

[x]Part of this paper was presented at the 59th Annual Meeting of the American Chemical Society, Montreal Oct. 1974

ISBN 0-12-169060-1

I. INTRODUCTION

Many attempts have been made for isolating the black caramel colour substances from several products as sugar molasses, coffee a.o. Practically all were concerned with one or other partial fraction of the caramel colours, depending on the isolation method used (1-15).

As earlier communicated (16, 17) a method was elaborated permitting isolation of the entire black caramel colour substances of all caramelised pure carbohydrates and carbohydrate-rich caramelised foods, such as coffee, chicory, the breadcrust, molasse etc. This method is based on a common universal property of all caramel colour substances. All black caramel colour substances as far tried, are isolated by special solvent dehydration as nearly creamy-white or white pulvers -called here F total-, reversibly colouring black-white on solvatation-desolvatation with many polar solvents as f.i. water and acetic acid, as shown in figure 1.

All these black caramel colour substances seem to originate similarly by polymerisation from small molecules, -called here Fsm-, heat degradation products of lower sugars. Indeed, on evaporation of these small molecules (Fsm), eluted

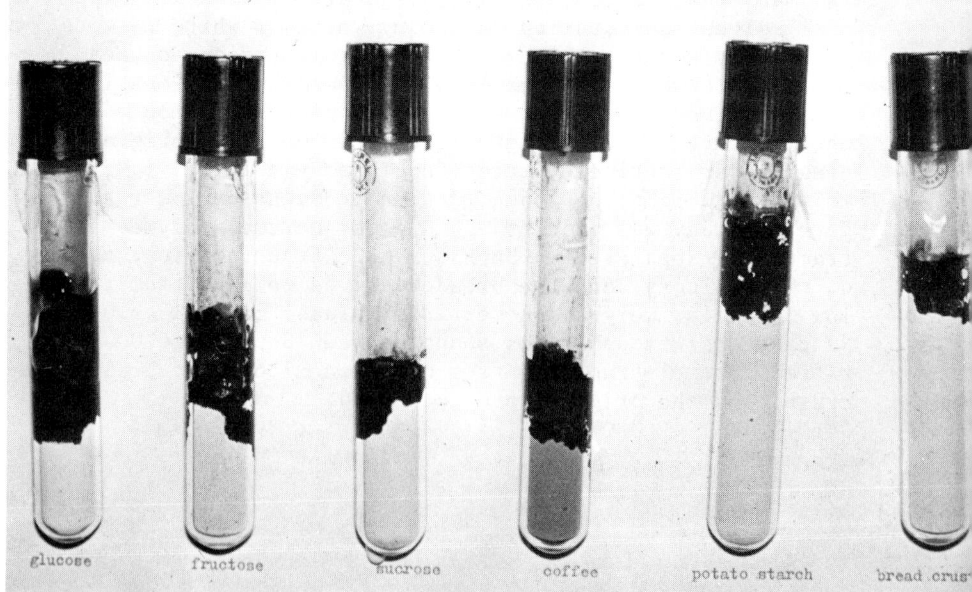

FIGURE 1. Top rehydrated, white-black reversible caramel colour substances of heat degraded carbohydrate products.

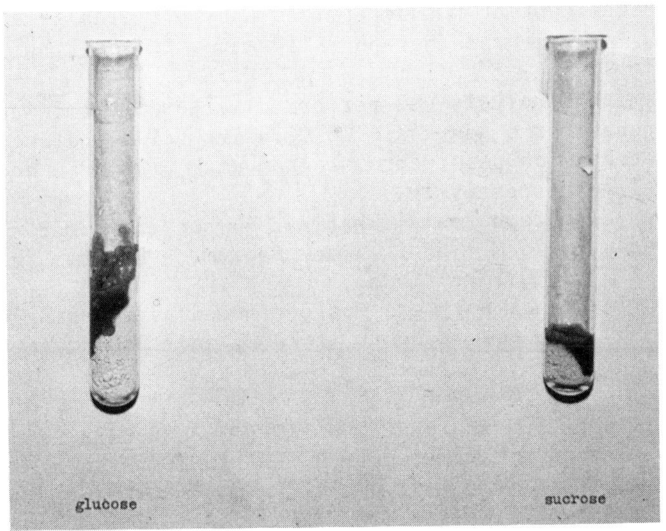

FIGURE 2. Top rehydrated, white Fsm fractions of some caramelised sugars.

from Sephadex G25-fractionated F total, a high polymer is obtained.

All these small brown-yellow molecules show the same white-black reversibility as the polymeric F total fraction, as shown in figure 2.

II. MATERIALS AND METHODS

Sugar caramels were prepared by dry oven heating at 180-200°C for 1-2 hours.

Wheat breadcrust and potato starch were heated at 240°C for 30 minutes.

For coffee, chicory and sugarbeet molasse, normal trade products were used.

Molecular sieves of Sephadex G25, LH20 and G200 column fractionation was used for fractionation and molecular weight estimation, absorbances being read at 420 nm.

Nitrogen content was determined by the Kjeldahl-method, sugars being gaschromatographically estimated as T.M.S. derivatives (18).

Polymer hydrolysis was performed with 90% formic acid in sealed tubes for 6 hours at 100°C, amino acids and sugars, being detected on paper-chromatogram with ninhydrin and aniline oxalate respectively.

Paper electrophoresis (Whatman 4) was performed at 800V., using Borax (0.05 M - pH 9.2) and Sodium carbonate-bicarbonate (0.1 M - pH 10.5) buffers.

Acetylation of polymers was performed by stirring with acetic anhydride for 72 hours at 20°C, and 10 hours at 80°C.

III. ISOLATION AND FRACTIONATION OF PURE CARBOHYDRATE CARAMEL COLOUR SUBSTANCES

A. Isolation

All caramel colour substances were isolated in accordance with the following outline, slight modifications being added for polysaccharide containing products (fig. 3).

To sugar caramels of Glucose, Fructose, Sucrose, Galactose and Xylose, dissolved in the smallest possible amount of water, glacial acetic acid was added, up to slight precipitate formation.

Desolvatation by adding dry aceton (20 Vol/Vol) changes this black solution into a faint yellow supernatant, the caramel colour substances being entirely precipitated as white or creamy white pulvers, called here F total.

From the yellow supernatant the low molecular weight and very actively polymerising Fsm fraction is precipitated by dry diethylether (10 Vol/Vol). Both fractions are vacuum dried, acetic acid and water being previously completely eliminated by repeated washing with dry aceton (only for F total, not Fsm) and diethylether. Both fractions show black-white colour reversibility on solvatation-desolvatation with several polar solvents as f.i. water, acetic and formic acid, pyridine a.o. (fig. 4). They are insoluble in nonpolar solvents, the highest molecular weight polymers being however insoluble in acetic acid too.

For caramelised potato starch, the evaporated waterextracts, previously freed from higher dextrins by tentative ethanol fractionation, were dissolved in glacial acetic acid, a few insolubles being discarded by centrifugation. White F total and Fsm pulvers were isolated by aceton and diethylether precipitation.

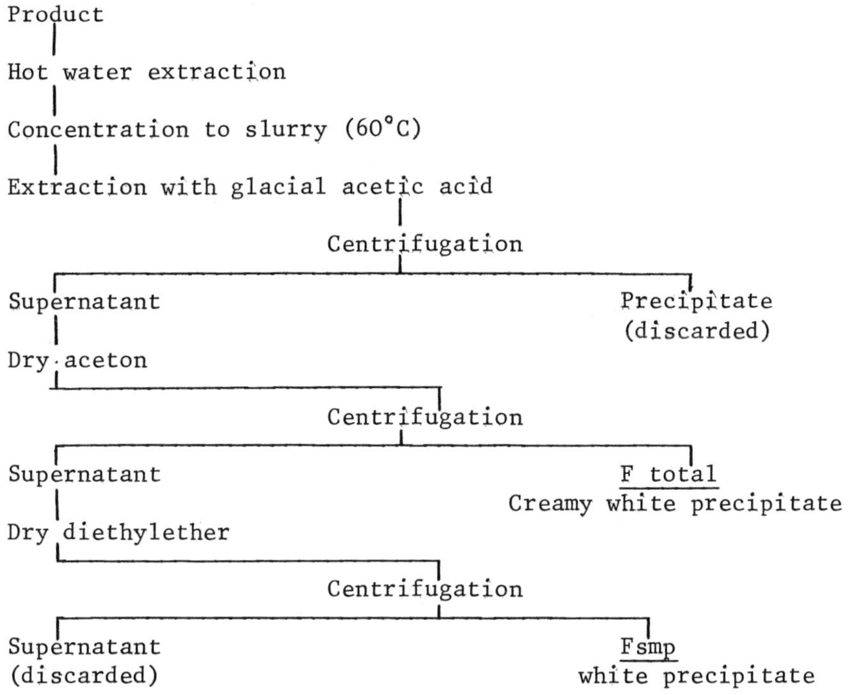

FIGURE 3. General outline for isolation of all black caramel colour substances as F total and Fsm pulvers.

B. Fractionation and Polymerisation Study of the Caramel Colour Substances Using Sephadex G25

Evolution of the F total caramel pulver of Glucose, Fructose, Sucrose and roasted potato starch, all developed two coloured bands on a Sephadex G25 column, caramels of protein containing products usually giving more bands (fig. 5).

All F total fractions show two clear distinct elution peaks, one due to the quick moving high polymer FI, and one to the slowest moving coloured fraction (Fsm), only a few lower polymeric materials usually being present between the FI and the Fsm fractions, as shown in the figures 6-7-8 and 9.

Hence, some Fsm seems to be coprecipitated by aceton through adsorption on the precipitated higher polymers of the F total.

Attempting to purify the slowest moving coloured fraction (Fsm) the hot evaporated Fsm eluates -called Fsme- reeluted on the same Sephadex G25 column showed the presence of high polymeric FI besides some lower polymers and the used Fsme

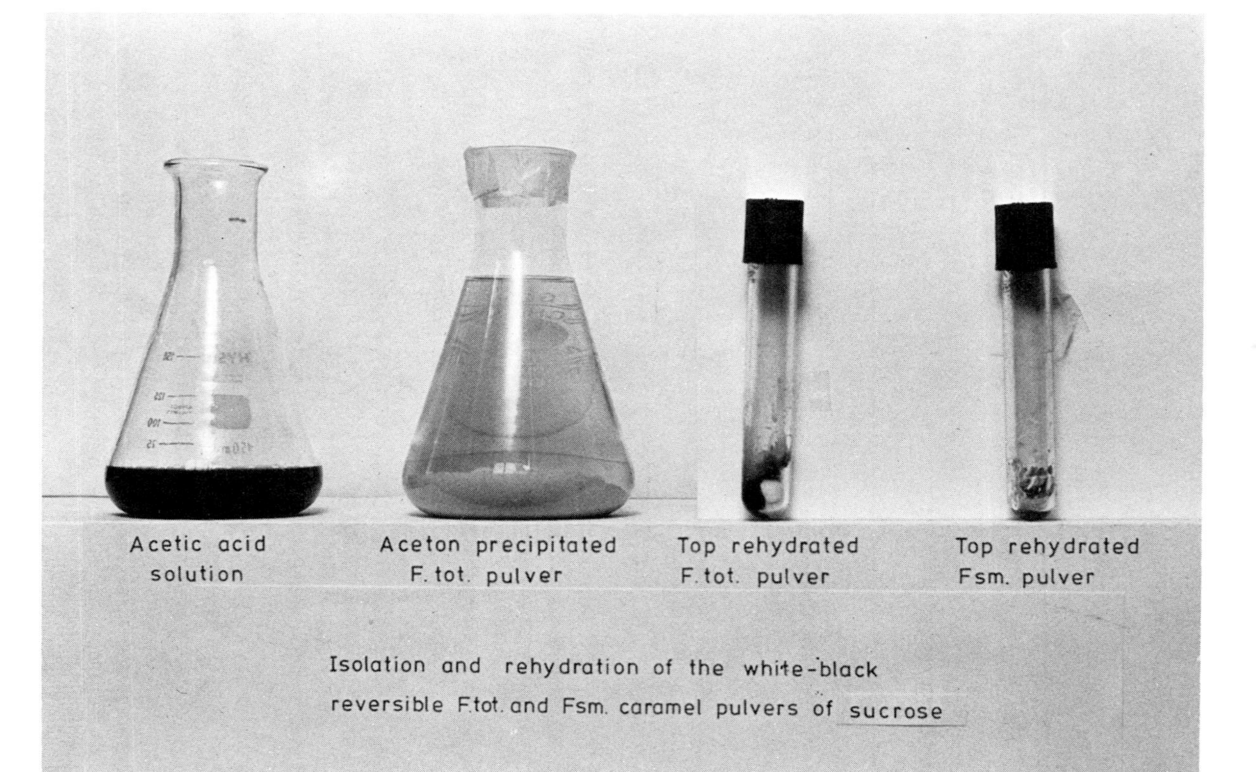
Acetic acid
solution
Aceton precipitated
F. tot. pulver
Top rehydrated
F. tot. pulver
Top rehydrated
Fsm. pulver
Isolation and rehydration of the white-black
reversible F.tot. and Fsm. caramel pulvers of sucrose

FIGURE 4.

fraction. This indicates the polymerising character of this smallest coloured fraction.

The same polymerisation was obtained on hot evaporation of the Fsm fraction, precipitated with diethylether from the aceton supernatant -called Fsmp- as shown in the figures 6-7-8 and 9. Similar results were obtained on hot evaporation of the eluted intermediate polymers between FI and Fsm.

Sucrose caramel shows two low molecular weight fractions, found also in the reeluted polymerised Fsm diagram. So clearly the same type of elution diagram is obtained with all pure carbohydrates, showing also the same type of diagram for the polymerised Fsm.

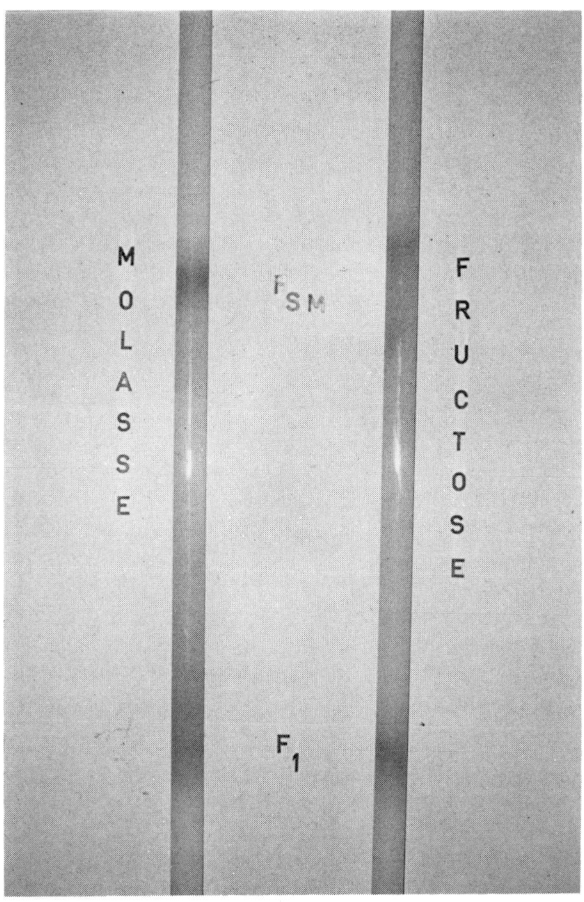

FIGURE 5. Elution of fructose and molasse F total pulver on G25 Sephadex.

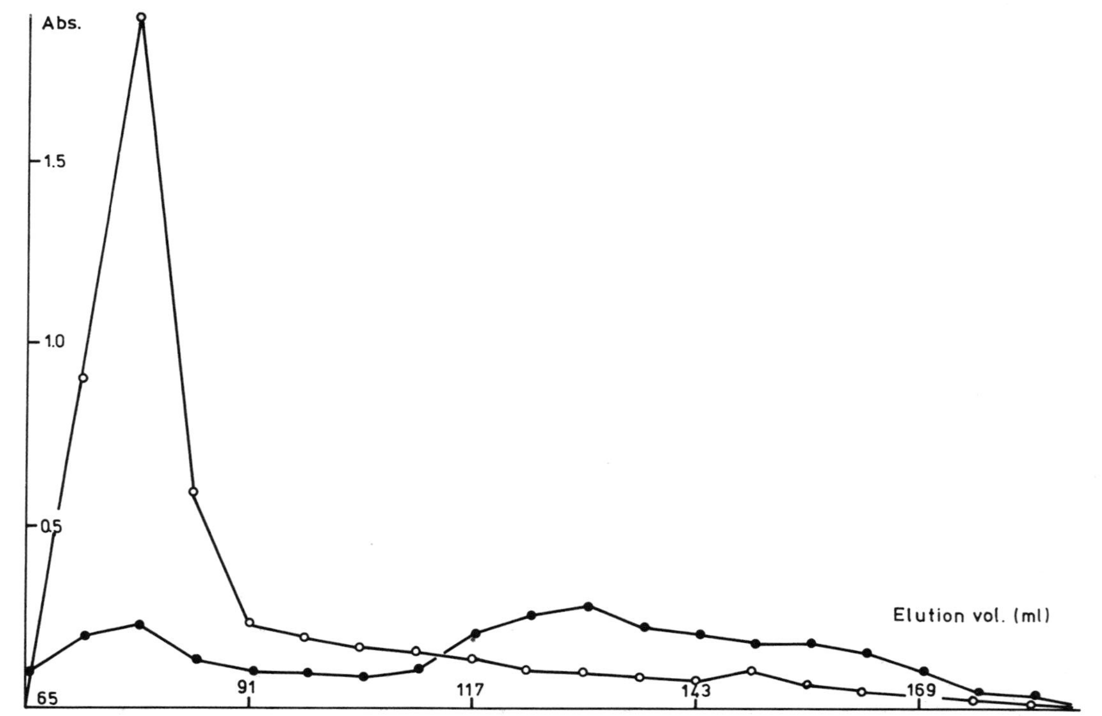

FIGURE 6. Elution diagram of fructose F total and polymerised Fsm caramel pulver in Sephadex G25 (column 77 x 1.5 cm) F total —o—o—
Fsme —●—●—

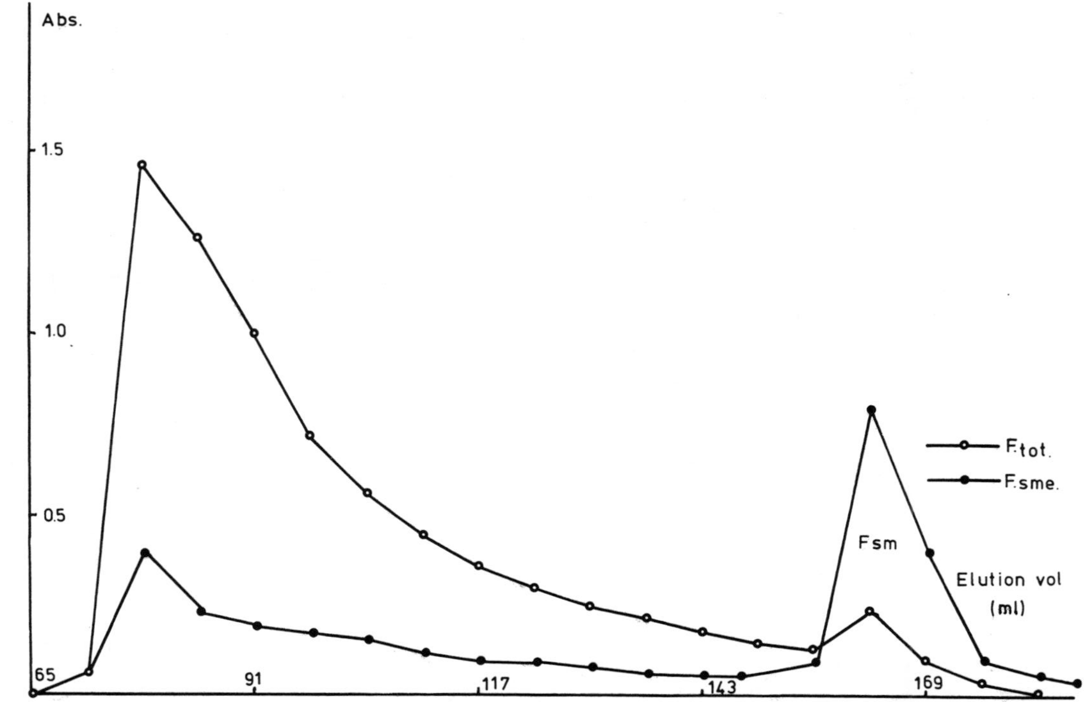

FIGURE 7. Elution diagram of glucose F total and polymerised Fsm caramel pulvers on Sephadex G25 (column 77 x 1.5 cm) F total —o—
Fsme —●—

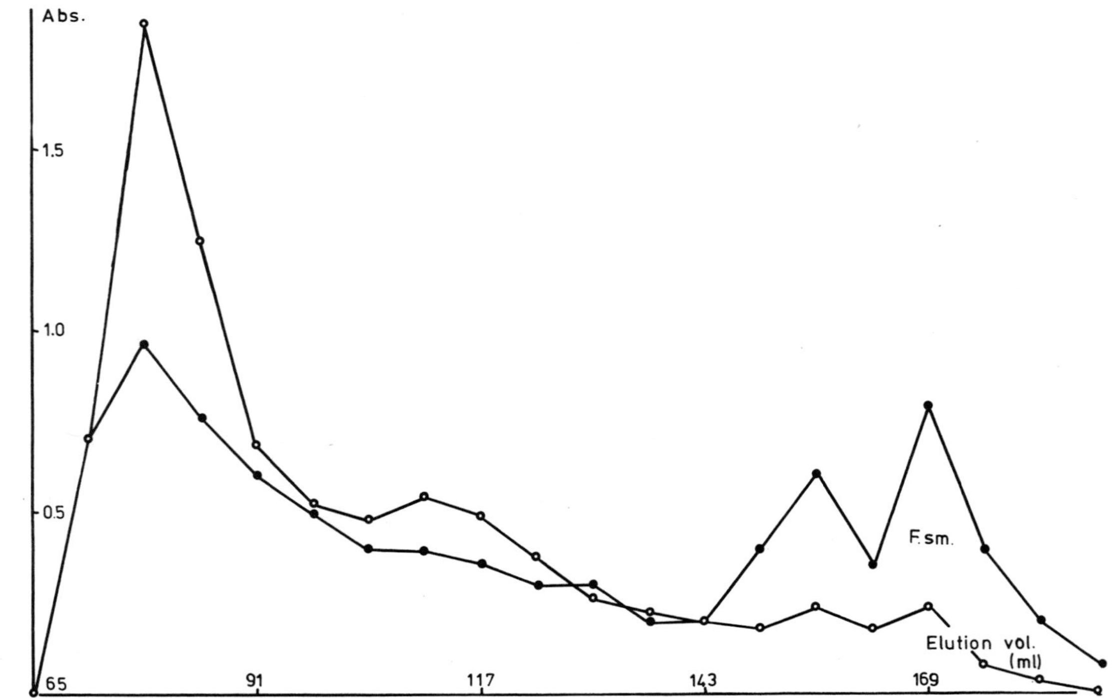

FIGURE 8. Elution diagram of sucrose F total and polymerised Fsm on Sephadex G25 (column 77 x 1.5 cm) F total —o—
Fsmp —●—

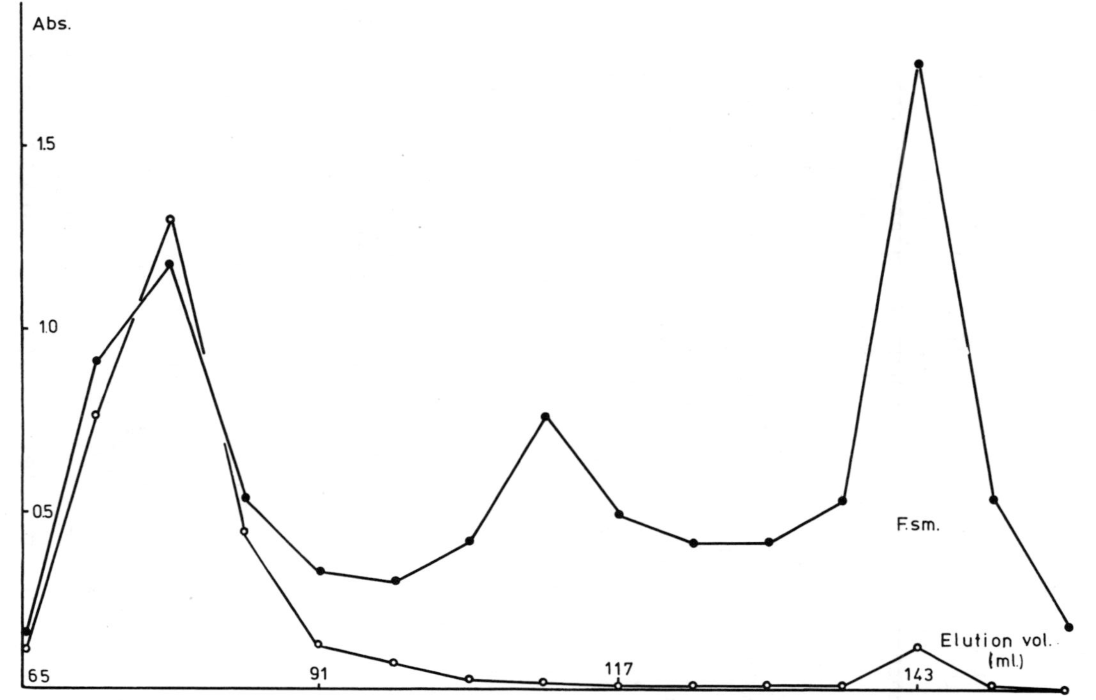

FIGURE 9. Elution diagram of caramelised potato starch F total and polymerised Fsm on Sephadex G25 (column 77 x 1.5 cm) F total —o—
Fsmp —●—

IV. ISOLATION AND FRACTIONATION STUDY OF THE CARAMEL COLOUR SUBSTANCES OF SOME CARBOHYDRATE RICH FOODS AS BREADCRUST, COFFEE, CHICORY AND MOLASSE

A. Isolation

All caramel colour substances were isolated in accordance with the above mentioned outline (fig. 3).

Molasse (23 g) thoroughly freed from sucrose and most water by repeated oozing through hot (70°C) stirred ethanol (1000 ml, 96%), was dissolved in glacial acetic acid. As for sugar caramels, a white-black reversible F total and Fsm pulver were obtained on addition of a surplus of aceton and diethylether as shown in figure 10.

The warm (400 ml, 40°C) waterextracts of 20 g breadcrust were freed from higher dextrins by tentative ethanol fractionation after concentration (30 ml). The evaporated extracts give a black solution in glacial acetic acid, a few insolubles being discarded by centrifugation. Here too, a white-black reversible F total and Fsmp pulver is obtained by aceton and diethylether precipitation as shown in figure 11.

Similar results are obtained using the same isolation procedure for starch and other polysaccharides.

The evaporated (60°C) boiling water extracts (400 ml) from

Fig. 10

Isolation and rehydration of the white-black reversible caramel pulver and Fsm of molasse

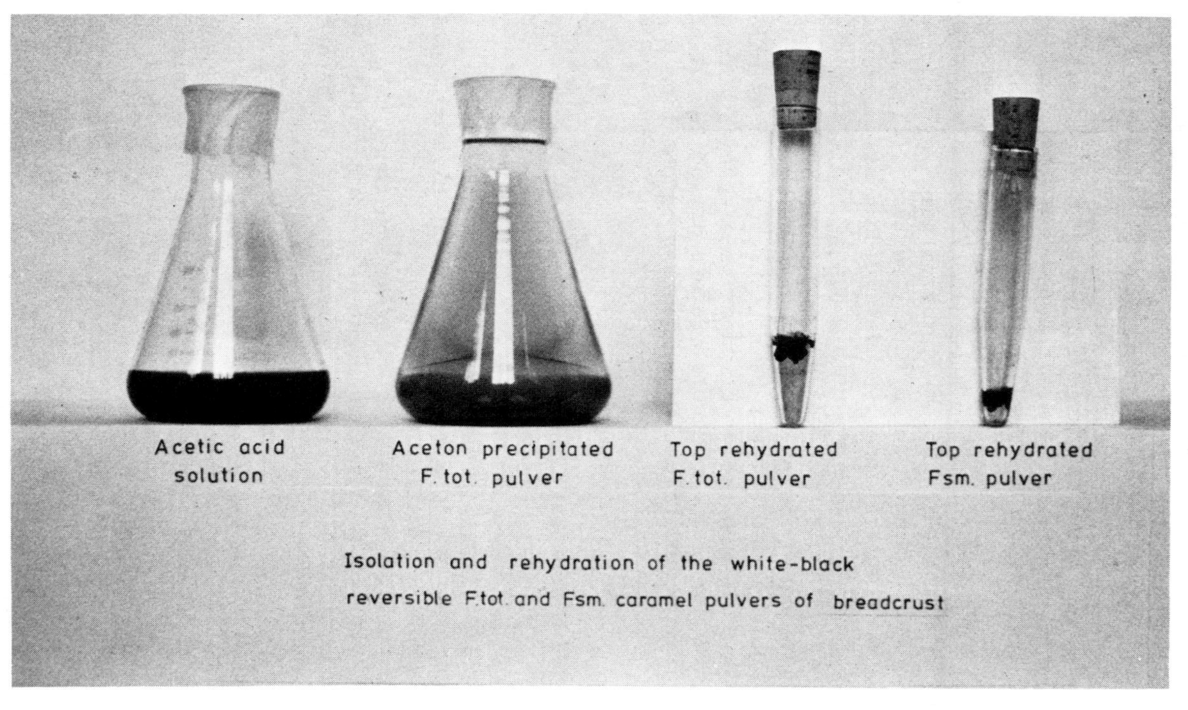
Acetic acid
solution
Aceton precipitated
F. tot. pulver
Top rehydrated
F. tot. pulver
Top rehydrated
Fsm. pulver
Isolation and rehydration of the white-black
reversible F.tot. and Fsm. caramel pulvers of breadcrust

FIGURE 11.

20 g coffee and chicory, washed with some dry ethanol, give a black solution in glacial acetic acid, a few insolubles being discarded by centrifugation. Similar results are obtained with hot glacial acetic acid extraction of coffee (20 g) previously washed with aceton and chloroform for water and pigment elimination.

Here the aceton precipitated F total, as well as the Fsm, are slightly brown coloured for coffee and goldish for chicory, due to adsorbed colour from the respective red-brown and goldish supernatant. Freed from most adhering colours by repeated precipitation, the ether dried F total and Fsm -although slightly coloured- show also "white-black" reversibility. However, this repeated precipitation eliminates the bulk of the Fsm from the F total fractions (fig. 12 and 13).

As mentioned in the Sephadex fractionation study, the coffee F total contains one black coloured fraction the FI -the highest molecular weight one- hardly decolouring by desolvatation and drying, although the brown FII and others do. For all white pulvers being extremely hygroscopic, drastic elimination of water or vapor is essential for preventing recoloration of the products. Before drying the white pulvers, the colour-inducing acetic acid, must be completely eliminated by aceton extraction.

B. Fractionation and Polymerisation Study of the Caramel Colour Substances

By fractionation on Sephadex G25, each of the F total pulvers divide into a major fraction of high molecular weight, and one or two smaller fractions of lower polymers, as seen in figure 14. So more elution peaks are found here as with lower sugar caramels. This might be due to protein copolymerisation, as the purified FI polymers were shown to contain protein and sugars too in the molecule.

Although practically all Fsm was lost from the F total of coffee and chicory by elimination of the adhering colours, the same polymerisation phenomenon of the Fsm was observed as with the pure carbohydrates, as shown in the elution diagrams of molasse, chicory, coffee and breadcrust F total pulvers (fig. 15-16-17-18).

As with pure sugar Fsm, here too, the amount of high and intermediate polymers build from the Fsm, depends on the repeating hot evaporation of these fractions.

In contrast to all brown coloured fractions of any other caramel source the highest molecular weight fraction FI from coffee F total is a deep-black coloured substance, the lower polymers however being brown coloured too, and easily decolorised.

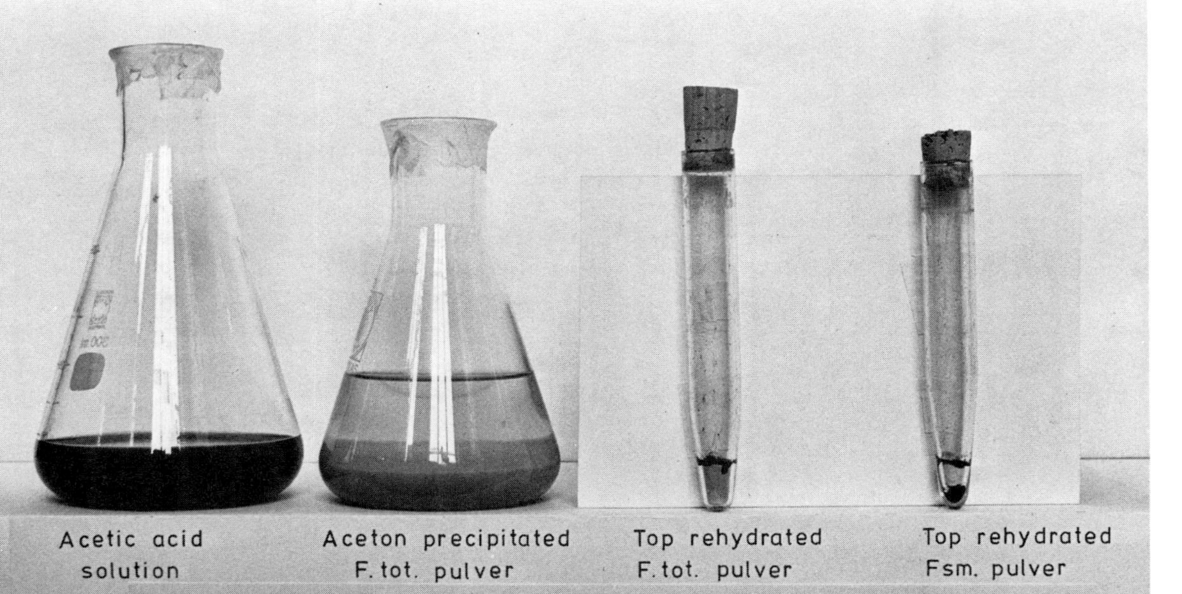
Acetic acid
solution
Aceton precipitated
F. tot. pulver
Top rehydrated
F. tot. pulver
Top rehydrated
Fsm. pulver
Isolation and rehydration of the white-black
reversible F.tot. and Fsm. caramel pulvers of chicory

FIGURE 12.

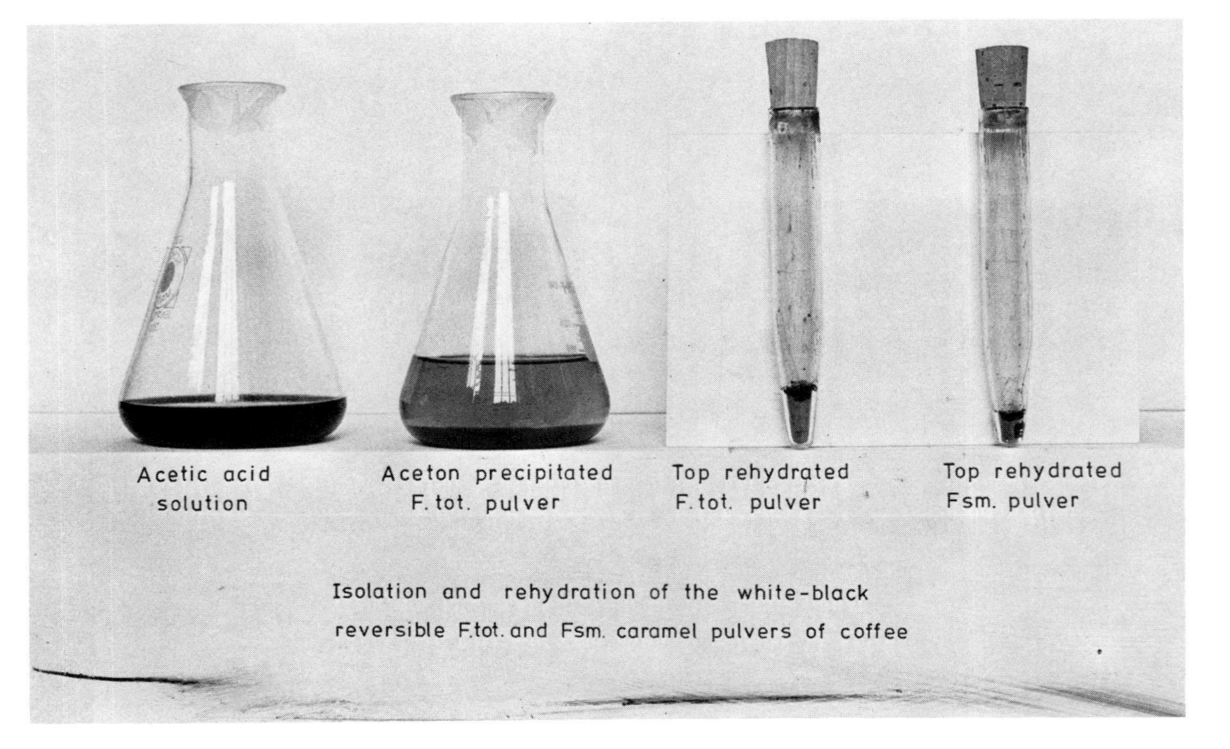

Isolation and rehydration of the white-black reversible F.tot. and Fsm. caramel pulvers of coffee

FIGURE 13.

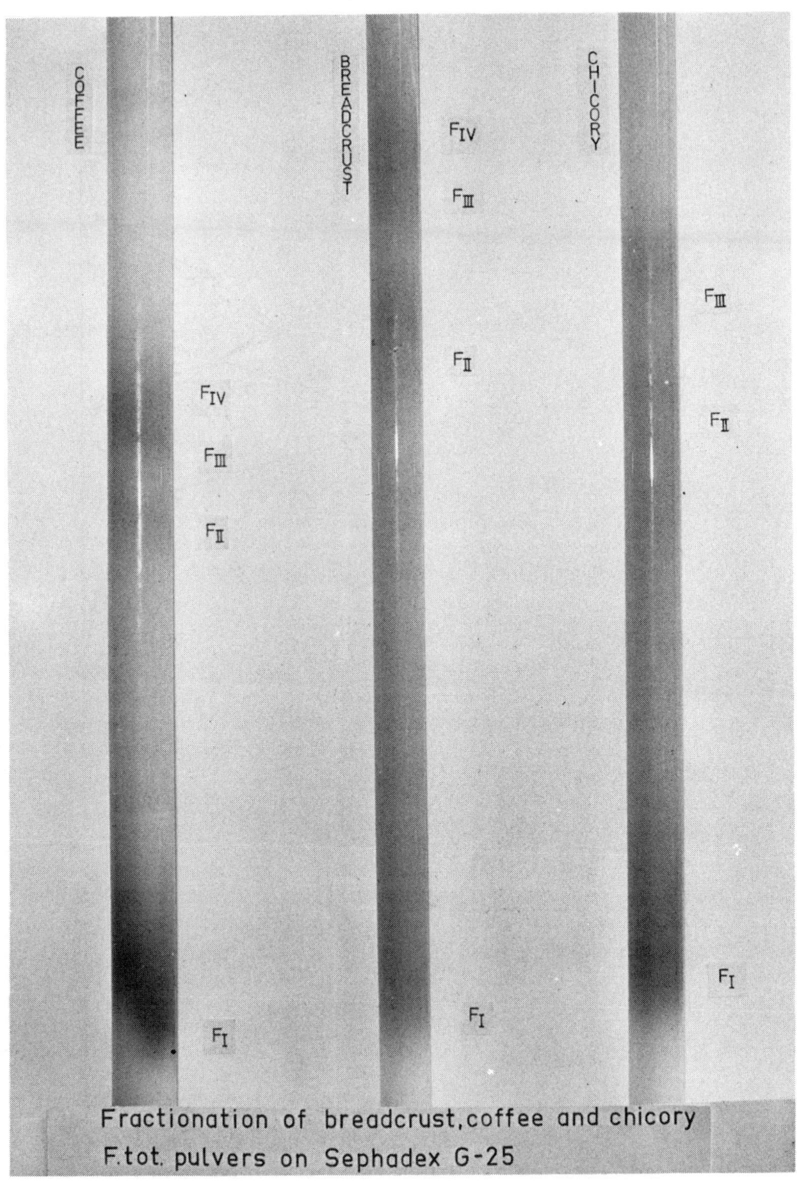
COFFEE
BREADCRUST
CHICORY
F_{IV}
F_{III}
F_{II}
F_{I}
F_{IV}
F_{III}
F_{II}
F_{I}
F_{III}
F_{II}
F_{I}
Fractionation of breadcrust, coffee and chicory
F.tot. pulvers on Sephadex G-25

FIGURE 14.

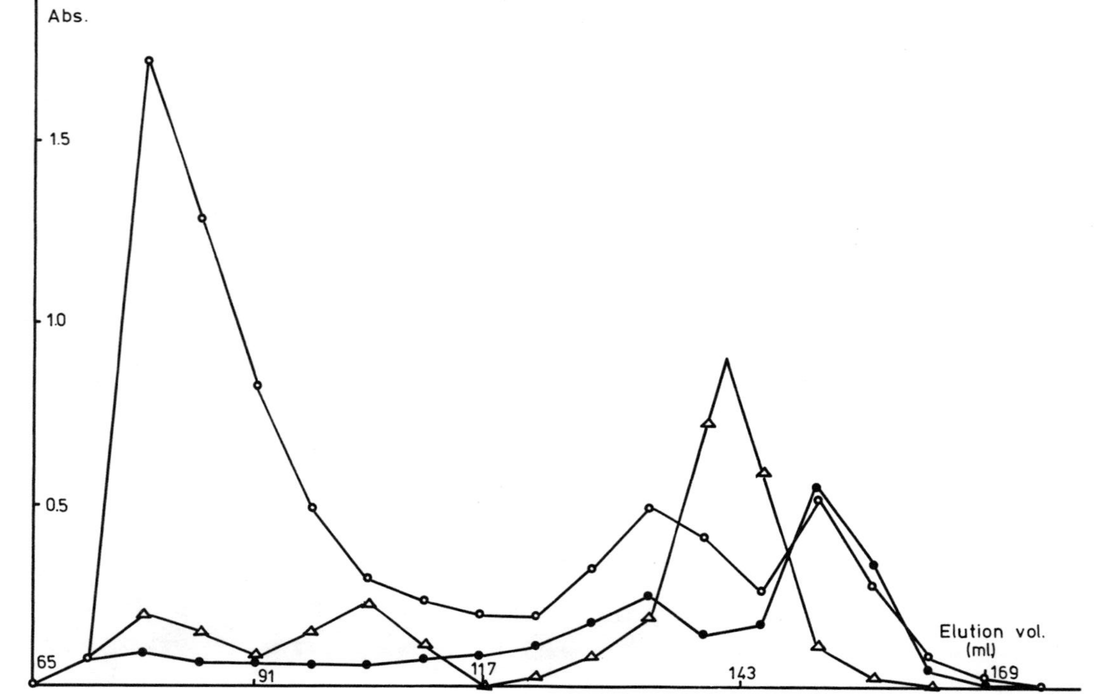

FIGURE 15. Elution diagram of molasse F total and polymerised Fsm pulvers on Sephadex G25.
F total —o— Fsme —△— Fsmp —●—

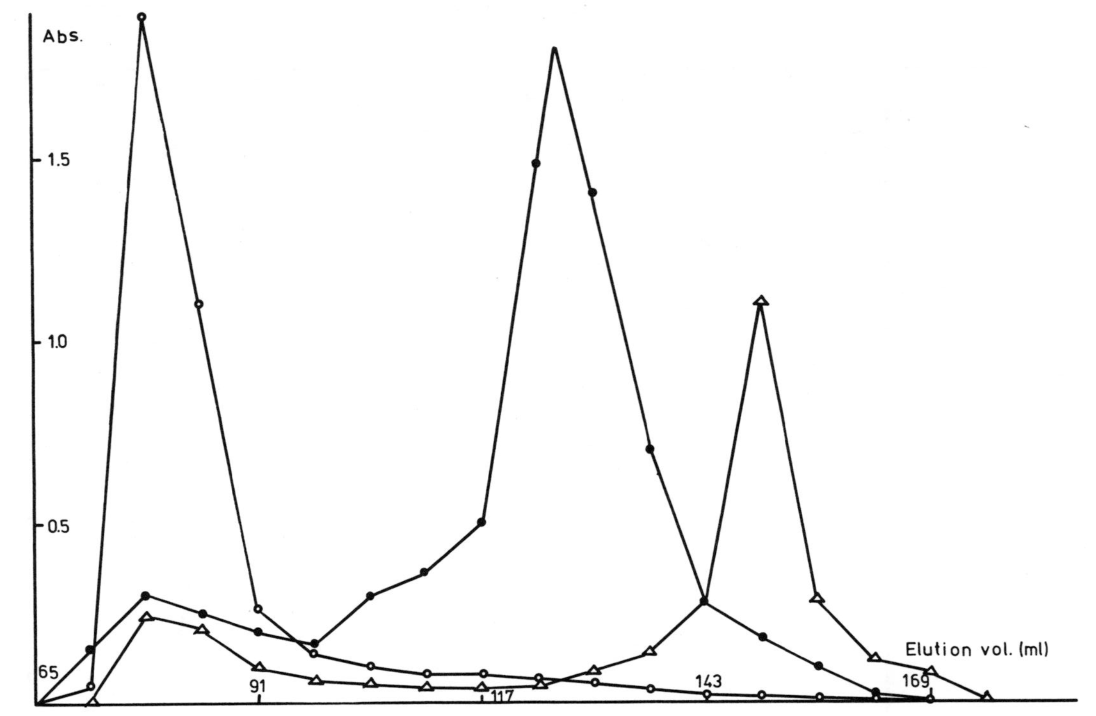

FIGURE 16. Elution diagram of chicory F total and polymerised Fsm pulvers on Sephadex G25.
F total —○— Fsmp —△— Fsme —●—

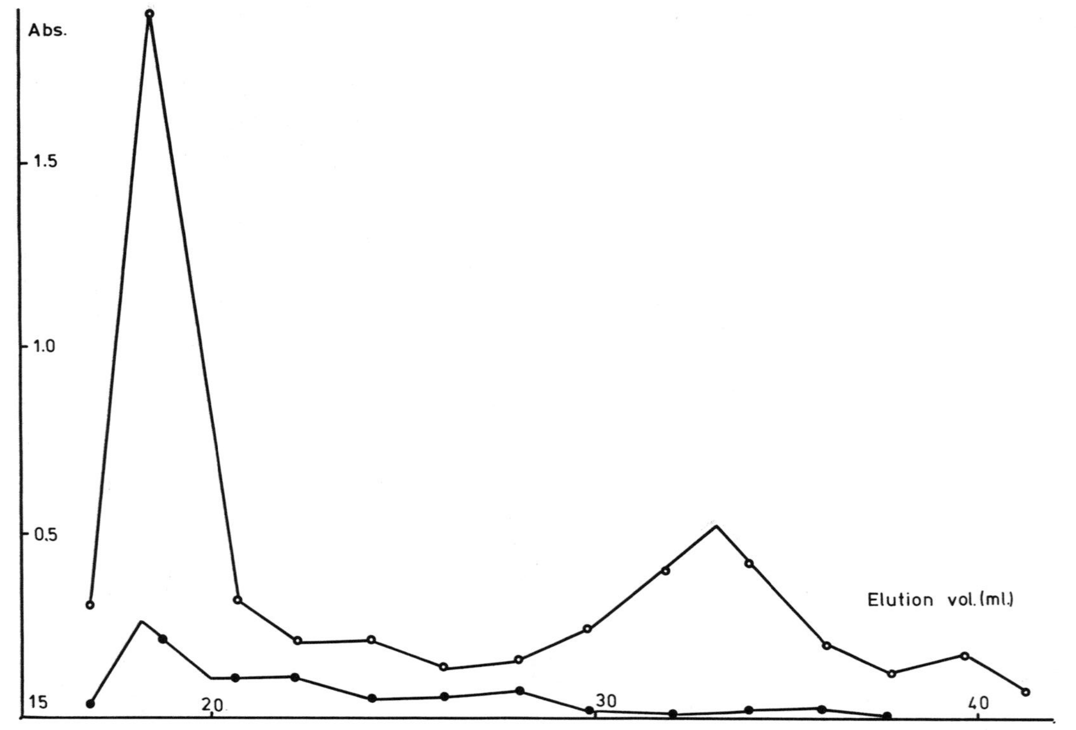

FIGURE 17. Elution diagram of coffee F total and polymerised Fsm pulvers on Sephadex G25.
F total —o—
Fsm —●—

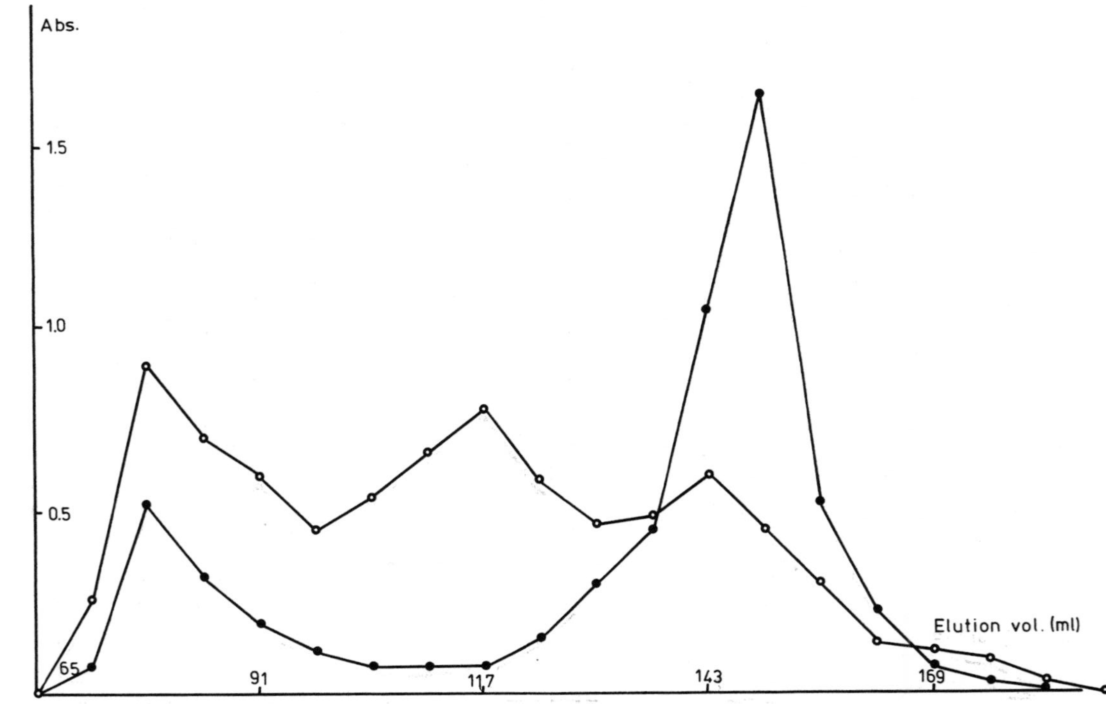

FIGURE 18. Elution diagram of breadcrust F total and polymerised Fsm pulvers on Sephadex G25.
F total —o— Fsme —●—

This FI seems to be the only caramel fraction hardly or not decolorising by the aceton desolvatation and the ether drying process, although the dried product is clearly much lighter in colour than the hydrated form, complete dehydration being very difficult to achieve. Coffee Fsmp, partially polymerised in the drying process, is slightly coloured too.

The FI fractions of coffee, chicory and the breadcrust, purified by repeated fractionation on Sephadex G25, were eluted on Sephadex G200 (fig. 20). Chicory FI separated into two fractions : coffee and breadcrust FI did not split up furtheron (fig. 19).

The fast running fractions leave the column at the exclusion limit, together with dextran blue, indicating molecular weights approaching or over 200.000. However, as indicated by Sephadex G100 fractionation, the FI still contains a substantial fraction of polymers lower than 100.000 molecular weight.

All caramel colour substances used, were obtained by glacial acetic acid extraction of the caramelised products or the evaporated water extracts, followed by aceton precipitation. Higher molecular weight polysaccharides, being insoluble in glacial acetic acid, normally should not be present in the extracts. Hence, sugars found to be present in the Sephadex G200 purified FI fraction, normally should not originate from co-eluted, non degraded, polysaccharides, but might be due to smaller pieces of heat damaged degraded polysaccharides, incorporated into the high polymeric FI fraction.

V. COMPOSITION OF THE SEPHADEX G200 PURIFIED FI FRACTIONS

The polymer obtained by polymerisation of pure molasse Fsme, freed from sugars and amino acids by Sephadex G25 fractionation -as shown by paper chromatogram of the eluates- is resistant to alkaline and acid hydrolysis.

However, on polymerising this Fsme in the presence of amino acids, the acids are incorporated in the polymer and eliminated by acid hydrolysis, although no sugar is detected in the hydrolysate of Sephadex G200 purified FI from glucose caramel. Here, the hydrolysates of the mentioned Sephadex G200 purified coffee-, chicory-, and breadcrust FI colour fractions show the expected presence of amino acids, but also the presence of sugars.

The composition of the FI G200 fractions is shown in table I, sugars being estimated by G.L.C. (18), amino acids being detected by ninhydrin on paper chromatogram.

As caramels and caramelised carbohydrate rich products have a specific caramel taste, all fractions eluted from the

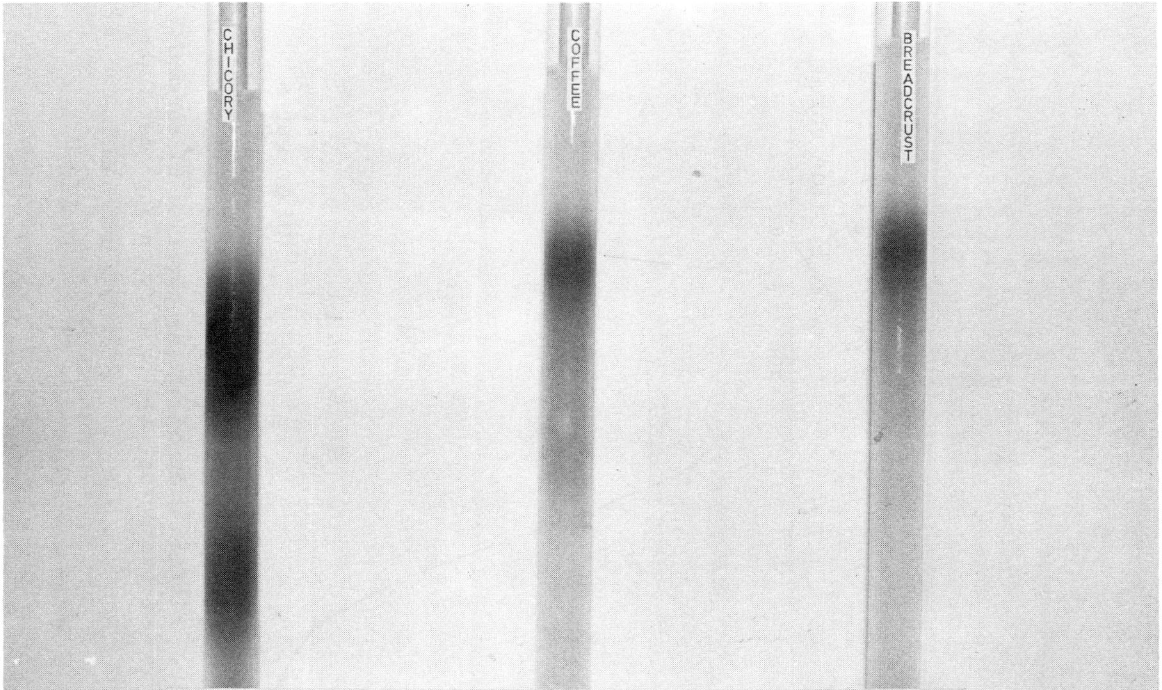

Fractionation of the F_I caramel colour fraction of Breadcrust, coffee and chicory on Sephadex G-200

FIGURE 19.

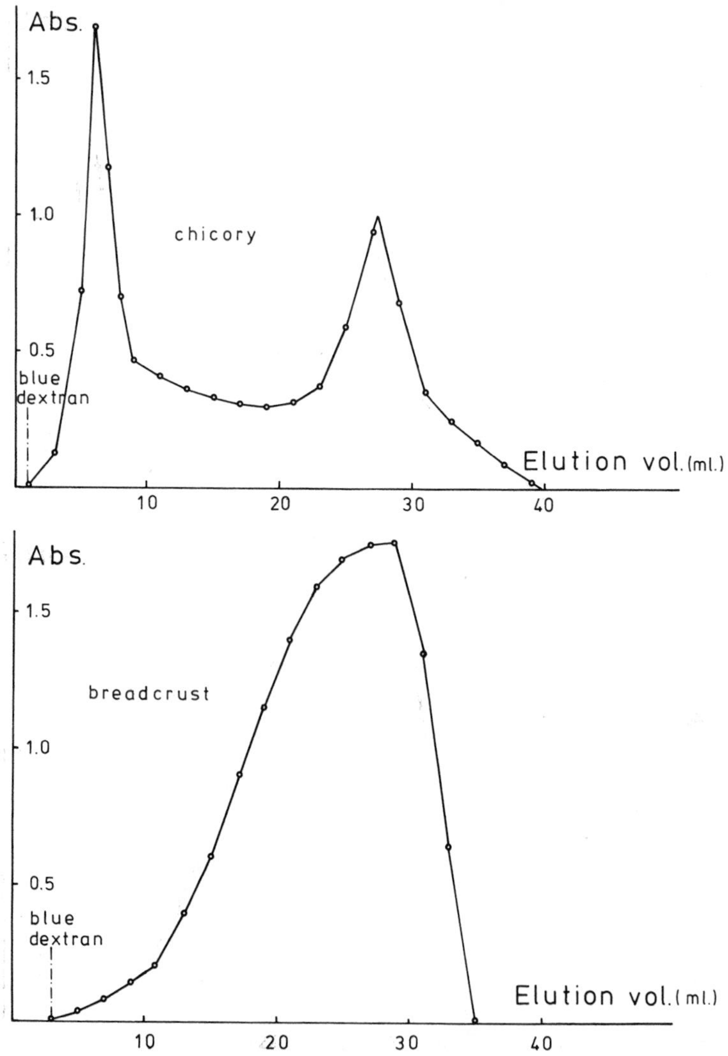

FIGURE 20. Elution diagram of the purified FI caramel colour fraction of breadcrust and chicory on Sephadex G200 (2 x 33 cm).

TABLE I. Composition of the Sephadex G200 fractionated FI of breadcrust, chicory and coffee

	breadcrust	chicory	coffee
% N2	3.91	2.27	2.38
% Protein	22.48	14.20	14.85
% Sugars total	38.34	46.68	24.58
Ara.	-	2.11	3.98
Xyl.	trace	-	-
Fru.	-	19.10	-
Gal.	-	-	12.36
Glu.	38.34	25.47	trace
Man.	-	-	8.24
% Ash	-	5.60	9.38
% Colour substance	39.18	33.52	51.19
Specific taste	none	none	none

Sephadex G25 and G200 column were tested for taste by a panel of six persons, trying to state taste relationship to one of the original products; coffee, chicory and breadcrust. Not one fraction could be attributed to its original product, by any of the panelmembers, proving that specific taste is not linked to the caramel colour substances. However, most taste and smell relationships were clearly recognised in the evaporated diethylether extracts of the coffee, chicory and breadcrust filtrates.

VI. PHYSICAL AND CHEMICAL PROPERTIES OF FSM (MOLASSE) AND CARAMEL COLOUR POLYMERS

The molasse Fsme, being eluted from the Sephadex G25 column between xylose and furfural, might have a molecular weight of 100-150. As no furfural is obtained from pure Fsme, or from polymerised pure Fsme, on boiling with strong concentrated acid, the furanose or pyranose structure must be lacking.

Fsm polymerises very actively and unreversibly, even at lower temperatures (20°C). So always some polymer is present in the Fsme, rendering elementary or structural analysis of the original product very difficult or nearly impossible. However the stable, not polymerising acetate can be obtained by letting the Fsme eluate, drop from the Sephadex G25 column, straight into refluxing acetic anhydride, the acetate being purified by ethanol (90%) fractionation on Sephadex LH20. The acetates, as well as the benzoates of Fsm and the polymers, all brown coloured, do not show white-black reversibility any more nor polymerisation. These acetates, waterinsoluble, but soluble in many nonpolar solvents, are purified by precipitation with petroleum ether from $CHCl_3$.

Deacetylated by diluted NaOH, the white-black reversibility of the Fsm, as well as the polymerising activity are completely recovered. The Fsme acetate, deacetylated and polymerised, gives the same type of elution diagramm as the original polymerised Fsme (fig. 21). Hence, original undamaged Fsm is recovered after the acetylation and NaOH treatment, indicating the chemical stability of the Fsm molecule.

Both Fsm and the polymers, negatively charged at alkaline pH do not move electrophoretically at acidic pH, indicating a weak acidic function. This is confirmed by a negative reaction with IO_3^-/I^- reagent.

As all eluted fractions are slowly decolorised by $NaBH_4$, indicating a carbonyl function, the weak acidity might be due to an enolic function. Decoloration of all fractions is also obtained with other reducing agents as f.i. sodium ethanolate, indicating for Fsm as well as for the polymers the possibility of conjugated double bounds. All fractions which slightly reduce ammoniacal silvernitrate, give also a positive test with $FeCl_3/K_3[Fe(CN)_6]$ (19-21).

At higher temperature, polymerisation progresses faster, giving polymers of higher molecular weight. Continued heating results in decreasing solubility in water and acetic acid, giving finally insoluble polymers. It reduces esterification ability too, rendering it finally impossible. Perhaps this might be due to higher polymerisation, rather than to further

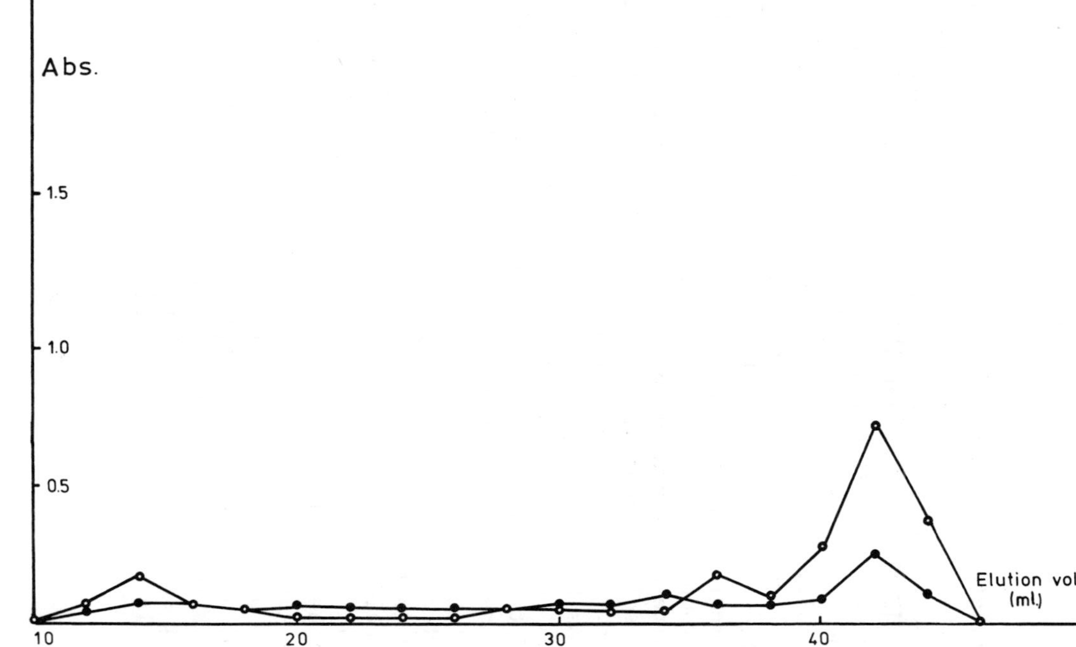

FIGURE 21. Elution diagram of polymerised molasse Fsme and polymerised deacetylated Fsme acetate on Sephadex G25. Original polymerised Fsme —o—
Polymerised deacetylated Fsme acetate —●—

heat degradation or dehydration. As caramel obtained from different sugars have somewhat different solubility in glacial acetic acid, it is rather doubtfull that the same Fsm is obtained from all sugars.

DISCUSSION

The solvatation-desolvatation colour reversibility being lost on blocking the hydroxylgroups by esterification, a tautomeric reaction might be involved, inducing conjugation of double bounds through irreversible stabilisation of the enolic form. This colour fixation, applies as well to watersoluble high polymer as to Fsm. As the polymer acetate is easily soluble in $CHCl_3$, the polymer must be a polyhydroxy compound. Hence, the loss of polymerising capacity of Fsm and polymers by esterification, might involve elimination of a polymerising tautomeric form, rather than a blocking of participating hydroxyl group(s).

The white unsolvatated form of Fsm-, water-, acetic acid-, and aceton soluble compound is precipitated by diethylether from aceton, indicating a polyfunctional low molecular weight compound. As F total acetate of molasse, ethanol fractionated on Sephadex LH20, shows the same fractionation pattern as the original F total on Sephadex G25, all higher molecular weight fractions are not merely agglomerates of Fsm, but covalent bound polymers.

Caramel colours, obtained by heating sugars or higher carbohydrates in the presence of mineral or organic bases -as f.i. amino acids in coffee-chicory and the breadcrust- show equally well the same black-white reversibility phenomenon, and the higher mentioned physical and chemical properties as the pure sugar caramels.

In view of the fact that all fractions of all mentioned caramel colours, are highly similar in all physical and chemical properties studied, these products might belong to a homological series of organic compounds, differing only by minor details as f.i. molecular weight.

ACKNOWLEDGMENTS

We wish to acknowledge our debt to the American Association of Cereal Chemists permitting publication of this paper, partially communicated at the 59th Annual Meeting of the A.A.C.C., Montreal 1974.

REFERENCES

1. Parker, K.J., Cookson, D., Williams, J.C. Proc. Tech. Session Cane Sugar Refinning Res. 103 (1970).
2. Parker, K.J., and Williams, J.C. Proc. Tech. Session Cane Sugar Refinning Res. 117 (1968).
3. Parker, K.J. Tate & Lyle Ltc. Res. Centre. Ann. Rept. 99 (1967).
4. Farber, L., and Carpenter, F.G. Proc. Tech. Session Cane Sugar Refinning Res. 145 (1970).
5. Farber, L., Carpenter, F.G., Mc Donald, E.J. Int. Sugar J. 73:8 (1971).
6. Smith, N.H. Proc. Tech. Session Cane Sugar Refinning Res. 84 (1966).
7. Smith, N.H. Proc. Tech. Session Cane Sugar Refinning Res. 105 (1968).
8. Binkley, W.W.Z. Zuckerind. 20 (6):291 (1970).
9. Gross, D. Int. Sugar J. 69:323;360 (1967).
10. Bugaenko, I.F., and Mukhamed, M. Isvest. V.U.Z. Pishch. Tekhnol. 1:161 (1972).
11. Houssiau, J., Waegeneers, R., Gurny, J. Communic. 14° assemblée générale de la C.I.T.S. Bruxelles 1971.
12. Tu, C.C., and Degnan, M. Int. Sugar J. 74 (9):259 (1972).
13. Binkley, W.W. Int. Sugar J. 62:187 (1960).
14. Binkley, W.W. Int. Sugar J. 62:36 (1961).
15. Motai, H., and Inque, S. Agr. Biol. Chem. 38 (2):233 (1974).
16. Casier, J.P.J., and Ahmadi Zenouz, A. Communication nr 145 at 59th Annual Meeting of the A.A.C.C., Montreal, Canada 1974.
17. Ahmadi Zenouz, A. Agricultura 25 (3):356 (1978). Doctor thesis.
18. Percival, E. Carbohydrate Res. 4:441 (1967).
19. Barton, G.M., Evans, R.S., Gardner, J.A.F. Paper chromatography of phenolic substances. Nature 170:249 (1952).
20. Gillio-Tos, M., Previtera, S.A., Vimercati, A. J. Chromat. 13:571 (1964).
21. Wagner, H., Hürhammer, L., Nufer, H. Arzneimittel Forsch. 15:453 (1965).

CHARACTERIZATION OF SYNTHETIC SUBSTANCES IN FOOD FLAVORS BY ISOTOPIC ANALYSIS

Jacques Bricout, Joseph Koziet

Institut de Recherches Appliquées aux Boissons
Créteil, Val de Marne (France)

INTRODUCTION

The flavor of many food products is directly imparted by the used ingredients or is a consequence of the technological process. Attempts were made to impart new flavor to traditional foods or beverages. This result can be obtained by the addition of flavoring substances. These were first obtained by extraction of herbs or spices. As our knowledge of the chemical composition of aroma is improving, it is now possible to duplicate natural flavor by careful compounding of synthetic substances. However the use of synthetic substances can be limited by national or international regulations: in some countries like France food aromas are classified in two main sections:

- natural flavouring obtained from vegetable through physical processes
- artificial flavouring obtained by a chemical process. This term includes:
  - substances which exist in natural products
  - substances not present in natural products

The use of any artificial flavor must be labelled and it becomes necessary to develop analytical methods in order to detect the addition of synthetic substances in food aroma.

Artificial substances not present in any natural product can be easily detected by integrated gas chromatography-mass spectrometry but the problem seems more difficult when we are dealing with a flavouring substance which exists naturally and which is also manufactured by chemical process. This process, in some cases, leads to the presence of some impurities which do not exist in nature. But as the technology of purification

ISBN 0-12-169060-1

of organic compounds improves, the detection of characteristic impurities becomes more tedious and time consuming. So, we have to consider the flavouring molecule as a whole and in order to gain some insight of its origin we have investigated the natural abundance levels of the isotopes of the atoms which constitute the flavouring molecule. This idea was supported by precedent results which have shown that isotopic analysis can help in the determination of the origin of ethanol (1-2).

EXPERIMENTAL PROCEDURES

1. Determination of the Radiocarbon Activity

The organic material (3-5 g) was burnt in an oxygen atmosphere to give CO_2 and the CO_2 was converted into acetylene via lithium carbide. Acetylene was cyclised to benzene over a catalyst. The benzene was dissolved in a scintillation cocktail and counted in a scintillation spectrometer. The activity of the sample was compared with 0.95 times the NBS oxalic acid (3).

2. Determination of the $^{13}C/^{12}C$ Ratio

The organic sample (5-10 mg) was burnt in an oxygen atmosphere in a closed system in which the gases are circulated by thermal convection through a vertical cobalt oxide oven maintaine at 600°C. After complete conversion of organic substances into CO_2 and H_2O, the oven is connected to a trap immersed in liquid nitrogen. CO_2 and H_2O are frozen out and oxygen is removed by vacuum. The trap is heated in dry ice and the CO_2 recondensed in a sample tube. Isotopic ratios on the CO_2 were measured on a mass spectrometer (VG Micromass model 602^{2}C) equipped with a double inlet system and a double collector. The relative difference between mass 45/mass 44 ratio from the sample and from the standard was calculated and corrections were made for ^{17}O contribution at mass 45 (4). The international standard is carbonate from the fossil skeleton of Belemnitella american (PDB) from the Pee Dee formation of South Carolina (5). Results are reported in terms of $\delta\ ^{13}C$.

3. Determination of the $^{2}H/^{1}H$ Ratio

After complete combustion of the organic material and elimination of CO_2 and O_2 water was recovered by distillation in a sample tube. This water was reduced by passage over metallic uranium turnings at 800° and the hydrogen evolved was collected in a sample tube using an automatic Toepler pump.

The hydrogen was analyzed on the same isotopic mass spectrometer, but in this case the source is specially designed to reduce H_3^+ ions formation which is cancelled electronically. In this case, the reference gas is the hydrogen obtained by reduction of Vienna SMOW water (6).
Results are expressed in term of $\delta^{2}H$

The method of ^{2}H measurement in water was calibrated using Standard Water of known isotopic composition according to the recommendation of Gonfiantini (7).

RESULTS AND DISCUSSION

1. Characterization of Synthetic Substances Derived from Coal or Petroleum

The radioactive carbon isotope ^{14}C provides a direct means of determining the products which are derived from fossil fuels. Due to their extreme age such fuels no longer contain ^{14}C as the ^{14}C half life is about 5730 years.
But all recent plant organic matter has a well established ^{14}C content as the source of carbon is atmospheric CO_2.
This ^{14}C activity of CO_2 before 1950 was about 13.56 dpm/g C, but this activity has been increased up to 26 dpm/g C in 1964 due to ^{14}C production in nuclear weapon test. Actually the ^{14}C specific activity of atmospheric CO_2 is about 18 dpm/g C. These observations were applied in order to characterize synthetic citral (Table I).

TABLE I. ^{14}C Radioactivity of Citral

Origin	^{14}C dpm/g C
Cymbopogon citratus	20.1 ± 0.4
Litsea cubeba	17.6 ± 0.4
Synthetic	0.25

Thus ^{14}C analysis allow a good characterization of citral synthetized from fossil fuel. But we know now that many synthesis of fragrance material start with pinene, a natural constituent of turpentine.
Thus, ^{14}C analysis cannot be used. So we have investigated the possibility of using stable isotope analysis as a means of characterization of the origin of flavour material.

2. $^{13}C/^{12}C$ Ratios in Flavoring Substances

$^{13}C/^{12}C$ ratios were determined on the same citral samples of Table I. The results are indicated in Table II.

TABLE II. $^{13}C/^{12}C$ Ratios in Citral

Origin	^{13}C ‰ PDB
Cymbopogon citratus	- 11.6
Litsea cubeba	- 26.7
Synthetic	- 27.4

This analysis does not allow a good characterization of synthetic citral which has a $\delta^{13}C$ similar to citral from Litsea cubeba but lemon grass citral shows a higher $\delta^{13}C$. This observation must be correlated to the phenomenon of carbon isotope fractionation associated with photosynthesis. The only

source of carbon for terrestrial plants is atmospheric CO_2 which shows a constant $\delta^{13}C$ of about - 7 ‰ PDB (8).
In most plants CO_2 is fixed by carboxylation of ribulose 1.5-diphosphate (9) to give 3-phosphoglycerate (Calvin pathway) and it was shown that carbon isotope fractionation does occur at this enzymatic step (10). Organic matter is depleted in ^{13}C by about 18‰. The other enzymatic reactions leading to the different organic substances of a plant do not imply large isotopic fractionation. The range of $\delta^{13}C$ for organic substances derived from Calvin plants is

In some plants atmospheric CO_2 is fixed by carboxylation of phosphoenolpyruvate (11) (Hatch and Slack pathway), an enzymatic reaction which occurs with a low carbon isotope fractionation (12).
Consequently organic substances derived from Hatch and Slack plants show a range of ^{13}C : - 15 $\delta^{13}C$ - 10% PDB
Few plants fix atmospheric CO_2 according to the Hatch and Slack pathway, among them: _Zea mays_, _Saccharum_ and _Cymbopogon citratus_ (lemon grass) (13).

In fact, we can consider that carbon 13 analysis can be useful only to characterize organic substances derived from non Calvin plants. Very few aromatic plants are non Calvin. _Vanilla planifolia_ represents a special case: this Orchid can fix CO_2 according to a third pathway called Crassulacean Acid Metabolism.
Carbon assimilation occurs in the night by carboxylation of phosphoenolpyruvate (PEP) and malic acid accumulates during the night. This acid is decarboxylated in the day and the liberated CO_2 is fixed by ribulose-1.5 diphosphate (RUDP).
But in these plants direct carboxylation of RUDP by atmospheric CO_2 can also occur during the day. The proportion of CO_2 fixed in the day by carboxylation of RUDP or in the night by carboxylation of PEP varies under the influence of environmental factors and can be estimated by carbon isotopic composition (14).
We have investigated the $^{13}C/^{12}C$ ratio in vanillin extracted from Vanilla beans and in synthetic vanillin obtained from wood lignin or from gaiacol.
The results are indicated in Table III.

TABLE III. $^{13}C/^{12}C$ Ratio in Vanillin

Origin	^{13}C ‰ PDB
Vanilla planifolia	
Madagascar	- 21
Mexico	- 20.8
Java	- 20.9
Vanila tahitensis	- 20.8
Synthetic	
From gaiacol	- 31.5
From lignin	- 27.8

The $\delta^{13}C$ of synthetic vanillin is characteristic of Calvin plants. Vanillin from Vanilla beans shows a $\delta^{13}C$ value which is intermediate between $\delta^{13}C$ value of Calvin plants and of Hatch and Slack plants. These results suggest that Vanilla planifolia assimilates CO_2 in the day and in the night and that ratio of these two carbon flows remains constant in the different countries of cultivation of Vanilla planifolia. In this particular case ^{13}C analysis appears as the method of choice to characterize synthetic vanillin in any food aroma.

III. $^{2}H/^{1}H$ RATIO IN FLAVORING SUBSTANCES

As the $^{13}C/^{12}C$ ratio is useful in very few cases, we have investigated the possibility of using $^{2}H/^{1}H$ ratio. This ratio can vary according to the class of substances investigated in a plant (15). So it was necessary first to investigate $^{2}H/^{1}H$ ratios in different classes of natural flavoring substances, in order to evaluate the importance of ^{2}H fractionation factor during biosynthesis of volatile substances. This study was first limited to terpenoïd and phenylpropanoïd compounds (16).

We have observed that the volatile plant substances can be classified according to their δ D value:

Phenylpropanoid	δ D > - 100 ‰ SMOW
Monoterpenes	-300< δ D < - 250 ‰ SMOW
L-menthol and its acetate	δ D < - 350 ‰ SMOW

These differences must be related to biosynthetic pathways. Phenylpropanoïd compounds derive from sugars by the shikimate pathways. Most of the hydrogen atoms of these phenolic substances were originally present in the sugar molecule and the δ D value of phenylpropanoïd is in the same range than δ D value of sugar. Terpenic compounds are highly depleted in deuterium as compared to sugar. This was also observed in lipidic substances (15). The origin of low deuterium value in lipidic substances lies mainly in the mechanism of deuterium fractionation associated with the biosynthesis of acetyl coenzyme A.

$^2H/^1H$ ratio appears as a sensitive indicator of biosynthesis pathway and we can expect that deuterium fractionation associated with the biosynthesis of a molecule does not occur in the same manner during chemical synthesis of the same molecule.

In order to check this hypothesis, we have measured the $^2H/^1H$ ratio in citral, linalol and menthol obtained either from different plants growing in different environment, either from synthesis. The results are indicated in Table IV.

TABLE IV. $^2H/^1H$ Ratio in Flavouring Substances Extracted from Plants or Synthetized

Origin	δ D ‰ SMOW
Linalol	
Thymus vulgaris (France)	- 257
Coriandrum sativum (France)	- 269
Petit grain (Paraguay)	- 244
Synthesis	- 170
Citral	
Citrus aurantifilia	- 258
Cymbopogon citratus	- 276
Litsea cubeba	- 251
Synthetic	- 174
Menthol	
Mentha piperita (France)	- 394
L-menthol natural	- 358
synthetic Europe	- 196
synthetic U.S.A.	- 242

δ D values of natural terpenic substances appear lower than δ D values of their synthetic equivalent.

Finally we have investigated the possibility of deutereium analysis for the characterization of synthetic anethole. The results are indicated in table V.

TABLE V. $^2H/^1H$ Ratio in Trans-Anethole

Origin	δ D ‰ SMOW
Foeniculum vulgare var.dulce (France)	- 86
Foeniculum vulgare var.vulgare (France)	- 91
Illicium verum (China)	- 96
Illicium verum (Nord Vietnam)	- 84
Synthetic (U.S.A.)	- 45

We observe that synthetic anethole has a higher δ D value than natural anethole. This difference seems at first surprising as synthetic anethole is derived from natural estragole present in turpentine.
We have measured the $^2H/^1H$ ratio in natural estragole from Foeniculum vulgare (δ D = - 70‰ SMOW). This value is higher than δ D of natural anethole from the same plant (δ D = 91 ‰ SMOW). This observation can be related to the different biosynthetic pathways of anethole and estragole (17). The phenylpropanoid skeleton of anethole derives entirely from cinnamic acid but the biosynthesis of estragole involves decarboxylation of cinnamic acid. So, higher δ D value of synthetic anethole seems to result at least partially from the higher δ D value of the starting estragole.

CONCLUSION

The preliminary results obtained show that isotopic analysis can be considered as a new way of characterizing synthetic flavouring substances. This method can be applied to chemical isolates or essential oils, but the weight of

substance necessary for a stable isotopic analysis (5 mg) is a real limitation of the method as the isolation of such an amount of a volatile substance from a food can be very time consuming. We can expect that this limitation can be reduced in a near future with technological improvement of isotopic analysis. Particularly a great deal of effort are made in order to monitor isotope ratios from gas chromatographic effluent (18): compounds are burnt as they emerge from the gas chromatographic column and combustion products are directly introduced in a computer controlled isotope ratio mass spectrometer.
These future developments can allow the determination of isotope ratios on microgramm amount of aromatic substances and thus can help in a correct assignment of the origin of food flavor as isotopic ratios can be considered in some cases as a built-in label of a molecule.

ACKNOWLEDGMENTS

We are grateful to G. Marien and C. Pachiaudi (Laboratoire du Radiocarbone de l'Université de Lyon I) for ^{14}C measurements.

REFERENCES

1. Guerain, J., Tourliere, S., Industrie Alimentaire et Agricole, 92:81 (1975)
2. Bricout, J., Fontes, J.C., Merlivat, L., Industrie Alimentaire et Agricole, 92:375 (1975)
3. Broecker, W. S., Olson, E. A., Radiocarbon, 3:176 (1961)
4. Craig, H., Geochim. Cosmochim. Acta, 3:53 (1957)
5. Craig, H., Geochim. Cosmochim. Acta, 12:133 (1953)
6. Craig, H., Science, 133:1833 (1961)
7. Gonfiantini, R., Nature, 271:534 (1978)
8. Craig, H., Keeling, C., Geochim. Cosmochim. Acta, 27:549 (1963)
9. Calvin, M., Bassham, J. A., in "the Photosynthesis of carbon compounds", Benjamin, New York, 1962
10. Whelan, T., Sackett, W. M., Benedict, C. R., Plant Physiol., 47:380 (1971)
11. Hatch, M. N., Slack, C. R., Biochem. J., 101: 103 (1966)

12. Reiback, P. H., Benedict, C. R., Plant Physiol., 59:564 (1977)

13. Smith, B. N., Epstein, S., Plant Physiol., 47:380 (1971)

14. Osmond, C. B., Allaway, W. G., Troughton, J. H., Queiroz, O., Lüttge, U., Winter, K., Nature, 246:41 (1973)

15. Smith, B. N., Jacobson, B. S., Plant and cell physiol., 17:1089 (1976)

16. Bricout, J., Merlivat, L., Koziet, J., C. R. Acad. Sci. Paris, 277:885 (1973)

17. Mannito, P., Monti, D., Gramatica, P., Tetrahedron Letters, 17:1567 (1974)

18. Matthews, D. E., Hayes, J. M., Third Conference on Stable Isotopes Oak Brook (Illinois), May 23-26 (1978)

ANALYSIS OF AIR ODORANTS IN THE ENVIRONMENT OF A CONFECTIONS MANUFACTURING PLANT

M.M. Hussein
D.A.M. Mackay

Life Savers, Inc.
Research & Development
Port Chester, N.Y.

I. INTRODUCTION

It is usual for a person walking by our manufacturing plants to guess, with some accuracy, what is being manufactured at the particular time. It is, of course, fortunate that these experiences are pleasant and enjoyable, but the emanation of odors from the manufacturing facilities shows that flavor losses are constantly occurring from the natural and synthetic flavorants utilized in confectionery products.

Identification and measurement of volatiles within and without the manufacturing plant is of interest for a number of reasons. An economic reason, for example, entails monitoring losses of certain costly flavorants. A research reason is to measure odorants at about their organoleptic thresholds.

The procedures developed for representative sampling and accurate analysis of odorants in air are described with emphasis on the comparison of various sampling devices. The unusual technique of sampling and trapping trace volatiles in very large volumes of air through the use of very long, large-bore coated (LBC) columns is also introduced and described.

Successful analysis of highly diluted complex mixtures in large volumes of air depends on concentrating the trace volatiles such that final transfer onto the analytical column is optimal i.e. all components are equally and maximally concentrated in a single band of the shortest possible length at the head of the chromatographic column.

ISBN 0-12-169060-1

Prevention of peak broadening is probably the most important factor in determining the least detectable amount of substance, quite apart from considerations of loss of chromatographic resolution power. There is always some maximum volume of directly sampled air or headspace such that the benefit of larger sample weight is offset by peak broadening resulting from the larger volume. With packed columns, this volume is usually about five ml though for any one specifically desired component, volumes perhaps ten times greater might be employed at the expense of compromising the rest of the analytical separation. For this reason, on the assumption that other operating parameters have already been maximized, sample concentration is always necessary at some stage to increase overall sensitivity.

This concentration stage is achieved through the use of sampling traps which, of course, are selection devices rather than sampling devices. The concentration attained for volatiles in air depends upon the selectivity of the trap for those volatiles. Validity of the analysis depends upon understanding of the selectivity process. Achieving optimal separation and resolution of the trapped volatiles depends on the mode of sample transfer from the sampling trap onto the analytical column; this can be done directly or via another trap affording another stage of concentration or selectivity.

Air sampling has been reported by many researchers using a variety of techniques (1,2). Dravnieks et al. (3), Jennings et al. (4), and Zlatkis et al. (5) introduced the use of porous polymers as sampling devices for organic volatiles. Such sampling techniques have found wide acceptance and applicability, with the techniques of sample collection and analysis varied according to the need and ingenuity of the researchers.

Although probably the most widely used and successful techniques for concentration of air and headspace volatiles employ these porous polymer traps, they have certain disadvantages which do not seem to have been thoroughly investigated. Mackay and Hussein (6) touched briefly on some of these, such as artifact formation and loss in the more volatile constituents in sampling large volumes at high flow rates. Butler and Burke (7) carefully studied the efficiency and capacity of several porous polymers and concluded that none is universally applicable and each has its own merits depending on the situation and the problem at hand.

The use of large bore coated (LBC) columns as trapping devices for volatiles in air was suggested recently by Mackay and Hussein (6) who demonstrated the principle by utilizing 40 ft of ¼" diameter Tygon tubing for the sampling of large volumes of breath for its sulfur-containing compounds. This technique is developed further in this paper by using large bore (¼") metal columns coated with a stationary phase and utilizing this coated column to sample and concentrate organic volatiles in air, as well as in breath. The technique proved to be generally comparable, if not superior, to direct sampling on porous polymer columns. The advantages of the LBC columns are the ease of sampling (whereby minimal back pressure is encountered), a minimum or absence of artifacts, and ease of elution of trapped materials onto the chromatographic column for analysis. However, no system was found to be entirely without some problem.

Trapping on porous polymers is quite effective when the sampling volume is small, which means no sample loss from breakthrough, and when the amount of polymer in the trap is also small, which results in minimal artifact formation upon desorption.

However, certain porous polymers, notably Tenax-GC, have a low breakthrough volume, i.e. low affinity for low molecular weight alcohols. And if large amounts of porous polymers are used for large volumes of highly dilute samples, considerable formation of artifacts will interfere in parts of the analysis, or the artifacts may be mistaken for constituents of the sample. The extent of artifact formation depends on the type and amount of porous polymer, the type and amount of sample, and the trapping and desorbing temperatures. As expected, the lower the temperature, and the lower the amount of polymer, the smaller the number and amount of artifacts. Prolonged conditioning of the polymer does not eliminate the problem, and may in fact exacerbate it.

Also, the artifact problem becomes increasingly important as the dilution of the sample increases, and can be further aggravated by the resort to larger amounts of porous polymers in attempts to offset the breakthrough problem resulting from the need to scavenge increasing volumes of air for the minimum detectable weight of material. However, for routinely difficult analyses the artifact problem with porous polymers is relatively minor, so that porous polymers offer excellent opportunities for trapping and transfer of volatiles subsequent to their prior concentration by other means.

The use of activated charcoal in very small amounts as recently advocated by Clark and Cronin (8) resulted in relatively artifact-free analysis, but the problem of irreversible adsorption still seemed to persist even when desorbing at 300°C; more pronounced in large, more concentrated samples (>40 μg/L).

The use of standard coated gas chromatographic packings, as first advocated nearly 20 years ago (9) for enrichment of headspace volatiles, was also evaluated and found to suffer from low retention of the most volatile components. The problem of artifact formation from substrate breakdown also exists, but can be eliminated or minimized by proper choice of liquid phase. Low retention of volatiles can be minimized by sampling at sub-ambient temperature, provided the moisture content of the sample is low and the liquid phase remains fluid. However, it was an intent of the present study to compare trapping techniques at ambient temperatures, with obvious relevance to convenience of field studies.

When the constituents which are sought are of sufficient concentration to be found in a 300 ml sample of air, 200 mg of Tenax-GC in a 6 inch pre-column offers a quick and accurate analysis. But when the volatile organics of the air are at extremely low concentrations, trapping on large bore coated (LBC) columns followed by subsequent desorption of the sample onto the short Tenax pre-column offers a superb technique.

The only artifacts in LBC columns coated with SE-30 silicone rubber are two peaks with very short retention times which may interfere with only a very few highly volatile components. If such components exist, an identical sample may be collected simultaneously on another column (up to 2 feet long) packed with porous polymer. The matched samples are desorbed on two similar Tenax pre-columns and then analyzed. The porous polymer column gives no artifact in the region of the very volatile components.

Transfer of the originally trapped samples was done by desorption onto short Tenax-GC pre-columns (suitably configured for sample injection), since they are widely used and their desorption is easily and quickly achieved onto the analytical column. Desorption from other pre-columns containing charcoal or coated column packing was also investigated. Sample manipulation was also carried out successfully on LBC columns other than the one used for trapping. The use of substrates other than SE-30 for coating the LBC column will also be reported at a later date.

Other useful features of sampling on LBC columns are that organic volatiles in high-moisture samples (such as breath, or headspace above aqueous solutions, or moist air) can be concentrated with no fear of moisture condensation provided the LBC column is maintained at 40-50°C; minimal back pressure enables direct sampling of breath by blowing directly into the LBC column; and large samples of air (10 to 100 liters) are easily passed through 50' LBC columns at room temperature with the aid of simple pumps. There is no need for temperature control. Segmentation and serial examination showed that breakthrough in 50 ft LBC columns rarely occurs, with less than half the column being utilized in normal use. In one case of overnight running (16 hours, i.e. 96 liters) some breakthrough of high volatiles was observed.

Another feature is that trapped samples can be readily exchanged between LBC columns, leading to a strong concentration effect at each exchange. After a few exchanges the concentration becomes high enough to permit direct introduction, if desired, into the analytical unit. It should be noted that in this procedure the exposure of the sample to artifacts and contaminants need be no greater than is inevitable from the choice of the substrate used for the GC analysis. This "concentration pump" technique will be fully described in a forthcoming publication.

II. EXPERIMENTAL

A. Materials

1. Chemicals. All chemicals referred to in this work were obtained from commercial suppliers.

2. Pre-Column Packings and Column Coating Materials.

a. Tenax-GC (60-80 mesh) product of Enka, N.V., Holland, obtained from Applied Science, Inc.

b. Coconut activated charcoal (Anasorb), 90/100 mesh, Analabs, Inc.

c. Porapak Q, 80/100 mesh, Analabs, Inc.

d. SE-30 Silicone Rubber (100% methyl).

e. 20% Di(2-ethyl-hexyl) phthalate (DOP) on Chromosorb W (80/100 mesh), acid washed and DMCS treated; coated in the authors' laboratory.

f. 10% Tween 80 on Chromosorb W (80/100 mesh), acid washed and DMSC treated; coated in the authors' laboratory.

3. Pre-Columns for Sampling and Analysis. Short lengths of Pyrex glass tubing 15 cm x 0.6 cm o.d. (0.38 cm i.d.) were used containing two glass wool plugs not over 9 cm apart to ensure that the packing material was well within the hot zone of the tube heating oven used as an accessory sample injection port.

B. Packing and Conditioning of Pre-Columns

The pre-column was packed tightly, using suction and vibration, to a depth of 9 cm. With 5 mg activated charcoal, however, a bed only 1 mm thick was obtained between the two glass wool plugs. The weight of packing was obtained by difference.

The pre-column was conditioned for at least 3 hours with a nitrogen flow of 50 ml/min at the following temperatures: Tenax-GC (275°C), Porapak Q (250°C), DOP (150°C), Tween 80 (150°C), activated charcoal (300°C).

C. Large-Bore Coated (LBC) Columns

Large-bore aluminum tubing (50 ft. long and ¼ inch o.d.; 0.186 inch i.d.) was coiled to fit into the oven of a gas chromatograph used only for desorption purposes. The column was cleaned by repetitive aspiration of benzene and dried. To coat the column, it was first filled with 10% SE-30 in toluene, then capped and left to stand for 2 hours. The solution was then removed by gentle suction which was continued for 20 to 30 minutes until no more toluene odor was detected. The column was weighed before and after coating. The column was then conditioned at 275°C for 6 to 8 hours with a nitrogen flow of 50 ml/min.

The weight of SE-30 coating in a 50 ft column ranged from 1.8 to 4.5 g. Thicker coating was achieved by repetitive application of the SE-30 solution. A 3.0 g SE-30 coating in a 50 ft, 0.186 inch i.d. column is equivalent to a thickness of approximately 0.014 mm, assuming uniform thickness.

D. Sample Collection

1. Short Pre-Columns. The 300 ml sample was collected directly on the pre-column at a flow of approximately 75 ml/min using suction from a 100 ml gas-tight syringe.

When a large sample was collected suction was obtained with a peristaltic pump at a flow of 100 ml/min.

Sub-ambient temperature sampling was accomplished by wrapping the center section of the pre-column with plastic tubing filled with salt solution, pre-cooled to -10 to -15°C in a freezer.

2. LBC and Long Packed Columns. The sample was collected with the aid of a peristaltic pump for the LBC column, or a suction pump for the packed column; with the flow regulated at 100 ml/min in each case. Sampling was continued until the desired volume was collected as determined by elapsed time.

E. Sample Desorption

When using short pre-columns either for direct sampling, or for transfer from other columns, the sample was desorbed off the short pre-column directly into the analytical column by heating the pre-column for 6 minutes in the tube oven. The temperature was that indicated in section B for the packing used. In all cases nitrogen flow was 50 ml/min through the pre-column while desorbing the sample. The carrier flow through the analytical column remained constant at 50 ml/min.

After air sampling on long Tenax or on LBC columns, the collected sample was usually transferred onto a short Tenax pre-column by heating the sampling column in an oven for 12 minutes at 225°C while connected to a pre-column outside the oven, and flushing with 50 ml/min nitrogen during desorption. The transferred sample was then analyzed as just described for short pre-columns.

Gas chromatographic analysis (by temperature programming) was initiated at the same time the sample desorption process was begun.

F. Gas Chromatographic Conditions

1. Instrument. Perkin-Elmer model 3920 gas chromatograph with flame ionization detector.

2. Column. Single aluminum 8 ft x ¼ inch o.d. (0.186 inch i.d.) packed with 10% Carbowax 20 M on Chromosorb W 80/100 mesh (acid washed and DMCS treated).

3. Carrier Gas. Nitrogen at 50 ml/min measured at ambient temperature.

4. Column Oven Temperature. Programmed at 4°/min from 70 to 230°C.

5. Injector Temperature. 250°C.

6. Interface Temperature. 250°C.

7. Recorder. 5 mv at ½ inch per minute chart speed.

8. Integrator. Vidar model 6230.

III. RESULTS AND DISCUSSION

Air odorants in a confections manufacturing environment consist predominantly of flavor volatiles which vary in type and amount depending on the products being manufactured at the time. Identification of the majority of components is accurately achieved by use of retention times, since these are well known from frequent analysis of the flavor oils.

Figure 1 shows a typical analysis of odorants in an air sample taken from the flavor processing area of the manufacturing plant. It is seen in this case that a 300 ml sample collected on a short pre-column packed with Tenax-GC is sufficient. Most of the components seen are calculated to be in the ppm range. The result is judged to be a mixture of peppermint, citrus, wintergreen, spearmint and fruit flavors, all of which had been processed in the prior 4 hours.

To compare sampling on Tenax-GC pre-columns with other systems, pre-columns of similar dimensions were packed with other porous polymers, or with coated gas chromatographic packings, and used for sampling the same environment. Activated charcoal in small amounts (5 mg) was also evaluated.

To compare the efficiencies of these short pre-columns as sampling traps each was used to trap the identical amount of a standard mixture of various flavorants used in confectionery manufacture. An exact amount of mixture was deposited directly onto the pre-column packing and the sampling process simulated by drawing 300 ml of air through the pre-column at 75 ml/min using a gas-tight 100 ml syringe. The pre-column was then analyzed and the individual peak areas of the components were compared to those obtained from direct injection of the same amount of mixture.

This procedure was followed in screening many packing materials. Table 1 shows comparative data obtained with some of the packings. It can be seen that appreciable loss of ethyl formate occurred with Tenax-GC, and ethyl alcohol was completely lost. While retention of Porapak Q for ethyl formate, the most volatile constituent of the mixture, and for ethyl alcohol is better than that of Tenax-GC, recovery of the less volatile constituents of the sample is less. As expected, cooling the pre-column improved the performance of Porapak Q and gave good retention for ethyl alcohol and ethyl formate on Tenax-GC. Sampling at ambient temperature using coated column packings resulted in severe losses in the more volatile components of the sample, up to and including ethyl butyrate.

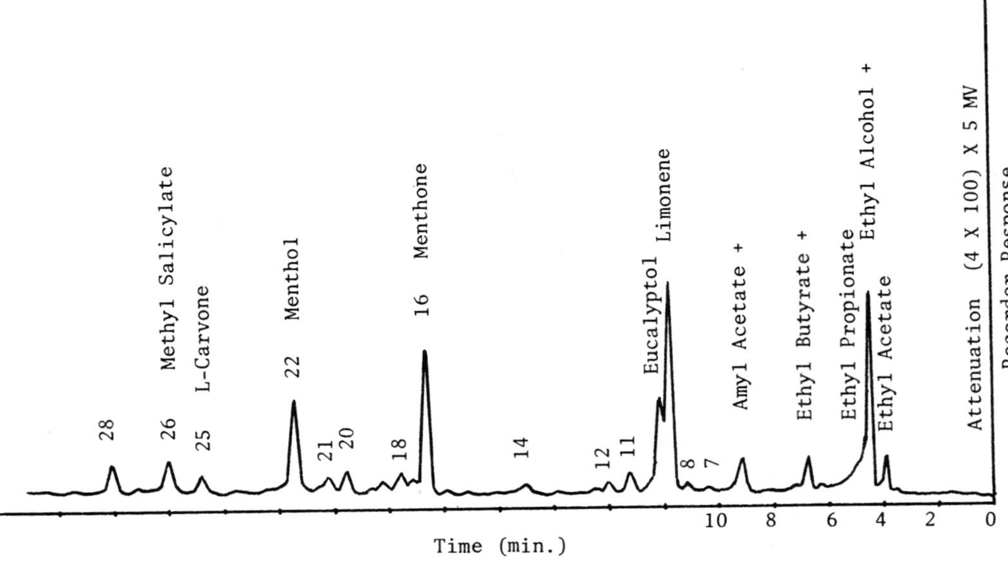

Fig 1. Analysis of 300 ml air in flavor processing area of manufacturing plant; sampled directly onto Tenax-GC pre-column, showing traces of flavoring oils from peppermint, wintergreen, spearmint, citrus, and fruit.

TABLE 1. Comparison of Sampling* Efficiency of Identical Volumes on Various Short Pre-Column Packings; Relative** to Direct Injection of Exactly Same Amount of Mixture

Sampling Temperature → / Compound ↓	Amount in 1.0 µl Sample	Tenax-GC (200 mg) R.T.	Porapak Q (345 mg) R.T.	Porapak Q (345 mg) -10°C.	20% DOP (430 mg) R.T.	20% DOP (403 mg) -10°C.	10% Tween 80 (418 mg) R.T.
		% Recovery					
Ethyl Formate	95.2 µg	72.0	86.7	93.5	0	0.1	0
Ethyl Acetate	103.2 µg	88.7	86.0	91.7	0	87.9	0
Ethyl Alcohol	95.4 µg	0	86.2	93.5	0	4.7	0
Ethyl Propionate	74.0 µg	83.8	85.7	92.0	0	88.0	0
Ethyl Butyrate	77.6 µg	84.7	83.9	90.3	17.1	87.5	0.7
Limonene	79.6 µg	86.0	75.9	82.1	72.4	88.1	89.0
Menthone	98.2 µg	90.9	69.7	82.9	73.8	97.1	85.5
Menthyl Acetate	70.3 µg	94.8	64.3	83.7	74.7	100.0	86.4
Neo-Menthol	92.0 µg	94.8	74.6	89.1	73.8	100.0	88.0
Menthol	94.3 µg	95.4	76.5	87.3	74.3	101.3	87.8

* Simulated sampling by depositing 1.0 µl standard mixture at end of pre-column followed by sweep of 300 cc clean air.

** Expressed as % of direct injection.

Cooling the pre-column markedly improved the performance, however. The DOP packing was found to be the most efficient of the gas chromatographic packings, although its performance was still inferior to that of the porous polymers.

Table 2 shows results of analyses of air odorants in 300 ml samples collected on short pre-columns packed with Tenax-GC, Porapak Q, activated charcoal, and DOP. The sample on the 20% DOP pre-column was collected at sub-ambient temperature. Due to the small amount of Tenax-GC in the pre-column, no appreciable artifacts from this source were noted. Artifact formation which interfered in the analysis for some of the sample components was noted with both Porapak Q and DOP pre-columns. The recovery values of the components (where no interference occurred) were comparable for Porapak Q, DOP and Tenax-GC. Interferences due to artifacts which also occurred in the activated charcoal analysis are noted in the table. Sampling on this small amount of activated charcoal, although not artifact-free, gave, for most components, results comparable to those obtained with Tenax-GC, provided the desorption temperature was fully 300°C.

Sampling on coated packings was found to be dominated by the characteristics of the liquid phase. The major drawbacks were the limitation on desorption temperature, the artifacts due to the bleed or thermal decomposition of the liquid phase, and low affinity for highly volatile constituents. The retention of sample components, as expected, depended on the volume of the air sample, the path length, and the amount of packing in the sampling trap, the type and loading of the liquid phase, and the sampling temperature. It should be noted, however, that the amount of active absorbent in the pre-column is only a fraction of that of the porous polymers. While the sampling efficiency of cooled pre-columns packed with coated column packing was generally adequate, problems may be expected from high-moisture samples due to condensation of water.

Artifact formation using Tenax-GC pre-columns was minimal but many interfering peaks occurred using Porapak Q, particularly of longer retention times. Of course, these results refer to a particular mixture of flavorants at a particular concentration, and extrapolation of results to other conditions may not be justified. Clearly the "truth" is not to be found by use of any one absorbent, but it may be approachable through inspection of all results with these different types of packings. By this method it was judged that the "true" total volatile content of this 300 ml air sample was about 16.0 µg, compared to results of 13.9, 12.2, 17.0 and 11.0 µg yielded by Tenax, Porapak, Charcoal, and DOP, respectively.

For concentration of traces of volatile organics in larger

TABLE 2. Comparison of Analyses of Odorants in Flavor Processing Area Using Various Short Pre-Columns. (Sample Volume is 300 cc Air)

Sampling Temp. →	(RT)	(RT)	(RT)	-10°C
Absorbent →	Tenax-GC	Porapak Q	Activated Charcoal	20% DOP
Weight →	(200 mg)	(345 mg)	(5 mg)	(403 mg) (a)
Compound	(μg)	(μg)	(μg)	(μg)
Ethyl Acetate	0.17	0.22	0.20	0.30
Ethyl Alcohol(b)	2.06	2.28	4.68	1.92
Ethyl Butyrate	0.45	0.28	0.39	0.35
Limonene+Eucalyptol	3.73	3.68	2.86	3.08
Menthone	0.38	*	0.33	1.56**
Iso-menthone	0.08	0.44**	0.41**	0.36**
Menthyl Acetate	0.13	0.34**	0.18	*
Neo-menthol	0.17	0.43**	1.06**	*
Menthol	0.66	0.66	1.25	0.83
Methyl Salicylate	2.92	3.16	3.25	3.02
Total Organic Volatiles in Sample	13.89	12.20	16.99	10.97

(a) i.e. Weight of DOP coating is 80.6 mg
(b) Value is inflated by a small amount of ethyl propionate which occurred as a shoulder following the alcohol peak.

* Could not be estimated due to appreciable artifacts.
** Value is inflated due to artifact interference.

volumes of air, sampling on LBC columns was found generally preferable to sampling on long Tenax-GC columns, which offered an alternative solution to the breakthrough problem. After two early artifacts (retention times 3.2 and 4.4 minutes) the LBC column technique using SE-30 is almost free of artifacts. In contrast, long Tenax-GC columns suffer from major interfering artifacts, the most prominent of which appear at 19 and 20 minutes. The total number of artifacts observed in an analysis of highly purified (molecular sieve and cooled activated charcoal) nitrogen exceeded 30, though these artifacts were mostly small in comparison to those noted at 19 and 20 minutes. Figures 2 and 3 show identical analyses performed on identical volumes of factory air (6 liters) sampled simultaneously at the same rate at the same location. One used a 50 ft x ¼ inch LBC column coated with 3.0 g SE-30 for the initial trapping; the other a 2 ft x ¼ inch column containing 1.75 g Tenax-GC. After transfer to the standard Tenax pre-column for analysis, both chromatograms show striking similarity. The main differences noted are the slightly lower concentration of ethyl acetate in the LBC column analysis and the higher concentration of the first peak, which is partially due to an artifact peak from the SE-30 coating. Other differences are the appreciable artifact peaks due to the Tenax (peaks 15a, 15b, 18, 20). Other instances of interference due to Tenax artifacts can be noted, but the effect is small.

Table 3 shows the amounts of individual components identified in the analyses represented by Figures 2 and 3. Apart from ethyl acetate and ethyl formate, where either interference or loss has occurred, recovery of all components is higher in the LBC column than in the long Tenax column.

It is possible to guess what products were manufactured at the time of these analyses, based on individual components and their relative abundance. The esters suggest fruit, the limonene suggests citrus, menthone and menthol suggest peppermint, carvone suggests spearmint, and methyl salicylate suggests wintergreen. All of these products were being manufactured that day.

Sampling large volumes of air over long periods tends to negate short term variation of the air-borne volatiles. This may or may not be advantageous.

As can be seen from these data, no single system is likely to be absolutely efficient and free from all interference for sampling of air volatiles, but a workable system can be found, depending on the situation and the sought components.

For sampling at high concentration, where a 300 ml sample can give sufficient information, a pre-column of Tenax-GC provides good results, especially if a back-up trap of activated

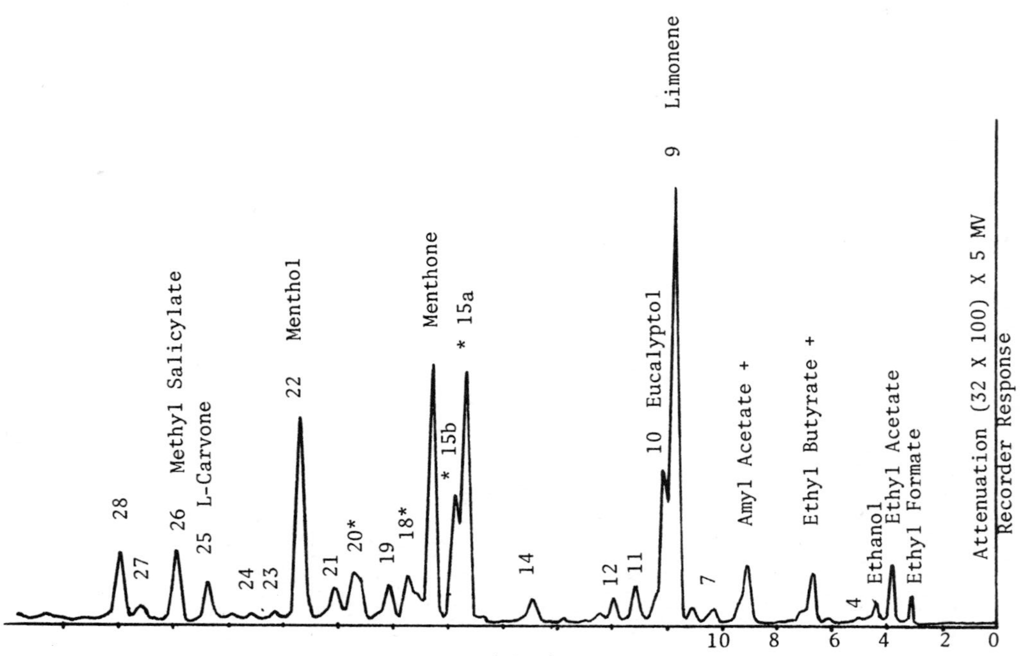

Fig. 2. Analysis of 6 liters air in flavor processing area of manufacturing plant; sampled directly onto 2 ft, ¼" o.d. Tenax-GC column (containing 1.75 g Tenax) at flow rate of 100 ml/min. Sample was desorbed for analysis onto short Tenax-GC pre-column.

* Peaks 15a, 15b are artifacts, while 18 and 20 are enhanced by artifacts.

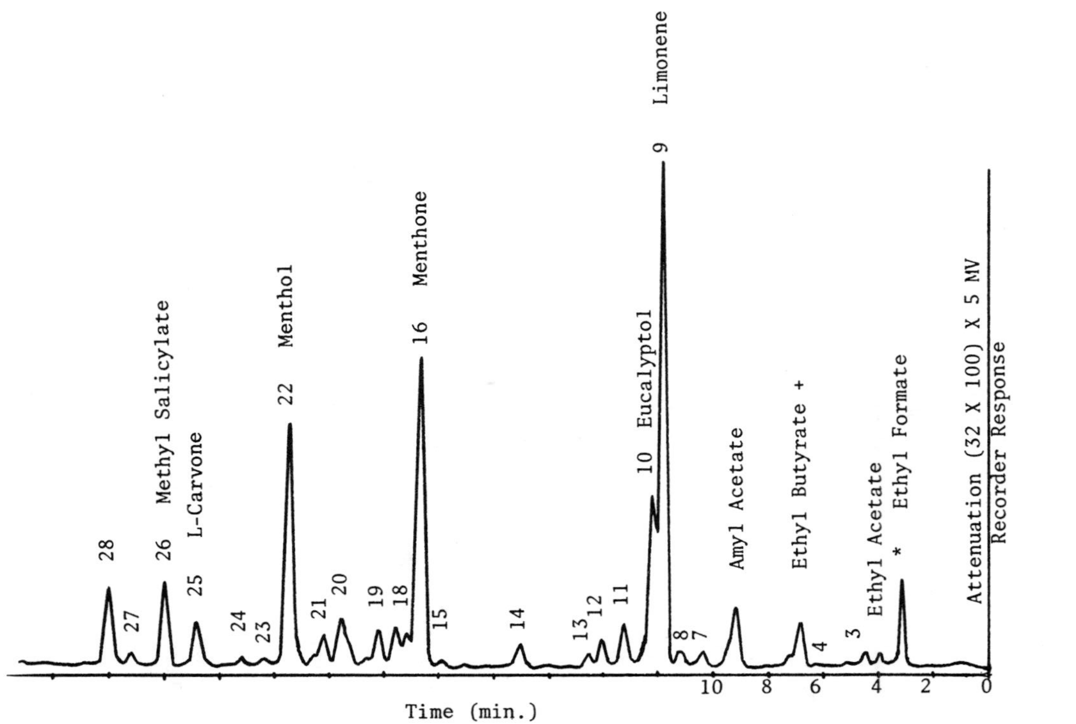

Fig. 3. Analysis of 6 liters air in flavor processing area of manufacturing plant; sampled directly onto 50 ft x ¼" o.d. LBC column (3.0 g SE-30 coating) at flow rate of 100 ml/min. Sample was desorbed for analysis onto short Tenax-GC pre-column.

* Enhanced by artifact.

TABLE 3. Comparison of Analyses of Odorants in Flavor Processing Area Using Long Tenax and LBC Columns (Sample Volume is 6 Liters Air)

Compound	Sampling On Long Tenax Column (μg)	Sampling On LBC Column (μg)
Ethyl Acetate	11.3	5.7
Ethyl Alcohol	4.8	6.8
Ethyl Propionate	1.9	3.6
Ethyl Butyrate	10.5	14.5
Limonene	37.2	44.3
Eucalyptol	13.0	15.6
Menthone	27.4	34.4
Menthofuran	3.5	3.3
Iso-menthone	4.9*	4.0
Menthyl Acetate	8.7*	6.9
Neo-menthol	7.2	7.8
Menthol	21.0	24.9
Carvone	4.6	6.6
Methyl Salicylate	13.2	16.6
Total Organic Volatiles in the Sample (including other components not listed above)	225.0	249.5
Concentration of Total Organics in Air	3.75×10^{-5}g/liter	4.16×10^{-5}g/liter

* Value is inflated due to artifact interference.

charcoal, or another Tenax pre-column, is used to contain the breakthrough problem.

For large air samples at very low concentrations LBC columns seem ideal. If the two early emerging artifacts interfere, simultaneous sampling on long porous polymer and LBC columns can be used, but should be rarely necessary.

To demonstrate the applicability of long columns (either LBC or porous polymers) to extremely low concentrations of volatiles, simultaneous samples of 10 liters air were taken from a location just outside the manufacturing plant and passed through separate trapping systems consisting of 50 feet LBC and 2 feet Tenax columns. At the time of sampling the temperature was -4°C and the odor was low. Analyses of the two samples after desorption onto short Tenax-GC pre-columns (Figure 4) show similar patterns of compounds except for the artifacts previously noted from Tenax at 19 and 20 minutes. Artifact formation was lower than usual in this analysis, which somehow may be due to the low sampling temperature, or to the low moisture levels.

TABLE 4. Identified Compounds in Air Outside Confections Manufacturing Plant

Known Compound	Sampling on Long (2 ft) Tenax-GC Column (μg)	Sampling on LBC Column (50 ft coated with 3.0 g. SE-30) (μg)
Ethyl Alcohol	0.30	0.34
Limonene	0.05	0.04
Menthone	*	0.06
Menthol	0.13	0.11
Methyl Salicylate	0.18	0.16

* Artifact interfered with analysis.

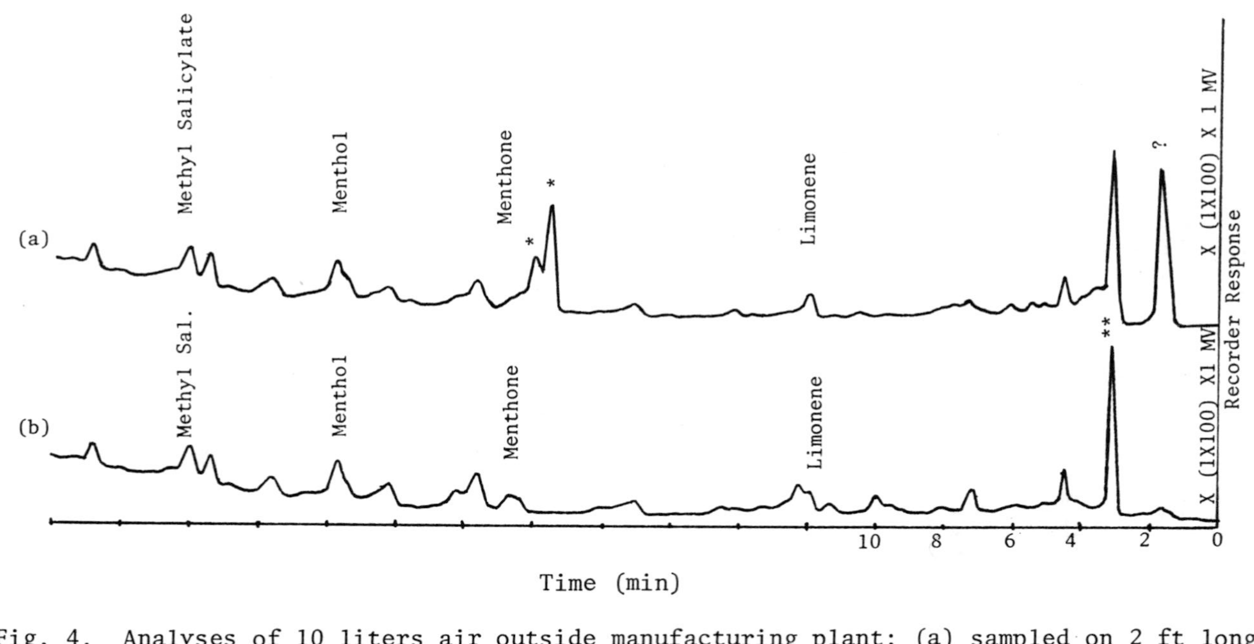

Fig. 4. Analyses of 10 liters air outside manufacturing plant; (a) sampled on 2 ft long Tenax-GC column (1.75 g Tenax), (b) sampled on 50 ft LBC column (3.0 g SE-30 coating; effective coating thickness is approx. 0.014 mm). Samples were collected at 100 ml/min over 100 minutes.

* Artifact.
** Contains artifact.

Table 4 shows comparative amounts of the few compounds definitely identified in this analysis. The calculated concentrations are found to be at or below odor threshold levels. For example, the concentration of methyl salicylate in the air sampled was calculated to be 1.7×10^{-8} g/liter, based on the known response of the F.I.D. to methyl salicylate. The published recognition threshold value for this chemical is 1.0×10^{-4} g per liter of air (10).

To demonstrate the applicability of LBC columns for analysis of organic volatiles in high moisture atmospheres, approximately 5 liters of breath were analyzed for peppermint oil components after ingestion of a 5 g peppermint candy. The sample was collected by blowing ten times directly into Teflon tubing connected to the LBC column which was maintained at 45°C to minimize condensation of moisture. The sample was then desorbed onto a short Tenax-GC pre-column and analyzed in the usual manner. The analysis is shown in Figure 5. The peppermit oil pattern is quite clear. The difference between this pattern and that of directly injected peppermint oil is only in the relative concentration of components. The most notable difference is the appreciable decrease of menthofuran in the breath, which may be due to this compound's tendency to oxidation. Other differences are the relatively lower concentration in the breath sample of the peppermint terpenes, such as the pinenes, limonene, and eucalyptol.

In addition to the peppermint oil, several other components, which must have originated in the breath, were noted. The weight of peppermint oil in 5 liters breath was about 0.06 mg, or 0.6% of the oil in the candy. In a previous analysis of peppermint flavored breath (6), 0.3 µg of peppermint oil was detected in a 10 ml sample. Collection and concentration of the larger breath sample in this analysis resulted in 200-fold concentration of the flavor compounds.

Quantitation of the volatiles in the sample was achieved by comparing the particular peak area in the sample to that of a known standard mixture at comparable concentration. The standard was injected directly and analyzed by the same GC conditions as the samples. The error encountered by this technique, i.e. comparing samples collected on pre-columns to direct injection, is minimal; less than 4% for the higher boiling components and even less for the more volatile ones. The areas were determined by electronic integration.

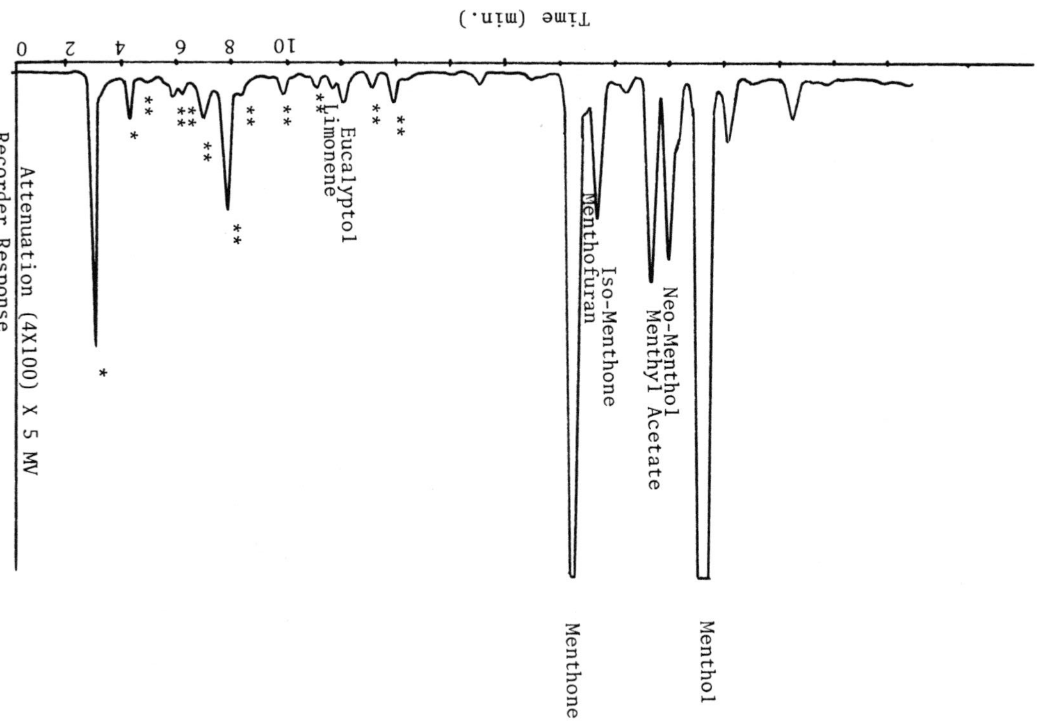

Fig. 5. Analysis for peppermint oil in 5 liters breath after ingestion of 5 g peppermint candy; sampled directly on 50 ft LBC column and desorbed onto Tenax-GC pre-column for analysis.

* Artifacts from sampling column.

** Components from breath not due to peppermint oil.

IV. CONCLUSION

Though no single system is of universal applicability for sampling and concentration of air volatiles, the use of LBC columns offers a unique analytical system with many advantages in examining large volumes of air or headspace for trace volatiles. In addition to efficiency of trapping, other advantages include low back pressure, and, when SE-30 is used as the coating phase, insensitivity to the moisture of the sample, and formation of very few artifacts.

ACKNOWLEDGMENT

The authors acknowledge the invaluable assistance of A.R. Pidel in performance of the analyses.

REFERENCES

1. Mieure, J., Dietrich, M.W. J. Chrom. Sci. 11, 559 (1973).
2. Russell, J.W., Environ. Sci. Technol. 9, 1175 (1975).
3. Dravnieks, A., Krotszynski, B.K., Whitfield, J., O'Donnell, A., Burgwald, T., Environ. Sci. Technol. 5, 1220 (1971).
4. Jennings, W.G., Wohleb, R.H., Lewis, M.J., J. Food Sci. 37, 69 (1972).
5. Zlatkis, A., Bertsch, W., Lichtenstein, H.A., Tishbee, A., Shunbo, F., Liebich, H.M., Coscia, A.M., Fleischer, N., Anal. Chem. 45, 763 (1973).
6. Mackay, D.A.M., Hussein, M.M., in "Analysis of Foods and Beverages-Headspace Techniques" (G. Charalambous, ed.), p. 283. Academic Press, New York, 1978.
7. Butler, L.D., Burke, M.F., J. Chrom. Sci. 14, 117 (1976).
8. Clark, R.G., Cronin, D.A., J. Sci. Fd. Agr. 26, 1615 (1975).
9. Mackay, D.A.M., Lang, D.A., Berdick, M., The Toilet Goods Association 32, 45 (1959).
10. Amerine, M.A., et al, "Principles of Sensory Evaluation of Food" p. 186. Academic Press, New York, 1965.

CAT AND HUMAN TASTE RESPONSES TO L-α-AMINO ACID SOLUTIONS

James C. Boudreau
Sensory Sciences Center
University of Texas at Houston
Houston, Texas

A quantitative comparison was made between human and cat taste responses to L-amino acid solutions. Neurophysiological measures from the cat (single unit recordings from the geniculate ganglion) were compared with psychophysical responses of humans. Amino acids elicit complex sensations from humans, although the sensations of sweet, bitter, and sour predominate. Of the three functional chemoresponsive neural groups identified in the cat geniculate ganglion, only groups I and II are discharged by amino acid solutions. Group I units are discharged by those amino acids that elicit a human sour sensation. Group II units are inhibited by the amino acids that elicit a strong bitter sensation and are excited by those tasting sweet or sweet-bitter. The neural and psychophysical responses to the amino acid solutions can be partly interpreted in terms of the chemical properties of the side chains. Group I stimuli and sour solutions contain amino acids with acidic side chains (where an acid is defined as a Brønsted acid), particularly those with hydrogen bond acceptor-donor complexes. Group II unit inhibition and bitterness are directly related to the hydrophobicity of the amino acid side chain, where hydrophobicity is calculated according to Tanford's hydrophobicity scale.

INTRODUCTION

Amino acids exist in free form in many animal and plant tissues. Many of these free form amino acids have been shown to be highly taste active and of major importance in food flavor and thus food selection (1). The measures available for determining the taste activities of amino acids are largely limited to psychophysical (behavioral) studies on humans and neurophysiological measures from other species. In this report the human sensations to amino acid solutions are quantitatively compared to neural response measures from the cat to determine the stimulus properties of the solutions.

ISBN 0-12-169060-1

The two different taste response measures were obtained independently by two different laboratories. These stimulus properties are then compared to certain chemical properties of the amino acids to determine the physical chemical properties of taste active amino acids.

TECHNIQUES: MEASURES COMPARED

A. Human Psychophysical Responses

The most intensive investigations into human psychophysical taste sensations to amino acids have been performed by investigators (2-4) at the Central Research Laboratories of Ajinomoto Co. These investigators selected 100 of the most sensitive tasters from a group of 1000 individuals. The taste tests used for selection consisted of measuring their sensitivity to a variety of taste solutions and the reliability of their judgements. A selection of human subjects is necessary because most adults are relatively insensitive to chemical stimulation (especially by sweet tasting compounds) of the front part of the tongue (5-9). Solutions of L-amino acids were tasted in distilled water in two concentrations, with a distilled water rinse. The tasters estimated the total magnitude of the sensation (on a 0-10 scale calibrated on NaCl solutions), as well as the distinct types of sensations elicited. The types of sensations were classified into six categories: sweet, bitter, sour, salty, umami (delicious) and "other tastes" (including astringent). Each sensation was expressed as a proportion of the total sensation magnitude. Their values are presented in Fig. 1 in terms of average sensation magnitude for sour, sweet and bitter sensations. The sensations of salty and "other tastes" were not elicited in any appreciable magnitude by the amino acid solutions. The sensation of umami is maximal for monosodium glutamate, which is inactive in the cat; other amino acids elicit small umami sensations. The psychophysical values in Fig. 1 and in other figures in this paper have been adjusted for 50 mM concentrations by linear approximation on a plot of magnitude vs. concentration (3). A table of the actual values obtained by (3) together with extrapolated values is contained in the Appendix. In general the results of comparing the two species are similar regardless of whether actual or estimated values are used.

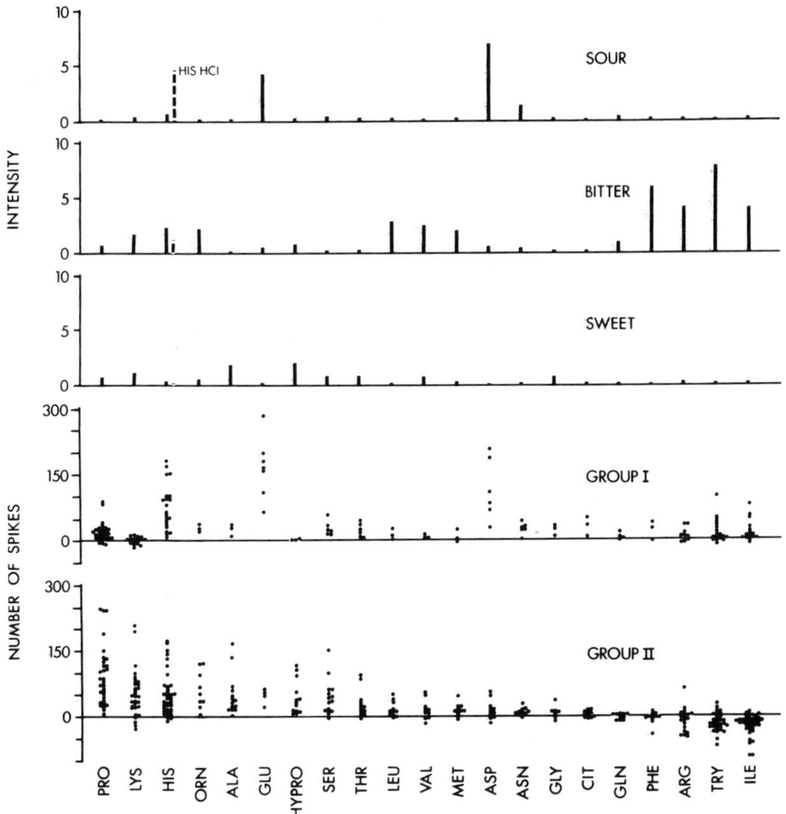

Fig. 1. Bottom two lines: cat single unit responses to amino acids, group I and group II units. Top three lines: human taste responses to amino acids. All values taken at or adjusted to 50 mM concentration in distilled water. Amino acids ranked on abscissa according to magnitude of group II response. HIS HCl only used in human.

B. Cat Neurophysiological Responses

The neurophysiological measures consisted of spike counts taken from single geniculate ganglion sensory neurons studied in anesthetized (pentobarbital) cats. These neurons innervate taste buds on fungiform papillae on the front part of the tongue. On the basis of various neurophysiological measures (9) these neurons have been divided into three separate functional groups. These neural groups were determined largely without regard to human taste systems (many compounds taste-active in humans were tested). Only the response measures from neural groups I and II will be reported here since group III neurons in general do not respond to amino acid solutions. The response of group I and group II units to a variety of amino acids are presented in Fig. 1. The basic measures represent spike counts taken during a 10 second stimulation period in which a 50 mM distilled water solution of a chemical was applied to the tongue. These spike counts were adjusted for the spontaneous activity level (unelicited spikes) for each cell by subtracting the spike count in the 10 second period preceding stimulation from the stimulated spike count. Negative spike counts were possible since some solutions decreased (inhibited) spontaneous discharge.

RESULTS (COMPARISON OF MEASURES)

Both the physiological and psychophysical measures (Fig. 1) indicate that the molecular complexity of the amino acids is matched by the response complexity. The psychophysical responses to the different amino acids are often complex in that different sensations may be elicited. The sensations most commonly elicited are those of bitter and sweet, although the sweet sensations tend to be of low magnitude (Fig. 1). Several solutions taste both bitter and sweet but, in general, bitter sensations are inversely correlated with sweet sensations in that strongly bitter compounds tend not to be sweet and vice versa (Fig. 2). Sour sensations are elicited by only a few amino acids and those compounds that are sour are not sweet (Fig. 2). The neural group most often affected by amino acid solutions is group II, to which the response may be one of excitation or inhibition.

There is a close similarity between the amino acids that excite group I neurons and those that elicit the sour sensation (Fig. 3). Although relatively few amino acids stimulate group I neurons, those that do also produce a sour sensation. The three most effective stimuli for group I neurons are aspartic acid, glutamic acid and histidine. The other amino acids that stimulate group I units can be ranked in this order: asparagine, serine, citrulline and ornithine with the remaining amino acids stimulating minimally. After aspartic and glutamic acids, the sourest amino acids are asparagine, histidine, glutamine, and serine. Thus the sour amino acids rank almost identically to those stimulating group I units.

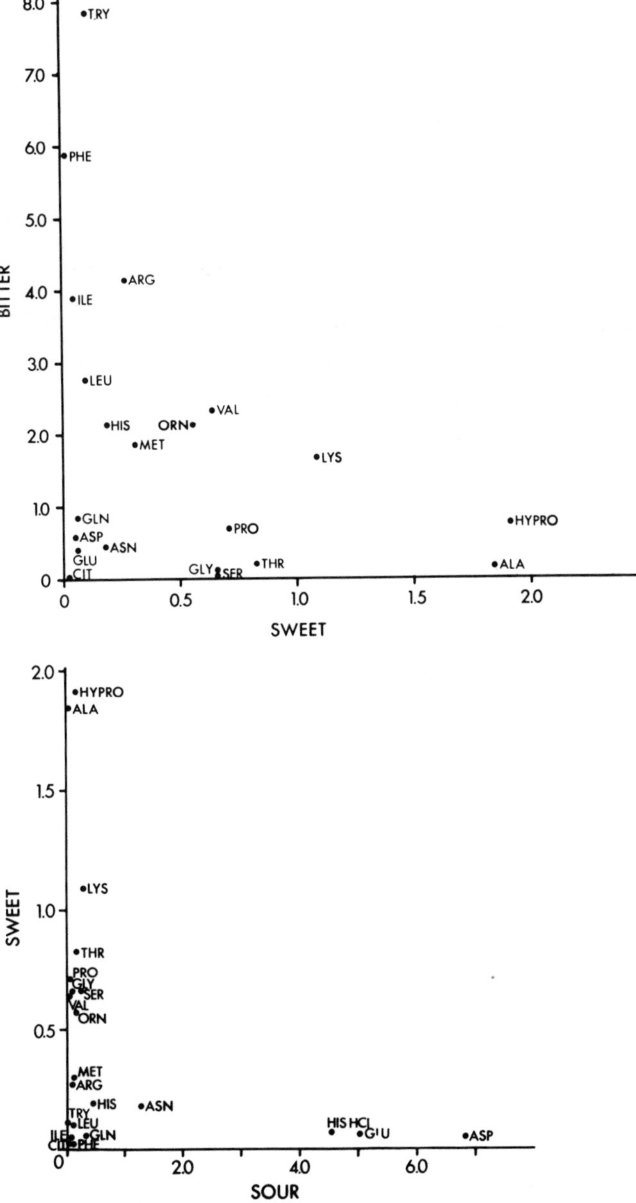

Fig. 2. Comparison of the magnitudes of the sweet, bitter and sour sensations elicited by 50 mM solution of amino acids.

The amino acid that clearly establishes a relationship between a group I stimuli and sour stimuli is histidine. In the cat a strong pH effect has been observed for histidine solutions with increased discharge to lower pH solutions (10). The sour sensation to HIS HCl (pH 4.2) was much larger than the sour sensation to the non-HCl histidine solution (pH 7.4). This pH effect has been interpreted for group I units in terms of protonation of the imidazole group (10).

The bitter sensation is elicited by many amino acid solutions. The amino acids evoking the largest bitter sensations are tryptophan, phenylalanine, arginine and isoleucine. Other amino acid solutions eliciting a bitter sensation with a magnitude greater than 1.0 include leucine, histidine, valine, ornithine, methionine and lysine. Other solutions elicit negligible bitter sensations. The sweetest amino acids, hydroxyproline and alanine, elicit negligible bitter sensations; and conversely, the bitterest amino acids, such as tryptophan, phenylalanine and isoleucine, tend to elicit

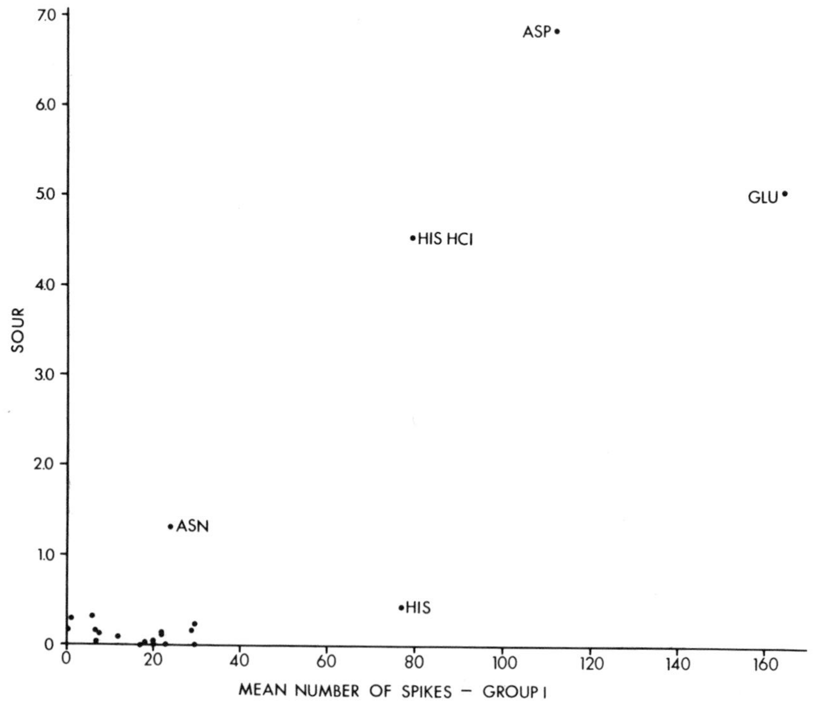

Fig. 3. Comparison of human sour sensations to cat group I response. No comparable value for the cat is available for HIS HCl, but similar pH effects have been oserved in the cat.

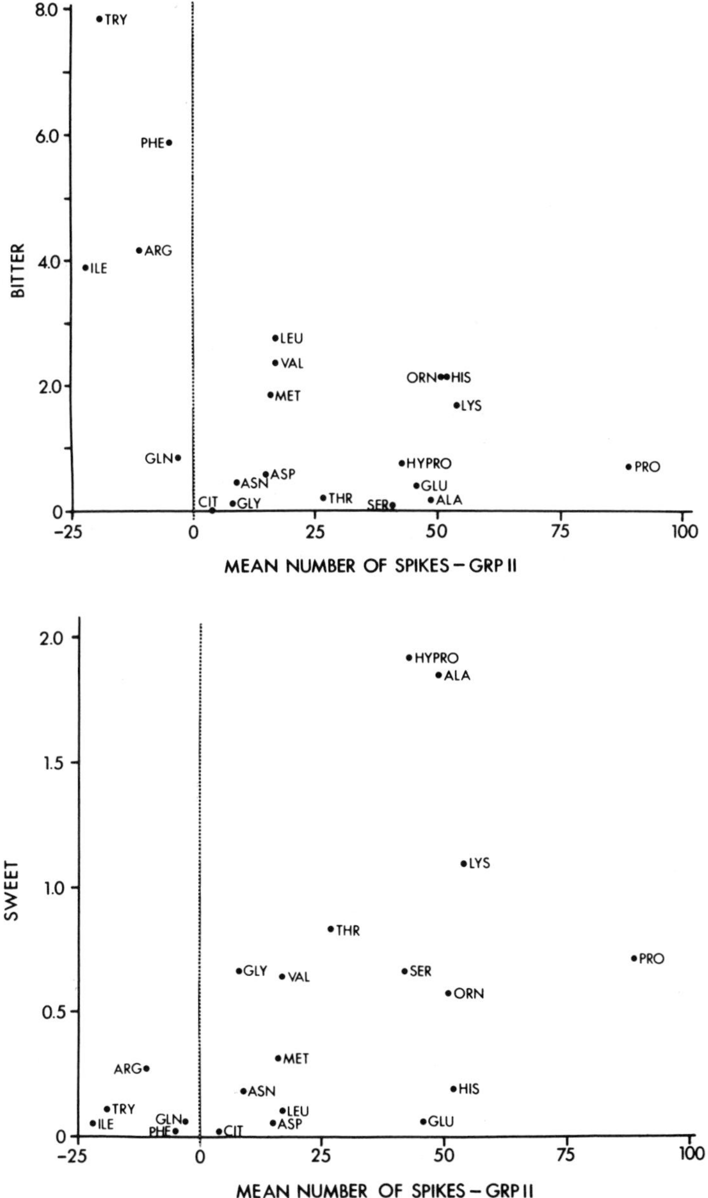

Fig. 4. Comparison of human sweet and bitter sensations and cat group II responses to 50 mM amino acid solutions.

sensations with only minor sweet components. Those amino acids yielding moderately intense bitter sensations, such as lysine and valine, are also accompanied by measureable sweet sensations. There seems to exist a group of sweet amino acids (alanine, threonine, serine, and glycine) that do not elicit bitter sensations.

The magnitude of the bitter sensation is inversely proportional to the magnitude of group II response (Fig. 4). Those amino acids evoking the strongest bitter sensations (tryptopan, isoleucine, phenylalanine and arginine) tend to inhibit group II unit discharge. Of the less bitter amino acids, leucine, valine and methionine elicit a group II response of about 15 spikes; and histidine, ornithine and lysine all elicit a discharge of about 50 spikes. Several amino acids, (alanine, serine, etc.) stimulate group II neurons but evoke little or no bitter sensation. Glutamine is the only amino acid that inhibits group II neurons but does not elicit a strong bitter sensation. The plot of bitter vs. group II response (Fig. 4) is similar in many ways to the plot of bitter vs. sweet (Fig. 2).

The sensation of sweet is elicited by a variety of amino acids, including those eliciting sensations of bitter (the sensations of sweet and sour being mutually exclusive -- Fig. 2). Thus the amino acids can be divided into those eliciting primarily a sweet sensation (hydroxyproline, proline, alanine, etc.) and those eliciting a sweet-bitter sensation, since very bitter amino acids have very little in the way of a sweet component (Fig. 2).

There is a fair degree of correspondence between sweet amino acids and those exciting group II units (Fig. 4). Those amino acids that consistently inhibit group II neurons are not sweet; those amino acids that excite group II units tend to taste sweet. All amino acids (except histidine and glutamic acid) eliciting group II responses greater than 25 spikes elicit a prominent sweet sensation. The sweetest amino acids (hydroxyproline and alanine) are not the best group II stimuli, however, and the best group II stimuli (proline, lysine, ornithine and histidine) are not the sweetest compounds.

DISCUSSION

Sensations represent an end product of considerable neural processing and as such are only indirectly related to any measure of peripheral neural activity. These factors and others not withstanding, it is apparent that, when viewed in terms of active stimuli, there is a close correspondence between certain human sensations and certain types of peripheral neural activity in the cat. To account for these similarities, a parsimonious explanation would be that the peripheral taste systems of the two species are quite similar and that at this neural level sour corresponds to group I excitation, bitter to group II inhibition and sweet (at least in part) to group II excitation. Such an explanation would be in accordance with data presented here and would also account for the interactions that have been observed between sweet and bitter, such as the inverse relationship between sensation magnitudes to amino acids (Fig. 2) and other interactions as reported by Békésy (11) and Birch (12).

To a great extent, the taste properties of the amino acids can be related to the chemical properties of the amino acid side chains, where the side chain is considered to include all of the molecule except the common glycine moiety. The side chains of the amino acids differ in that they may be composed entirely of hydrophobic hydrocarbon groups or consist of these and a reactive group.

The amino acids that taste sour or stimulate group I units possess side chains with a carboxyl group, an amide group or an imidazole group (Fig. 5).

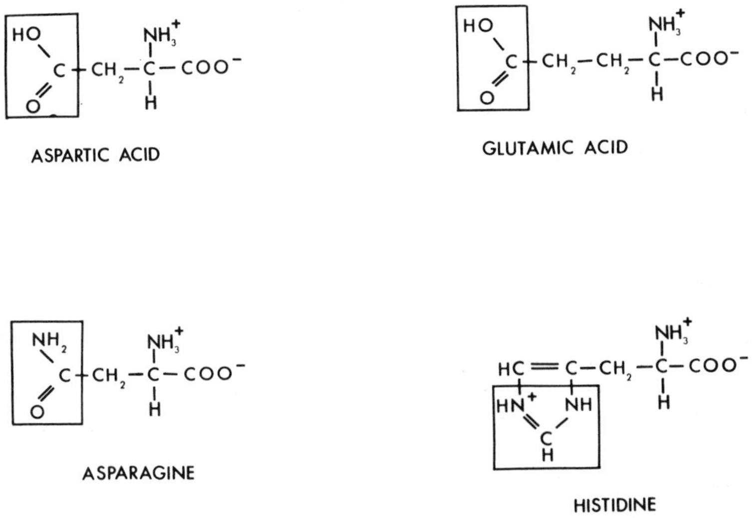

ASPARTIC ACID

GLUTAMIC ACID

ASPARAGINE

HISTIDINE

Fig. 5. Structural formulae for the best sour-group I stimuli. Amino acid side chains characterized by hydrogen bond acceptor-donor complexes as indicated by boxes.

These planar side chain groups have properties of both hydrogen bond donation and hydrogen bond acceptance (13) and have the common arrangement:

where X may be N or O. The carboxyl group and the imidazole group are maximally stimulatory when the carboxyl group is undissociated (sodium glutamate is nonstimulatory) and when the imidazole group is protonated. Being both hydrogen bond donors and acceptors, these molecules readily form hydrogen bonds with water molecules and with one another (especially in aprotic solvents (14), and are involved in rapid proton transfer reactions in aqueous solutions (15). Other compounds especially active on group I units (10) are those with phosphoric acid groups and the two heterocycles pyridine and thiazolidine. Strong acids are also active in low pH solutions (4.0 or less), where presumably the active molecular species is the hydrated hydrogen ion which has solution properties similar to the weak acids described above. In most foods the concentration of the hydrated hydrogen ion is insignificant and the active molecular species would include those with carboxyl groups, phosphate groups and a restricted set of nitrogen compounds. The introduction of nitrogen compounds extends the pH range of the sensory system since phosphate and carboxyl groups are largely dissociated and hence nonstimulatory at pH's above 5.0. That the common sour-group I stimulus usually contains a hydrogen bond acceptor-donator complex indicates that system activation entails considerable stimulus specificity.

The bitterness of amino acids has been related to the hydrophobicity of their side chains (16-18). In Fig. 6 the magnitude of the bitter sensation is plotted against the hydrophobicity of the side chains, where hydrophobicity is determined by Tanford's hydrophobicity scale and expressed in terms of Kcal of free energy of transfer from hexane to water (19,20) or from transformed surface tension measurements (21). The bitter sensation is directly proportional to side chain hydrophobicity, especially for those with purely hydrophobic side chains. Arginine and histidine are bitterer than would be predicted by hydrophobicity values alone, since the presence of nitrogen in the side chain tends to complicate these hydrophobicity measures (19, 22). The response of group II units to amino acids with purely hydrophobic side chains can be directly related to the hydrophobicity value for the side chain (Fig. 6). The most highly hydrophobic

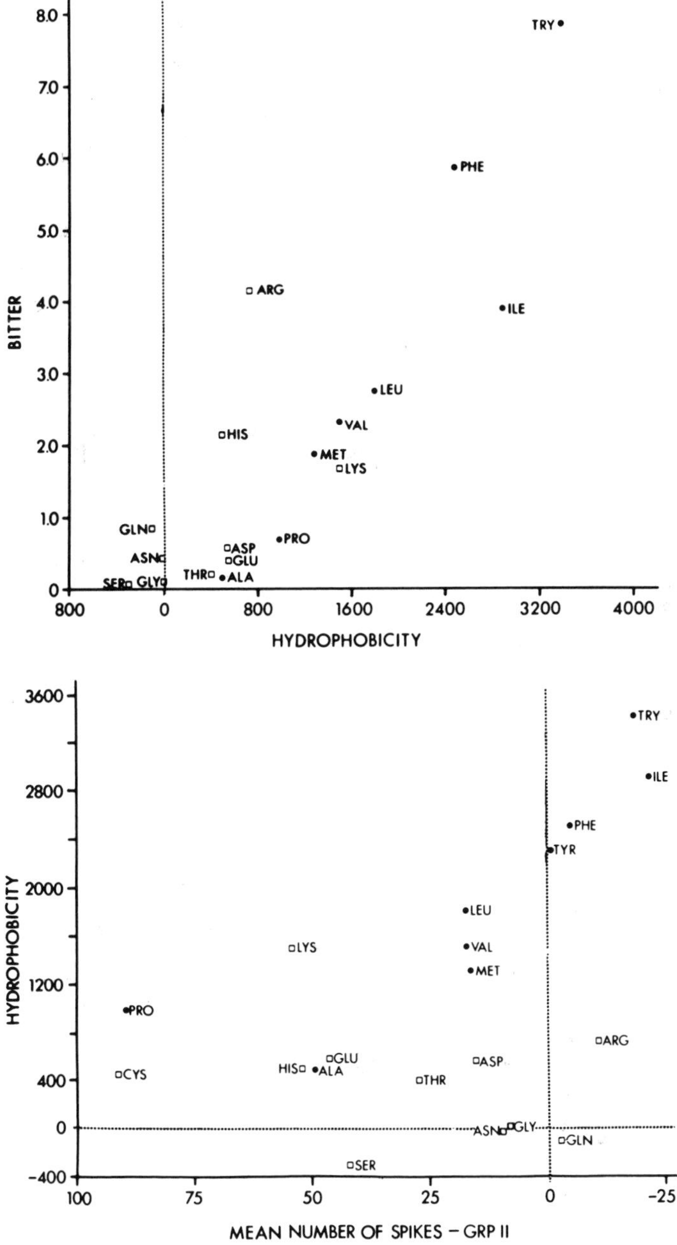

Fig. 6. Comparison of taste response measures with amino acid side chain hydrophobicity measures. Those amino acids with a hydrophyllic group indicated by an open square.

amino acids (tryptophan, isoleucine and phenylalanine) are all strong inhibitors. Those amino acids with moderately hydrophobic side chains (leucine, valine and methionine) elicit moderate group II responses, those low on the scale of hydrophobicity (alanine and proline) elicit large group II responses. Other amino acids with mixed side chains but high hydrophobicity values are the group II inhibitor tyrosine (hydroxylated phenylalanine) and lysine, a potent group II stimulus.

No common side chain property seems present among the amino acids that taste sweet, although, amino acids with acidic side chains (especially group I - sour stimuli) and those with strongly hydrophobic side chains are not sweet. Cat group II neurons are stimulated by a larger variety of amino acids (23) than those eliciting sweet sensations. In addition to sweet amino acids, those with mildly hydrophobic side chains and those with acidic side chains are also strong group II stimuli. There seems to be a fundamental difference in the stimulus dimensions for human sweet and cat group II stimuli. In fact the few L-α-amino acids that are effective in both human sweet and cat group II response constitute one of the few classes of sweet compounds effective in the cat. Many sweet compounds such as sugar, cyclamic acid, and saccharine are largely inactive neurophysiologically (24) and behaviorally (25, 26) in the cat. The taste activities of compounds similar to the L-amino acids may be different in the two species: D-tryptophan, strongly sweet, is inhibitory for cat group II neurones; and pyrrolidine, a strong cat group II excitor, tastes bitter. According to Wieser et. al (27), who investigated many amino acids and related compounds, sweet compounds require an amine group and a carboxyl group, whereas in the cat, optimum group II discharge is seen to simple heterocycles such as pyrrolidine, 3-pyrroline, and morpholine (23). It is apparant that although there are similarities between cat and man, there are also differences, no doubt reflecting the evolutionary history and nutritional requirements of the two species.

ACKNOWLEDGMENTS

J. Oravec, C. Madigan, W. Anderson, J. Watkins, N. Tran H. Ngyuen and J. Lucke participated in various aspects of this research. I thank T. Ninomiya and associates for reading and commenting on an earlier draft of this paper. This research financed in part by NSF Research Grants.

REFERENCES

1. Solms, J., in Gustation and Olfaction, an International Symposium (Ohloff, G. and Thomas, A. F., eds.) pp. 92-100, Academic Press, New York, 1971.
2. Kirimura, J., Shimizu, A., Kimizuka, A., Ninomiya, T., and Katsuya, N., J. Agr. Food Chem. 17, pp. 689-695, 1969.
3. Ninomiya, T., Ikeda, S., Yamaguchi, S., and Yoshikara, T., Rep. 7th Sensory Evaluation Symposium, JUSE, pp. 109-123, 1966.
4. Yoshida, M. and Saito, S., Jap. Psych. Res. 11, pp. 149-166, 1969.
5. Békésy, G.v., J. Appl. Physiol. 19, pp. 1105-1113, 1964.
6. Birch, G. G., Structure-Activity Relationships in Chemoreception, (Benz, G., ed) pp. 120-121, Information Retrieval, Ltd., London, 1976.
7. Dzendolet, E., and Murphy, C., Chem. Senses and Flavor 1, pp. 9-15, 1974.
8. Meiselman, H. L., and Dzondolet, E., Percep. & Psychophys. 2, 1967.
9. Boudreau, J. C. and Alev, N., Br. Res. 54, pp. 157-175, 1973.
10. Boudreau, J. C. and Nelson, T. A., Chem. Senses and Flavor, 2, pp. 353-374, 1977.
11. Békésy, G.v., Science 145, pp. 834-835, 1964.
12. Birch, G., CRC Crit. Rev. Fd. Sci. and Nutr. 8, pp. 1-56, 1976.
13. Vinogradov, S. N. and Linnell, R. H., Hydrogen Bonding Van Nostrand Reinhold Co., New York, 1971.
14. Davis, M. M., Acid-Base Behavior in Aprotic Organic Solvents, Nat. Bur. Stand. Monog. 105, 1968.
15. Caldin, E. and Gold, V., Proton-Transfer Reactions, John Wiley & Sons, New York, 1975.
16. Ney, K. H., Z. Lebensm. Unters. -Forsch. 147, pp. 64-68,
17. Ney, K. H., in Naturliche und synthetische Zusatzstoffe in der Nahrung des Menschen, C.I.I.A. Symposia (Ammon, R. and Hollo, J., eds.) pp. 131-143, Dr. Dietrich Steinkopf Verlag, Darmstadt, 1974.
18. Weiser, H. and Belitz, H.-D., Z. Lebensm. Unters. -Forsch. 159, 1975.
19. Nozaki, Y. and Tanford, C., J. Biol. Chem. 246, pp. 2211-2217. 1971.
20 Tanford, C., The Hydrophobic Effect, John Wiley & Sons, Inc., New York, 1973.

21. Bull, H. B. and Breese, K., Arch. Biochem. Biophys. 161 pp. 665-670, 1974.
22. Fendler, J. H., Nome, F. and Nagyvary J., J. Mol. Evol. 6, pp. 215-232, 1975.
23. Boudreau, J. C., Anderson, W., Oravec, J., Chem. Senses and Flavor 1, pp. 495-517, 1975.
24. Boudreau, J. C., unpublished research.
25. Carpenter, J. A., J. Comp. Physiol. Psych. 48, pp. 139-144, 1956.
26. Bartoshuk, L. M., J. Comp. Physiol. Psych. 89, pp. 971-975, 1975.
27. Wieser, H., Jugel, H. and Belitz, H.D., Z. Lebensm Unters.-Forsch., 164, pp. 277-282, 1977.

APPENDIX

Table 1. Taste Responses of the Human and Cat to Amino Acid Solutions

Amino Acid	Molarity[1]	Sweet	Sour	Bitter	Group I	Group II
	0.056	1.95	0.02	0.17		
L-ALA	0.561	3.22	0.32	0.24		
	0.050	1.84	0.02	0.16	23	49
	0.011	0.80	0.14	1.47		
L-ARG	0.057	0.28	0.11	4.30		
	0.050	0.27	0.11	4.15	8	-11
	0.001	0.10	1.73	0.62		
L-ASP	0.023	0.04	5.86	0.51		
	0.050	0.05	6.83	0.59	112	15
	0.067	0.21	1.56	0.53		
L-ASN · H_2O	0.200	0.23	2.58	0.71		
	0.050	0.18	1.32	0.45	24	9
	0.228	1.11	0.22	1.55		
L-CIT	0.457	2.23	0.36	1.61		
	0.050	0.02	0.003	0.02	30	4
	0.002	0.19	2.70	0.56		
L-GLU	0.136	0.05	3.99	0.31		
	0.050	0.06	5.09	0.40	164	46
	0.068	0.07	0.38	1.04		
L-GLN	0.205	0.82	0.68	0.53		
	0.050	0.06	0.32	0.85	6	-3
	0.133	1.59	0.22	0.28		
L-GLY	0.533	3.71	0.15	0.14		
	0.050	0.66	0.09	0.12	22	8
	0.064	0.21	0.50	2.45		
L-HIS	0.193	0.72	0.35	3.88		
	0.050	0.19	0.44	2.15	77	52
	0.038	1.61	0.15	0.65		
L-HYPRO	0.229	3.58	0.10	1.71		
	0.050	1.91	0.17	0.77	1.0	43
	0.046	0.05	0.06	3.72		
L-ILE	0.229	0.17	0.08	6.68		
	0.050	0.05	0.06	3.90	18	-22

Amino Acid	Molarity[1]	Sweet	Sour	Bitter	Group I	Group II
L-LEU	0.076	0.13	0.13	3.62		
	0.152	0.08	0.11	5.02		
	0.050	0.10	0.10	2.77	12	17
L-LYS-HCl[2]	0.022	0.73	0.20	1.13		
	0.164	1.69	0.28	2.48		
	0.050	1.09	0.29	1.69	1	54
L-MET	0.020	0.19	0.09	1.13		
	0.201	0.50	0.15	3.91		
	0.050	0.31	0.14	1.86	7	16
L-ORN-HCI	0.059	0.61	0.16	2.32		
	0.119	0.69	0.25	2.86		
	0.050	0.57	0.15	2.16	29	51
L-PHE	0.018	0.10	0.05	3.55		
	0.091	0.03	0.07	7.06		
	0.050	0.02	0.06	5.87	20	-5
L-PRO	0.217	1.78	0.08	1.75		
	1.216	3.52	0.05	3.05		
	0.050	0.71	0.03	0.70	20	89
L-SER	0.143	1.56	0.55	0.20		
	1.427	3.01	0.85	0.28		
	0.050	0.66	0.23	0.08	30	42
L-THR	0.168	1.75	0.33	0.42		
	0.588	2.25	1.20	0.24		
	0.050	0.83	0.16	0.20	22	27
L-TRP	0.015	0.06	0.24	3.75		
	0.034	0.10	0	6.61		
	0.050	0.11	0	7.84	17	-19
L-VAL	0.026	0.51	0.13	1.32		
	0.068	0.80	0.08	2.95		
	0.256	1.52	0.14	4.67		
	0.050	0.64	0.06	2.36	7	17

1. Human values at .05 extrapolated. All human values from (3)
2. Free base used in the cat.

AROMA ANALYSIS OF VIRGIN OLIVE OIL BY HEAD SPACE (VOLATILES) AND EXTRACTION (POLYPHENOLS) TECHNIQUES.

G. Montedoro, M. Bertuccioli, F. Anichini
Istituto Industrie Agrarie,
Universita Studi di Perugia, (Italy)

This paper presents the procedure and the results showing the organoleptic importance of the polyphenols and volatiles compounds. The original technique developed involves (1) the isolation and identification of volatile components by enrichment of "headspace" vapors on porous polymer (Tenax GC traps), their displacement and introduction through a closed system to the GC column (Bertuccioli *et al.*) coupled with MS and their quantitative evaluation; (2) the isolation and identification of the polyphenols by methanol-water mixture extraction and their evaluation by different techniques (TLC, calorimetric analysis). The evaluation of different compounds was carried out on the course of the olive oil fruits maturity and ripeness, examining also the influence of pigmentation. The results were undertaken to establish relationship between volatiles, which the GC-Sniff evaluation had determined to be "aroma significant," polyphenol levels, and intensity ratings of the aroma attributes. The importance of the ripeness, pigmentation rating, and oil extraction process on the major components of oil aroma are illustrated. The causes of the fluctuations in relative amounts of individual groups of constituents are discussed on the basis of possible pathways of products present in oil.

ISBN 0-12-169060-1

1- INTRODUCTION

Many papers have been published on the phenolic (1-8) and volatile (9-24) costituents of olive oil and other ones (26-30) It has to be pointed out that the pulp of the fruit is particularly rich in the first ones (from 2 to 5%) (31) and that a certain percentage of these is found also, in variable amounts, in the oil. According to recent study (7) the presence of some of those components in the oil affects very clearly its organolectical characteristics.

As far as the volatile constituents are concerned it that the results of studies have shown that there are many of those components in the olive oil, but only a few of them, which are found in an appreciable amount, are responsible of the "aromatic note" (9).

On this base we are studying, since several years, this lipid to examine both the phenolic (1,2,3,22,23,24,25) and the volatile constituents of the "head space (18,19,20,21). For each of these two groups of components we have developed a procedure the details of which will be given later.

Using this procedure it has been possible to characterize the above mentioned constituents, to follow their evolution during the ripening of the fruit on taking into account the degree of pigmentation of the same fruit; these studies have been carried out on different cultivars of the same or of different locations (20,22), clarifying the mechanisms, by which, these costituents accumulate in the droplet oil.

We have also studied the importance of the storage of the fruit and of the technological processes, in particular of the technique of mechanical oil extraction (milling, mashing, pressing) (21,24).

At last we tried to establish, for both of the above mentioned groups of constituents, a possible correlation between the qualitative and quantitative composition of the oil and this organolectic characteristics , considering both the olfactory (volatile constituents) and the tasteful (polyphenols) aspects (19,25).

2 - EXPERIMENTAL

A - Methods of analysis

A_1- Volatile Constituents

A detailed list of the different volatile costituents found in the virgin olive oils (9,11,14,18,20) is shown in Table I. Different techniques have been mainly used: Distillation (9,13,15, 17); entrainement (putling) of inert gases and concentration in deodorant chaorcal (10); "headspace" (18,20) . Althoug these methods allow a systematic estimate of the different volatile compounds, no one of them is surely free from the possible formation of artefacts; moreover they do not allow the possibility of doing, in a future prospective a qualitative and quantitative sound dosage to be correlated with the organolectic analysis.

As we mentioned above one of us (the authors) has developed a technique (18), derived from fermented beverages (32,33) which allows to test, by gas chromatography and in strictly defined conditions, only the constituents which are really present in the "headspace" and responsible of the olfactory and tasteful notes.

On the other hand the volatile "headspace" analysis have been used also by other Authors (10,11,14,15,21) for the systematic study of aromatic components of different other oils (26, 27,28,29,30).

Details of the technique are as follows (18,20):
20 g of oil are put into a syringe (of 5 lt volume) and then are kept at 25°C for 30'.

Successively they are adsorbed on a trap of Tenax CG. The relative "headspace" in analyzed directly by gas chromatography (Varian 1400 coupled with Data System CDS-111) under the following working conditions:
- Stailess steel column 3.2 M x 3.2 mm i.d.; coated with Carbowax 20 M 3% on Chromosorb W 80-100 mesh; Gas carried : He with flowrate of 18 ml/min; Column temperature: starting at 38°C for 10', then increasing by 2.7 °C /min untill 170°C; Injector temperature (Valco Valve) 250°C, detector 250°C.

For GC-Sniff evaluation, volatile effluent was splitted 1:1 and the aromas of the eluting components were sniffed.

An example of a chromatogram of the "headspace" whit the relative individual sniffed component sensation of an oil is shown in Figure I, while Table II shows the different identified compounds.

The quantitative evaluation, restricted only to some constituents, has been accomplished by calculating the curves of solubility adding those compounds in sequence to the oil and by te dosage of the latter ones in non-equilibrium conditions (19, 34).

TABLE I - Identified olive volatile components.

HYDROCARBONS

Isopentane (10)
2-Methyl-Pentane (10)
Hexane (10,18)
Octane (9,10,15,18)
Nonane (10)
Naphthalene (15)
Ethyl-Naphthalene (15)
Acenaphthene (15)
Aromatic Hydrocarbon (15)

ALIPHATIC ALCOHOLS

Methanol (15,17)
Ethanol (10,17,18)
Isopropyl alcohol (9,15)
1-Pentanol (10,18)
3-Methylbutan-1-ol (9,10,17,18)
1-Penten-3-olo (9,18)
cis-3-Hexen-1-ol (9,10,17,18)
trans-2-Hexen-1-ol (9,10,17,18)
1-Hexanol (9,10,17,18)
1-Heptanol (9)
1-Octanol (9,15,18)
1-Nonanal (9,15)
2-Phenylethanol (9,15)

OXYGENATED TERPENS

1,8 Cineole (9)
Linalol (9)
α-Terpineol (9,15)
Lovandulol (15)

ALDEHYDES

Acetaldehyde (9,10,17,18)
Propanal (17)
2-Methyl-butanal (9,18)
3-Methyl-butanal (9)
Butanol (15)
Pentanal (9,15,18)
trans-2- Pentenal (9)
Pentanal (cis-2?) (9,18)
Hexanal (9,10,17,18)
cis-2-Hexenal (9)
trans-2-Hexenal (9,10,18)
Heptanal (9,15)
2,4-Hexadienal (9)
Heptenal (cis-2?) (9)
trans-2-Heptenal (9)
Benzaldehyde (9,15,18)
Octanal (9,15,18)
2,4-Heptadienal (isomer A) (9)
trans-2-Octenal (9,15)
Nonanal (9,15,18)
trans-2-Nonenal (9,15)
2,4-Nonadienal (9)
trans-2-Decenal (9,15)
2,4-Decadienal (isomer A) (9)
2,4-Decadienal (isomer B) (9)
trans-2-Undecenal (9,15)

KETONES

Aceton (17)
3-Methylbutan-2-one (9)
3-Pentanone (9,10,17,18)
2-Hexanone (9)
4-Methyl-3-penten-2-one (18)
2-Methyl-2-hepten-6-one (9)
2-Octanone (9)
2-Nonanone (9)
Acetophenone (9)

ETHERS

Methoxybenzene (anisole) (9,15)
1,2-Dimethoxybenzene (9)

FURAN DERIVATES

2-Propyl furan (15)
2-n Propyl dihydrofuran (15)
2-n Pentyl-3-methyl furan (15)

THIOPHENE DERIVATES

2-Isopropenyl thiophene (15)
2-Ethyl-5- hexilthiophene (15)
2,5-Diethylthiophene (15)
2-Ethyl-5-hexyldihydro-thiophene (15)
2-Ethyl-5-methyldihydro-thiophene (15)
2-Octil-5-methylthiophene (15)

ESTERS

Methyl acetate (10)
Hthyl-acetate (9,10,17,18)
Hthyl-propionate (9)
Methyl-butyrate (9)
Hethyl-2-methylpropionate (9)
2-Methyl-1-propylacetate (9)
Methyl-3-methylbutyrate (9)
Methyl-2-methylbutyrate (9)
Hethyl-butyrate (9)
Propyl propionate (9)
Methyl pentanoate (9)
Hethyl-2-methylbutyrate (9)
Hethyl-3-methylbutyrate (9)
1-Propyl-2-methyl propionate (9)
3-Methyl-1-butyl acetate (9)
2-Methyl-1-butyl acetate (2)
2-Methyl-1-propyl-2-methyl-propionate (9)
Methyl hexanoate (9,15)
Cis-3-hexenyl acetate (9,10,18)
Hexyl acetate (10,18)
Methyl heptanoate (9,15)
Methyl octanoate (9,15)
Ethyl benzoate (9)
Ethyl Heptanoate (9)
Ethyl Ocatanoate (9,15)
Methyl salicilate (15)
1-Octyl acetate (9)
Ethyl phenylacetate (9)
Ethyl nonaonate (15)
Ethyl decanoate (15)
Ethyl palmitate (13,15)
Methyl Oleate (13,15)
Methyl linoleate (13,15)

In Table III we have reported the calculated values concerning hexanal, trans-2-hexenal, 1-hexanol and 3-methylbutan-1-ol, which are the most representatives of the different volatile constituents of the "headspace" of the oil (manuscript in preparation) (18):

In this paper we expressed the concentrations in conventional units as given by the integrator.

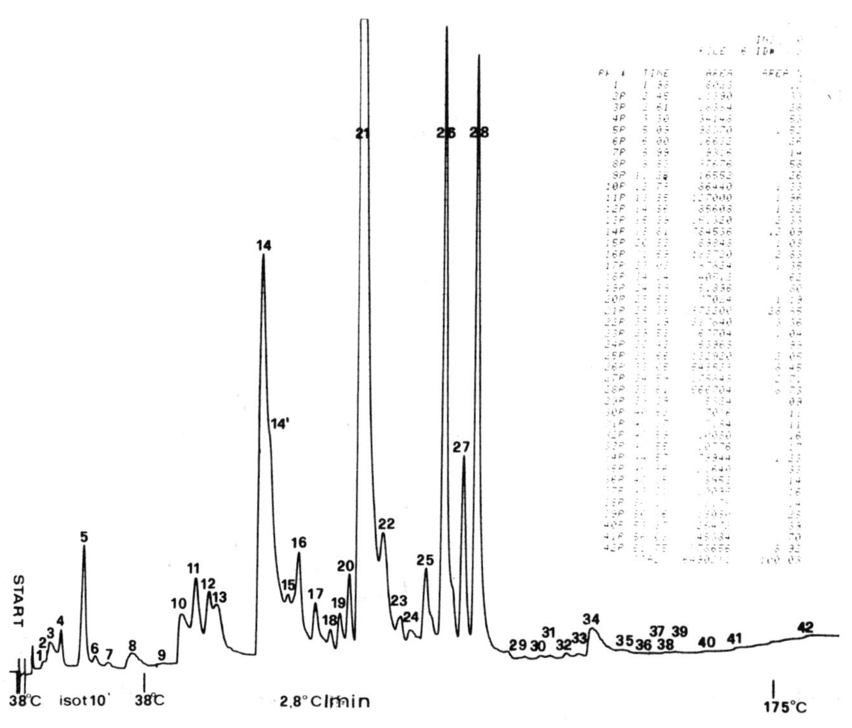

Fig. 1. GC profile of olive oil volatiles "headspace"

Peak code	Area %	Odor Description
1	12	odorless
2	33	"
3	28	pungent odor
4	53	odorless
5	1.52	"
6	26	"

(Continued Figure I)

7	14	ether like
8	58	alcohol
9	26	odorless
10	1.33	pungent odor,acrid
11	1.96	" " "
12	1.32	odorless
13	2.33	"
14	12.09	fruity odor,leaf
15	1.08	leaf odor
16	2.83	fruity,banana like
17	1.35	leaf odor,green
18	62	" " "
19	80	" " ,olive
20	1.19	pungent odor, (fusel oil)
21	28.85	leaf odor, green
22	3.36	fusel-like odor
23	1.04	fruity, olive
24	99	pungent odor
25	2.05	fruity, banana like
26	8.45	fruity odor,aromativ flav.
27	2.71	fruity green,leaf odor
28	8.73	fatty
29	09	floreal
30	11	fruity
31	11	orange-rose odor
32	16	almond
33	17	bread
34	1.23	odorless
35	33	fruity
36	14	terpene like
37	16	odorless
38	14	"
39	28	"
40	39	cooked olive
41	70	odorless
42	8.92	"

TABLE II - Compounds identified in olive oil headspace.

Components	Peak no	Identification r.t.	Identification m.s.
HYDROCARBONS			
n-Hexane	4	+	
n-Octane	5	+	
Ethylbenzene	19	+	+
ALCOHOLS			
Ethanol	8	+	
1-Penten-3-ol	17	+	+
3-Methylbutan-1-ol	20	+	+
1-Pentanol	22	+	+
1-Hexanol	26	+	+
cis-3-Hexen-1-ol	26'	+	
trans-2-Hexen-1-ol	27	+	+
1-Octanol	31	+	
ALDEHYDES			
Acetaldehyde	3	+	
2-Methylbutanale	8	+	+
Pentanal	10	+	
Hexanal	14	+	+
cis-2-Pentenal	14'	+	
trans-2-Hexenal	21	+	+
Octanal	24	+	
Nonanal	28	+	+
Benzaldehyde	32	+	
KETONES			
3-Pentanone	10	+	
4-Methyl-3-penten-3-one	18	+	+
ESTERS			
Ethylacetate	7	+	
Isobutyl acetate	13	+	
3-Methyl-1-buthyl acetate	16	+	
Hexyl acetate	23	+	
cis-3-Hexenyl acetate	25	+	

TABLE III
Linear regression analysis of different volatile components solubility in virgin olive oil.

Volatiles Components	tg	cor. coeff. (n = 19)	mg/Kg oil
3-Methylbutan-1-ol	94	0.980	0.1 - 5
1-Hexanol	88	0.964	0.1 - 2
Hexanal	96	0.959	0.5 - 5
trans-2-Hexenal	199	0.981	0.5 - 10
Nonanal	11	0.689	5 - 50

A_2- Polyphenols

A list of the different phenolic components found in the virgin olive oils is shown in Table IV (1,8).

TABLE IV
Phenolic compounds identified in olive oil.

Components	References	Concentration mg/Kg oil
(3-Hidroxyphenil)Ethanol	(4,8,5)	7 - 98
β(3,4-Hidroxyphenil)Ethanol	(4,8,5)	1 - 50
Oleuropein	(8)	tr
Oleuropeilaglycone	(8)	5 - 77
(α-p-m)Coumaric acids	(5,1)	
Caffeic acid	(1,8,5)	0.5 - 8
Hydrocaffeic acid	(1)	
Sinapic acid	(1)	
Protocatechuic acid	(5,1)	
Vanillic acid	(5)	
Homovanillic acid	(5,1)	
Gentisic acid	(1)	
Quinic acid	(1)	
Shikimic acid	(1)	
p-Hydroxybenzoic acid	(5,1)	
Syringic acid	(1)	
Apigenin	(5)	
Luteolin	(5)	

The extraction techniques are derived from the one we developed (1), which has been, as reported later, recently modified (22). This modification was suggested by re-examinating the different techniques used by these authors (4,5,8), who all add to the watery methanol (1) selective solvents.

Since there is no important quantitative improvement when the different techniques are compared with each other, as shown bu the results reportated in Table V, but they are quite more laborious, and considering also, (Table VI), that the phenolic compounds are bound in a certain manner to the lipoproteic membrane by whic the oil drop is circumscribed (35,36), the extraction was carried out additioning a tensioactive substances whic eliminates the above mentioned membrane (37, 38,39).

TABLE V

Phenolic compounds extraction by means different solvents.

Solvents	Total Phenols mg/Kg oil
Methanol : water 80:20(v/v) (1)	99.5
Petroleum Ether + Oil	
Methanol : water 60:40(v/v) (4)	108.0
Hexane + Oil	
Methanol : water 60:40(v/v) (5)	123.5
Methanol : Chloroform 95:5 (v/v) (30)	114.0

TABLE VI

Effect of enzymatic preparations and tannin phenolic adsorbent agent treatments on the phenolic oil content.

Test	Total Phenols mg/Kg oil
+ Cellulase	75.3
+ Ac. Protease	96.8
Test	28.4
+ Methil Cellulose	26.4

The importance of this addition is illustrated by the results reported in Table VII.

TABLE VII - Effect of tensionactive on the oil phenolic compounds extraction.

Solvents	Total Phenols mg/Kg oil
Methanol : water = 80:20 (v/v)	90
+ Tween 20	134.4

The experimental method may be summarized as follows (22):

- 50 g of oil have been homogenized twice with 50 ml of methanol-H_2O (80:20 v/v), previously added to Tween 20 of 2%. The two extracts cleared from the oil by centrifugation have been collected and decanted during 24-48h at 20°C.

In the supernatant, free from impurities (lipids or lipoproteins), the quantity of the polyphenolic components have been determined with Folin-Ciocalteu reagent (40).

The values have been expressed as gallic acid.

B - Materials

The oil was extracted by pressing process (disk press) (24) both with previous milling and mashing of the paste and with direct extraction of fruit. The oil before the analysis was dehydrated on sodium sulphate.
Scheme of the industrial oil extraction process by single pressure is shown in picture following:

```
                                    Sansa  →  Solvent  →  Olive
                                  ↗ (Cake)    Extraction   Oil
Olive→Milling→Mashing→Pressing
                                  ↘
                                    Oily→Centrifugation→Virgin Olive
                                    Must                     Oil
```

In the laboratory we have proceeded also to direct extraction from fruit:

Olive → Pressing → Oily Must → Centrifugation → Virgin Olive Oil

The results are referred to oil obtained by industrial process.

This has been sampled from the fruits harvested, in given periods, in different cultivars and in different locations.

The evolution of the phenolic and volatile constituents of the "headspace" has been studied both considering the above mentioned classification and on all the olives sampled.

3 - RESULTS

A - Influence of the fruits ripening.

A_1- Role of the pigmentation

The fruits of each sample have been divided into classes of pigmentation.This classification has been based upon a preliminaryresearch (41) in wich the values of the D.O. 625 nm/ /D.O. 520 nm ratio of a methanolic solution obtained by lixiviation of the fruits for 48 h, have been considered. Those values allowed us the fruits into green ones, semi-black and black ones . The evolution of the total volatile and total polyphenolic constituents, as well as the distribution of the different fruits, cultivar Moraiolo, in relation to their degree of pigmentation over time are reported in Figure II. As can be seen, the degree of pigmentation is very important for both the analytical parameters studied.

The concentration of the different constituents increase of the degree of pigmentation untill a stage is reached beyond which an inversion of this relationship is observed. During the semi-black stage the fruit is submitted to a deep physiological transformation. Indeed, while the volatile constituents are increasing considerably and continuously over time, the phenolic compounds are decreasing .
As far the quantitative and qualitative variations are concerned some volatile constituents, the results allow the following remarks (Figure III):

- With the increase of the degree of pigmentation there is an increase of aldehyds and hydrocarbons and, in the contrary, a decrease of alcohols and parts of esters.
- After the stage of completed pigmentation the aldehydes and hydrocarbons decrease, while the alcohols and esters tend to increase.
- In the different classes of pigmentation and over time the aldehyds and the alcohols show always opposite patterns of variation.
- The aldehyds are always found in higher quantitaties than the other constituents studied, as has been shown also by other authors (9). The values we found lied between 50% (green

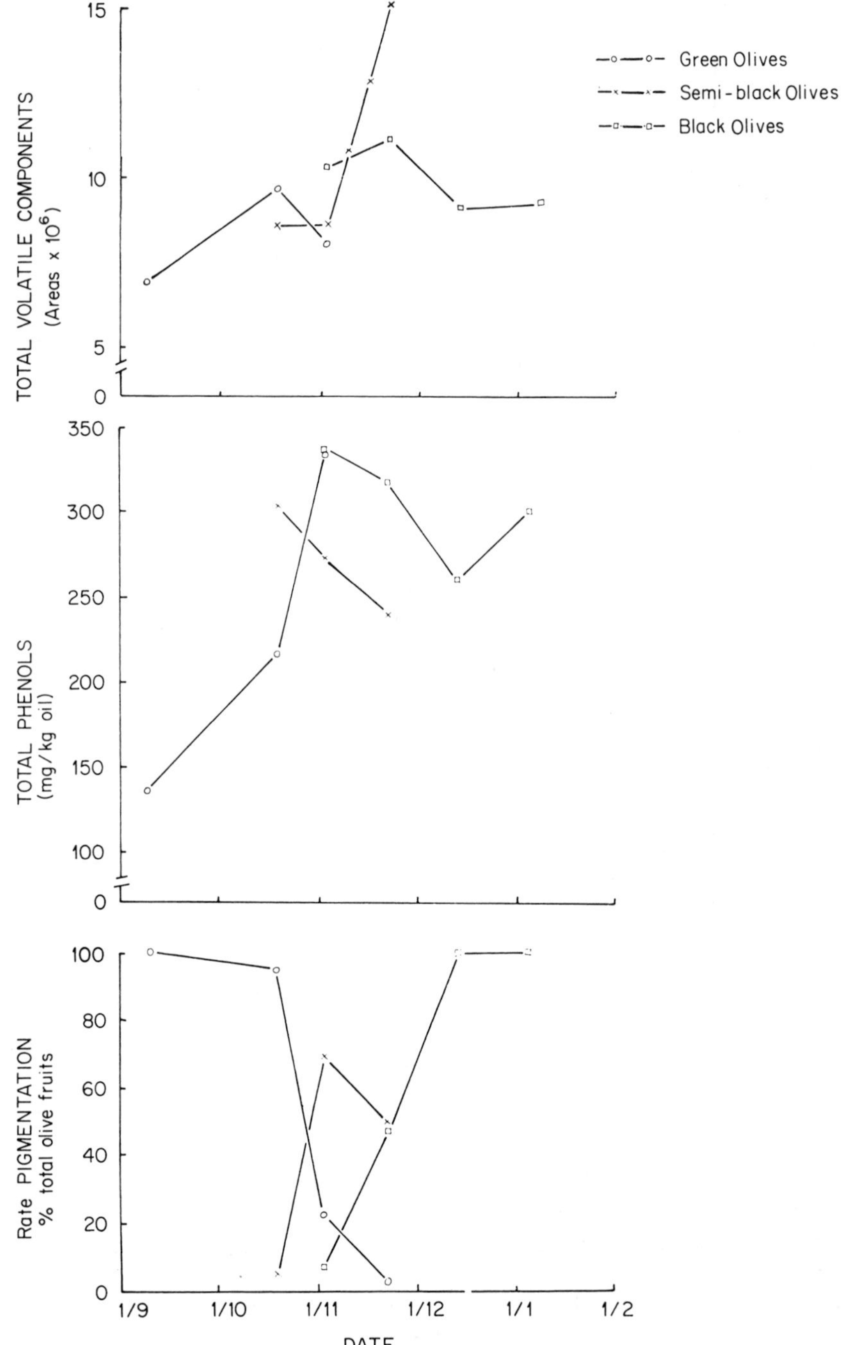

Fig. 2. Grade pigmentation of olives and evolution of oil volatile and phenolic costituents during the fruits ripening.

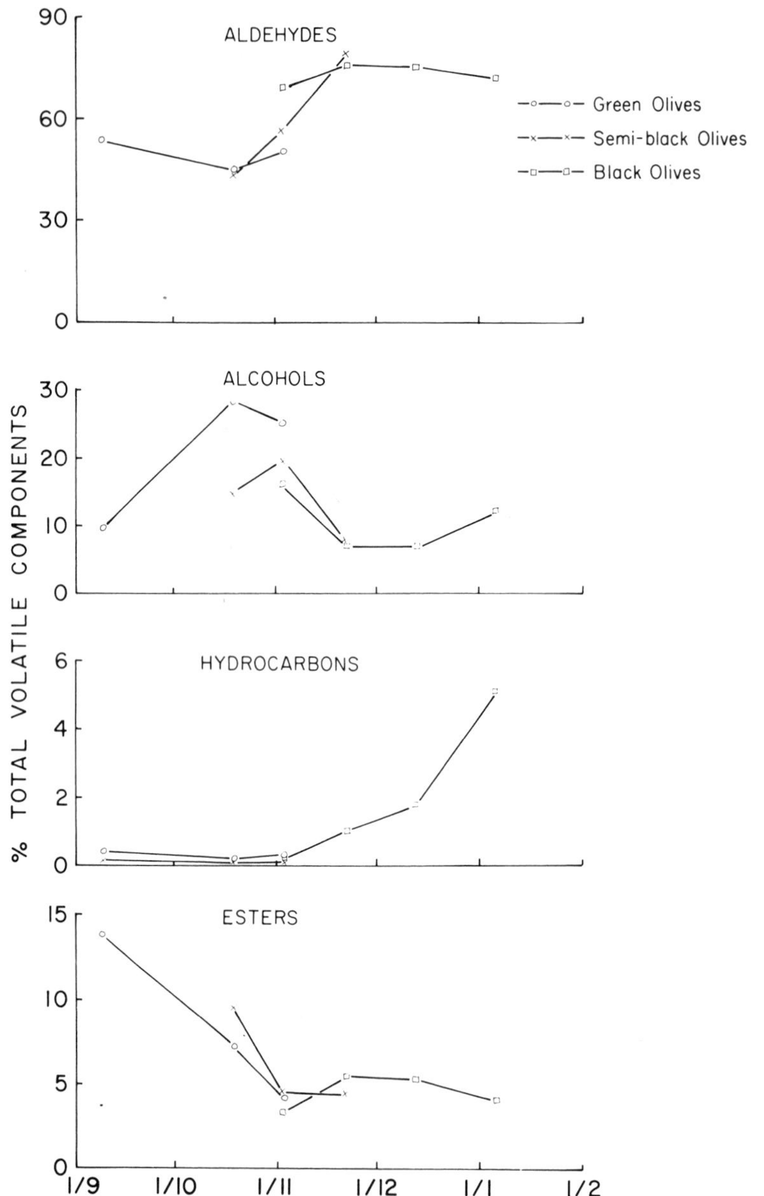

Fig. 3. Evolution of different volatile costituents of oil during the fruits ripening in relation to olive grade pigmentation.

olives) and 75% (black olives) of the total amount of the constituents (9). Among these costituents which are, as will be shown later (Fig. IV), the most significative ones as to their quantity and organoleptic role, are the 2-hexenal (85-90% of the aldehydes), the trans-2-hexen-1-ol (16% of the alcohols) and the cis-3-hexenyl acetate (20% of the total amount of esters).

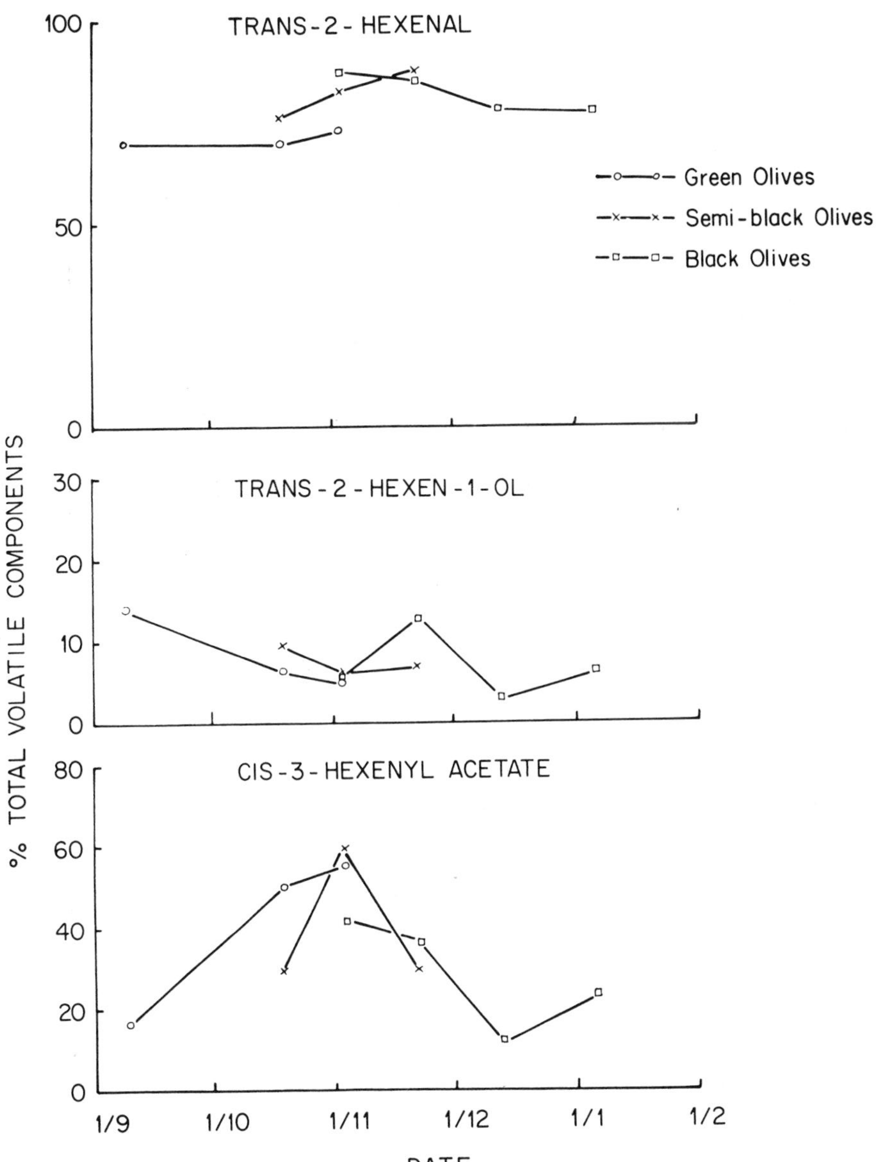

Fig. 4 . Evolution of single volatile costituent of oil during the fruits ripening in relation to olives grade pigmentation.

In the Figure V the results of thin layer chromatography of methanolic extracts of olive oils ,obtained from the fruit of different pigmentation and harvested in different date of the same cultivar are reported.These results shows that non significant qualitative differences have been observed for the phenolic costituents.

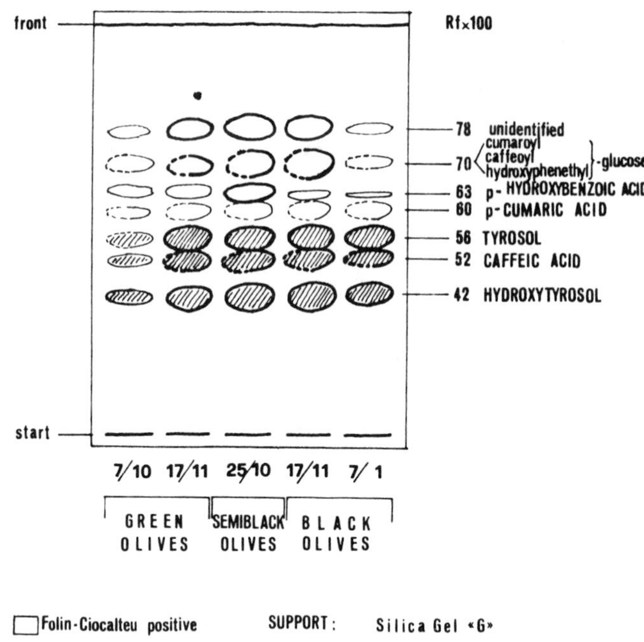

Fig. 5 . TLC of oil phenolic compounds,extracted from olives of different grade of pigmentation,during their maturation.

A_2 - Role of the harvesting period

This variable has been studied for the two different groups of constituents with respect to oil obtained from olives which have been harvested during a wide period (1/10to 1/3) without a previous division into classes of pigmentation.

The values concerning the different constituents "in toto" (Figure VI), support widely the observations on the role of the pigmentation, namely that the highest concentrations of both costituents are found during the phenological period between the semi-black olives and the complete black olives.

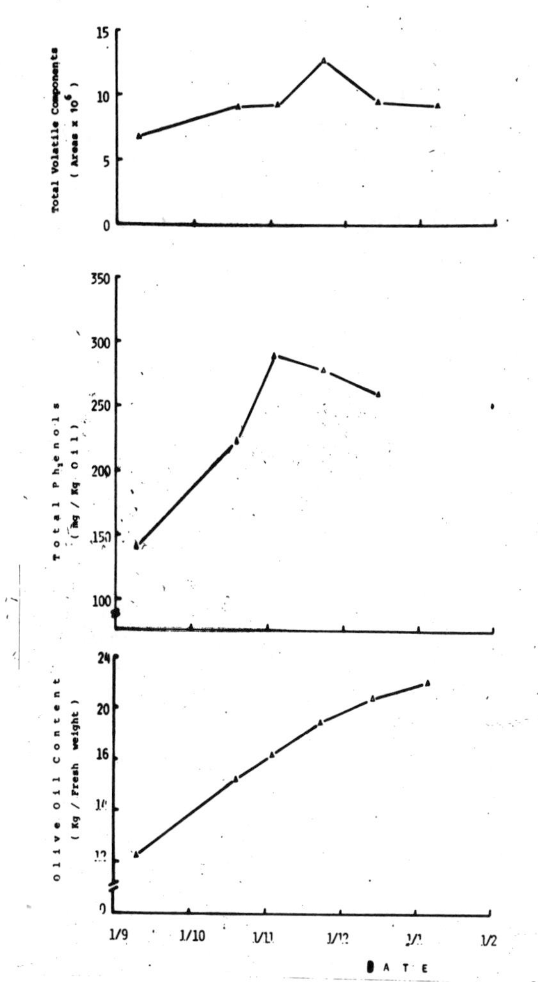

Fig. 6 . Evolution of oil content and oil volatile and phenolic costituents during the olives ripening.

On the other hand this period corresponds to the one during which the oil concentration reaches nearly its maximum and the aromatic characteristics of "fruity" is the highest (20).

A_3- Role of the cultivars

The evolution over time of the two constituents has been studied not only on the cultivar Moraiolo grown in Umbria, but also on Canino and Frantoio which are grown in Lazio. The data show that there are quantitative differencies only among the volatile constituents. The highest values of the latter ones have been found in the cultivar Frantoio, the lowest ones in Moraiolo. On the contrary the patterns of variation and the concentrations of the phenolic constituents are quite homogeneous. Only the cultivar Canino shows slightly lower values than the other two cultivars that are very similar. ȯt is readily seen in Figure VII.

The results related to single classes of volatile components (Figure VIII) show that the percentage distribution of the different components varies homogeneously over time, regardless of the absolute quantity.

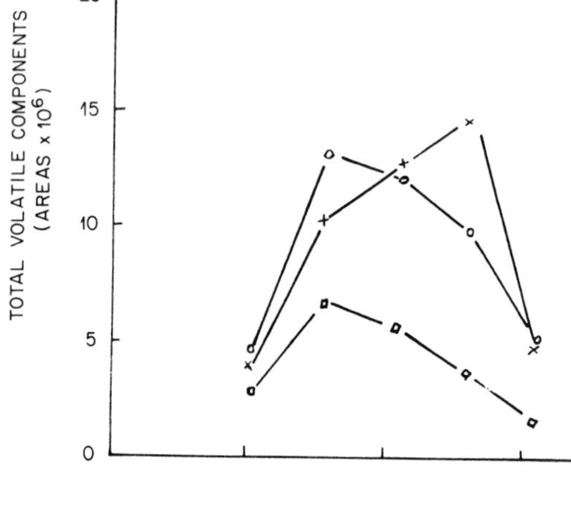

Fig. 7 . Evolution of oil volatile and phenolic costituents during the olives ripening in relation to the different cultivars.

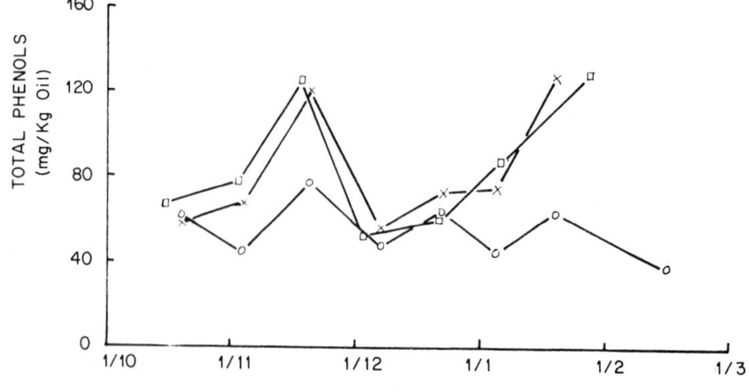

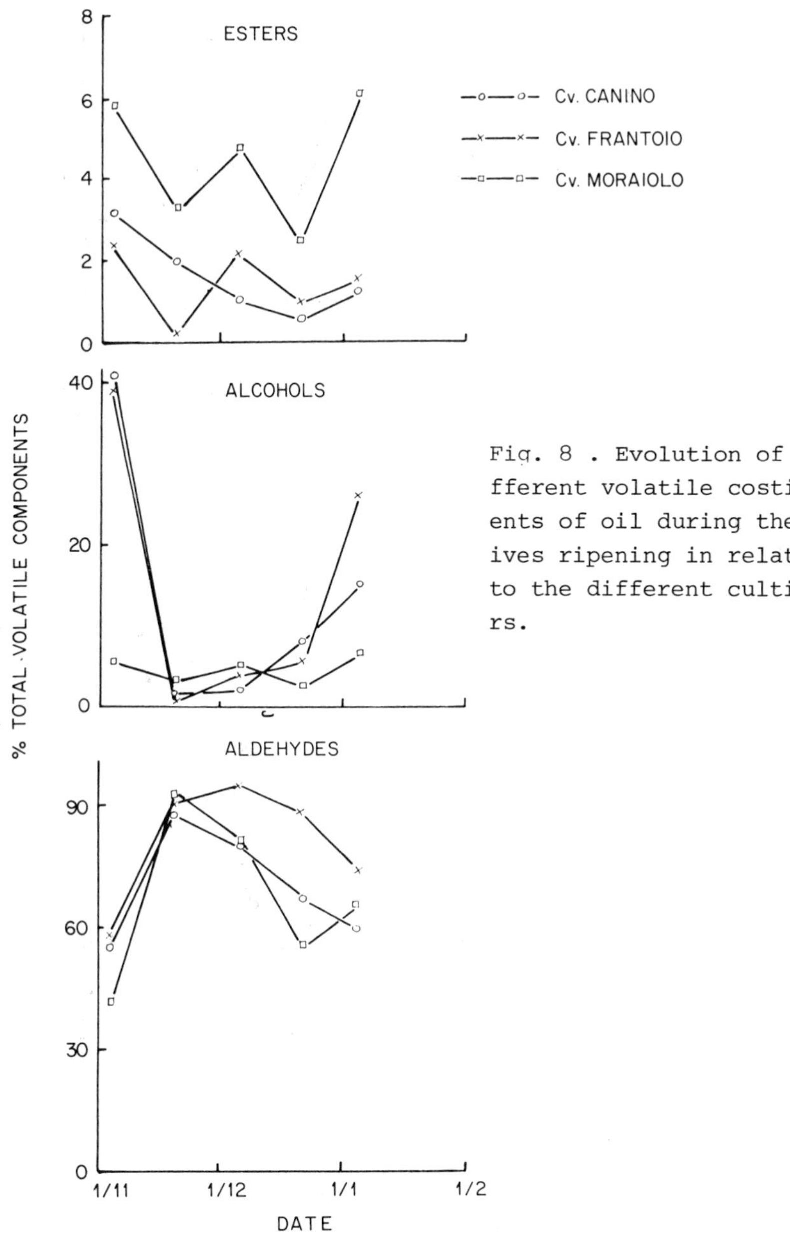

Fig. 8 . Evolution of different volatile costituents of oil during the olives ripening in relation to the different cultivars.

A_4- Role of the pedoclimatic conditions

The role of the climatic conditions has been studied on the cultivar Frantoio grown in two different regions (Umbria and Lazio).

The graphics in Figure IX show that for both the total phenolic constituents there are no significative differences of the pattern of variation over time between the two re -

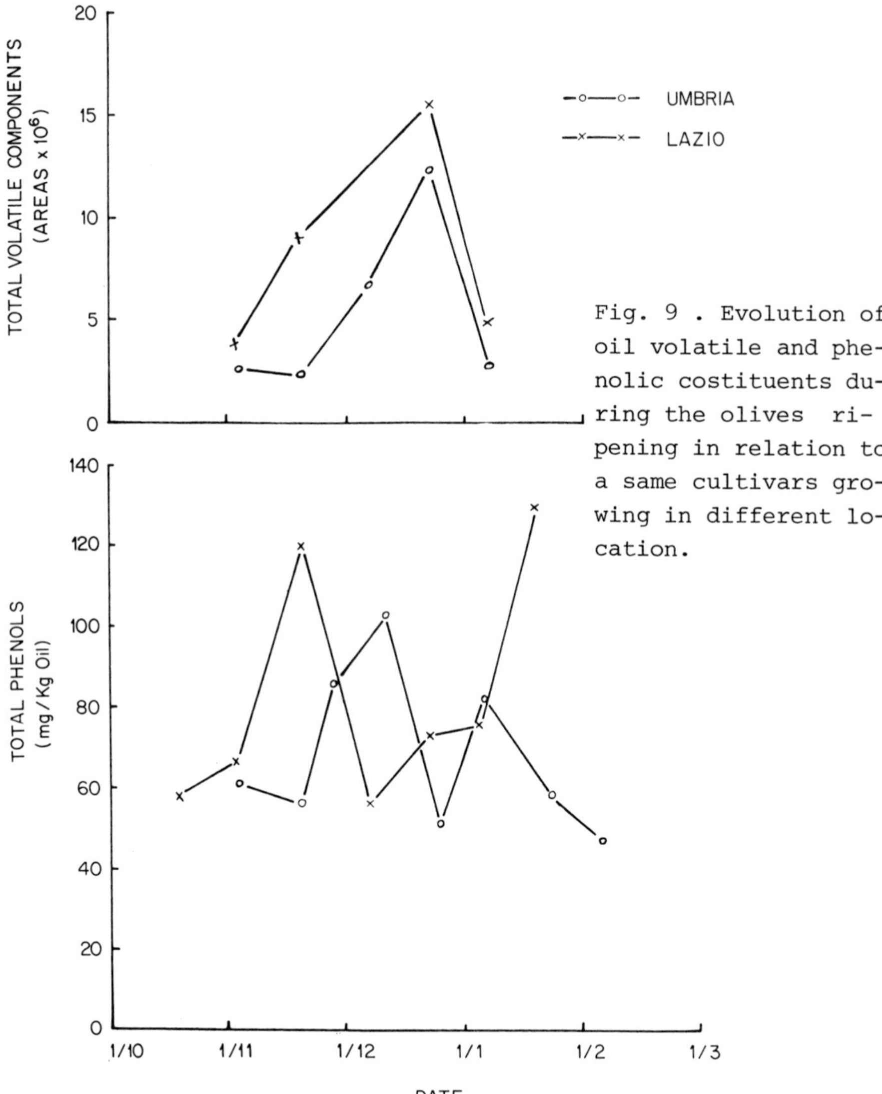

Fig. 9 . Evolution of oil volatile and phenolic costituents during the olives ripening in relation to a same cultivars growing in different location.

gions, while there are clear quantitative differences when the same harvesting period is considered.

The slightly higher concentrations found in the cultivar Frantoio grown in Lazio and the highest and lowest values of this one, which are anticipated compared with the Frantoio grown more northwards show again the importance of pigmentation and ripening, if we consider that more southwards the plants are grown first they ripen. In accordance to the observations also in this case the differences are only in the distribution of the different volatile components (Figure X).

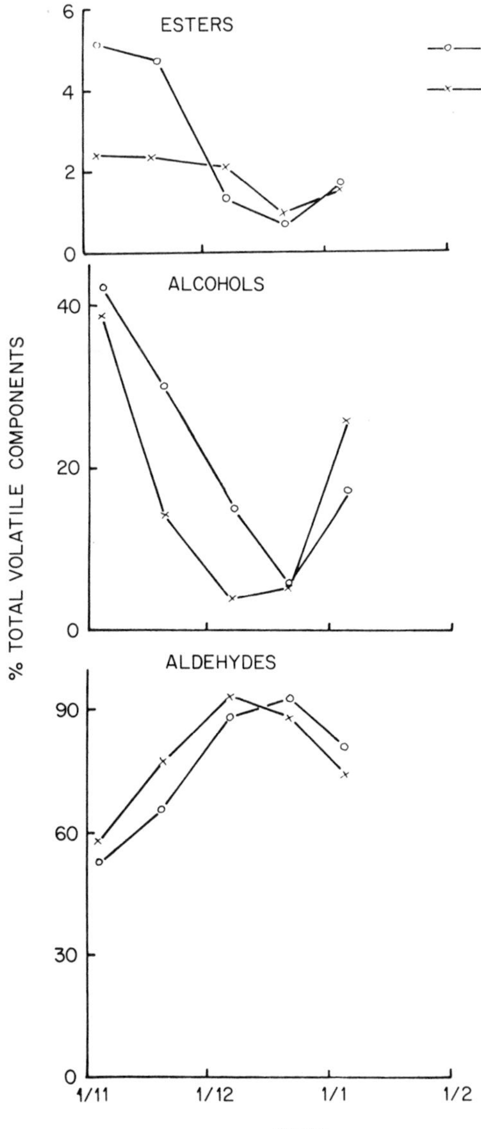

Fig. 10 . Evolution of different volatile costituents of oil during the olive's ripening in relation to a same cultivars growing in different location.

B - Genesis and Localization

B_1 - Volatile costituents

As already mentioned the most plentiful constituents are the trans -2-hexenal and hexanal, also the trans-2-hexen-1-ol is found in remarkable quantities. Their formation, are in connection with the distruction of the cell structure. The enzymatic processes involved in those mechanisms are hydrolyses and oxydations which proceed depending on the pH and on the temperature, with a amazingly high speed.

The pathway is that of linoleic acid that gives, by enzymatic oxidation (lipoxygenase), 13-hydroperoxy 9-11 octanic acid, which forms, with an aldehyd-lyase, hexenal .

This compound gives, by reduction (alcohol oxidoreductase) 1-hexanol. In a different pathway, the same hydroxyperoxyd gives , by its cleavage (aldehyd-lyase), cis-3-hexenal and trans 2-hexenal and by the reduction of the latter ones (alcohol oxidoreductase) respectively cis-3-hexen-1-ol and trans-3-hexen-1-ol and then the respective 1-hexenols (42,43,44,45,46).

Considering the above mentioned indications the presence of these constituents is widely explained also because the oil is extracted by milling of the fruits.

In Table VIII we have reported a number of results that show the relative concentrations of volatile constituents of oils extracted from fruits subjected to mechanical treatments of different intensities so that the degree of rupture of the cells was also different.

TABLE VIII - Effect of the Cellular Structures damage on the oil volatile components.

	Pressing			Milling Pressing		
	G	S	B	G	S	B
Volatiles Components:						
Total areas (x 10^6)	3.05	3.09	5.80	9.70	8.60	10.40
Aldehydes %	17.28	23.99	46.56	45.82	44.32	69.34
Esters %	23.56	10.45	6.38	7.20	9.49	3.50
Variation respect only pressing:						
Aldehydes %				+165	+ 84	+ 48
Esters %				- 69	- 9	- 45

Olive Fruits:
G = green; S = semi-black; B = black.

Those results (Table VIII) not only confirm the above mentioned mechanism, but show also that the degree of ripening, affecting the rheological state of the carpus, influences very much the intensity of the biochemical and chemical processes. Indeed the higher the degree of maturity (black fruits) is, the higher is the concentration of the aldehyds and the lower is the concentration of the esters (42,43,44,45,46) which are pre-existant in the fruit, when it is considered that the direct pressing is a mechanical procedure of a relative intensity. The process of milling by which the cellular structure is destroyed, causes a clear increase of both phenomenons, but in particular of the oxydative ones which are the intenser the greener is the fruit (42,43,44,45,46).

The volatile costituents are found in the oil because of the role played by the lipoproteic membrane of the oil drop. When the latter one is brought into contact with the chloroplasts on the which are located in the enzymes, the above mentioned components may be formed and kept enclosed in the lipid substrate (citoplasmatic) (19).

In addition to the above metioned mechanisms the lipophil and hydrophil characteristics of the different compounds may be of fondamental supplementary importance. Indeed the mashing, which is, in the normal diagrams of extraction, employed to favour the gathering of the oil drops by coalescence, causes, as shown in Table IX, an increase of the liposoluble components (aldehyds, esters and hydrocarbons) and a decrease of the hydrosoluble ones (alcohols etc.) because of solubilization mechanisms and partially of enzymatic oxydation (19).

TABLE IX - Effect of the paste's mashing on the volatile components of oil.

		Milling Pressing	Milling Mashing Pressing
Volatiles components:			
Total areas	10^6	6.70	9.06
Alcohols	%	8.15	7.32
Aldehydes	%	62.85	75.85
Esters	%	3.77	5.5
Hydrocarbons	%	7.30	1.07
Others	%	17.93	9.66

B_2 - Polyphenols

As we already pointed out, the components which are, according to some authors (7), organolectically the most interesting ones are the β-(3,4 dihydroxy-phenyl-)ethanol(hydroxytyrosol) that is found only in very good quality oils, the (3-hydroxy-phenyl) ethanol(tyrosol)and some phenolic acids (caffeic acid, cumaric acid, p-hydroxybenzoic acid), which are the only ones found in poor quality oils.

Some of these acids (hydroxytirosol and tyrosol) derive from the hydrolysis of the oleuropein (bitter glucosyde) of which the olives are very rich (47,48), while others (benzoic and cinnamic acids derivatives) (1,2,6) derive from the hydrolysis of flavonoids (anthocyans, flavones), which are found in considerable amounts especially at ripening (31,49,50).

The presence of these constituents in the oil is not due to the effect caused by extraction process (6,7) (milling and mashing) which, on the contrary, reduce their concentration as it can ben seen in Table X (51). It may be argued, on the contrary, that their transfer into the oil occurs already in the cell (25,52).

TABLE X - Effect of the paste's mashing on the phenolic compounds of oil.

Operations process	Total phenols mg/Kg oil
Pressing	616
Milling + Pressing	450
Milling + Mashing + Pressing	260

The above mentioned decrease is, as the results of Table XI show, not due oxydation processes (51) but it may be explained by the solubulizing effect of the vegetation water and particularily by their dissolved colloidal substances (proteins and polysaccharids), which bind the above mentioned components by adsorbation and/or by hydrogen and electrostatical bindings.

We have to keep in mind that all these components are hydrosoluble and in a limited fashion liposoluble (1,4) and moreover very reactive.

TABLE XI - Effect of the SO_2 addiction to the olive paste on phenolic compounds oil content.

Operations process	Total phenols mg/ Kg oil
Milling + Mashing + Pressing	260
+ $K_2S_2O_5$ 1g/Kg	228

C - Role of the technology

We have studied the effects of some operating phases of the process of mechanical extraction by a single pressure. We also reported the results regarding some biochemical treatments on the olive pastes during the milling phase (which increases the oil yield) (35,36), and of the olives storage.

C_1 - Extraction

The diagram of the process of mechanical extraction by single pressure (36) by been previonsly showed.

As far the role of the milling and the mashing is concerned we refer to the date shown in Table VIII,IX,X,XI.

Instead we report about the pattern of variation of the two different constituents during the pressing process under conventional conditions and after treatment of the pastes with tannic phenolic adsorbent agent and tannic acid (35,36). The analytical results reported in Table XII show that the concentrations of volatile and phenolic constituents of the different oil fractions increase progressively when the untreated pastes are considered (21,24).

The evolution of phenols (Table XIII), by effect of treatment with phenolic adsorbent agent and tannic acid present an inverse pattern of variation compared with that of untreated ones. That gives, as the data of Table XIV show, a same final concentration compared with the conventional extraction, contrary to what we observe for the volatile constituents (24).

This phenomenon may be explained considering the above mentioned studies on the genesis and the localization of these components (19,24).

The liposoluble volatile constituents emigrate into the oil just as the colloidal substances come into contact with the solid phase of the paste that contains the enzymes (Chloro phasts), while, on the contrary, the phenolic constituents are either adsorbed to the same colloids or emigrate into the vegetation water.

TABLE XII - Influence of the pressure intensity on the volatile components, total phenols, alcoholic insoluble substances of the virgin olive oil.

Time min.	Pressure atm.	Oil Kg/100	Volatile Costituents areas x10^6	Total Phenols mg/Kg oil	Alcoholic insoluble substances extracted g/100 Kg olives
10	0	9.70	6.82	146	738
25	16.6	14.60	6.92	146	966
45	66.6	17.78	-	160	1037
60	133.3	20.25	7.03	180	1050

TABLE XIII - Influence of the pressure intensity on the volatile components,total phenols, alcoholic insoluble substances of the virgin olive oil extracted from paste treated by tannin phenolic agent.

Time min.	Pressure atm.	Oil Kg/100 Kg olives	Total Phenols mg/Kg oil	Alcoholic insoluble substances extracted g/100Kg olives
10	0	15.70	172	873
25	16.6	18.96	114	1065
45	66.6	21.00	-	1088
60	133.3	22.05	104	1094

This is in agreement with one of our previous paper (36) in which the increase of the concentration of the colloidal substances in the vegetation water, is consequent to tannin phenolic adsorbent agent and tannic acid, treatment of pastes .

This increase has been found just in the first phase of extraction (25') (Table XIII) and is in agreement with the decrease of the phenolic constituents.

TABLE XIV - Effect of the treatments to the olive pastes by tannin phenolic adsorbent agent on the volatiles and phenolic costituents of the virgin olive oil.

Volatiles Costituents	Olive Pastes non treated	Olive Pastes treated
Areas x 10^6	5.11	6.55
Aldehydes %	38.22	38.71
Alcohols %	18.09	16.47
Esters %	4.53	4.68
Hydrocarbons %	9.30	6.13
Others %	29.86	34.01
Total Phenols (mg/Kg oil)	151.00	153.00

In this case the adsorbent agent and the tannic acid act also on the water determining ist desolvation (35,36), which favours the exchange micelles-oil. The results shown in Table XV concerning the effects caused by the homogeneization of the leafs in water and oil and in oil only are very indicative. These data shows indeed that the water doesn't enrich the oil with volatile constituents, formed by trituration of the leafs, in spite of their very good liposolubility (19,25).

TABLE XV - Effect of water on the enrichment of oil in volatile and phenolic costituents extracted from leaves by their homogenization.

	Oil	Oil H_2O Leaves	Oil Leaves
Volatile Costituents			
Areas 10^6	3.69	3.87	8.30
Aldehydes %	48.00	59.00	69.00
Alcohols %	15.00	18.00	20.00
Esters %	10.00	10.00	7.00
Others %	27.00	13.00	3.00
Total Phenols (mg/Kg oil)	109.60	84.00	144.00

As far as the phenolic constituents are concerned the mechanism is the same as above except for the additional role of the water which tends, in this case, to take them away from the oil because of their hydrosolubility (1,4).

C_2 - Storage

This practice, which is inevitable in the present day oil technology, causes over time, because of the hydrolytic enzymatic mechanism of the cell wall, by which the cell structure is degradet (this begins already during the ripening of the fruit Table VIII, IX), a loss of the different constituents of the oil; there are included also the liposoluble volatile constituents because of their low solubilization in the water and their successive enzymatic degadation (increase of the alcohols). Table XVI e XVII illustrates this mechanism. Table XVI refers to the composition of the oil obtained from olives that have been pressed immediatly after harversting and after 10 days of storage (2).

TABLE XVI - Effect of the olives storage on the volatile and phenolic costituents oil content.

	Oil Extraction	
	after wards harvesting	after 10 drays storage
Volatile Costituents		
Areas x 10^6	5.35	3.75
Aldehydes %	26.62	13.58
Alcohols %	17.14	25.77
Esters %	4.37	2.24
Hydrocarbons %	11.44	14.14
Others %	40.40	44.26
Total Phenols (mg/Kg oil)	104	89

Table XVII refers to test of solubility of some "standards" in the water, addition of colloids, precipitated from oily-must extracted by centrifugation of exactly alike pastes and treated with tannin phenolic agents and tannic acid (36). The concentrations of the tested volatile constituents in the "headspace", which have been examined respectively after one and three days after the addition, confirm the hypothesis of the solubilizing effect of the water on all constituents, included the liposoluble ones (19).

TABLE XVII - Effect of the alcohol insoluble substance of olive vegetation water on the some volatile compounds vapour pressure.

Volatiles Costituents	H_2O		H_2O+Alcohol insoluble substances			
			Olive pastes		Olive pastes + tannin phenolic adsorb. agent	
	Areas (mm^2)					
	a	b	a	b	a	b
Hexanal	350	50	328	52	380	48
trans-2-Hexenal	477	8	440	7	480	7
Hexanol	246	158	292	165	301	148

headspace analysis carried out: a = afterwards ; b = after three days.

Organolectic importance

As pointed out by other Authors (10,14,16,26,27) only a few of the volatile constituents present in the oils (also different botanical origin), are responsabile of their organolectic characteristics (9,26).

Also, for the phenolic constituents has been already reported about those of organolectic importance (7), even if we do not agree with authors (7). The results, reported in Table XVIII concerning an attempt to correlate the concentration of volatile and phenolic constituents with the quality index of an oil, shows that the phenolic constituents have, other than pointed out by the above mentioned authors (7) only a limited importance for the olfactory characteristics. Indeed these components have a sharp negative influence on the tasteful and on the organolectical characteristics when they are present above certain concentrations, where is present also the hydroxytyrosol. On the contrary the data of Table XIX show that there is no limited of the concentration of the "headspace" costituents. This is true in agreed of the very low values of organic aci-

TABLE XVIII - Volatile Costituents, progressive phenolic compounds contents and sensory evaluation of differents virgin olive oil.

		O			I			L			
		A	B	C	D	E	F	G	H	I	L
Volatile Costituents Areas x 10^6		6.92	9.72	15.24	9.05	10.40	6.60	6.20	7.70	9.25	5.81
Aldehydes	%	52.94	45.82	78.98	75.70	69.34	62.85	48.49	64.18	70.55	46.56
Alcohols	%	9.93	28.60	8.26	7.19	16.41	8.15	15.06	11.26	9.38	20.93
Esters	%	13.85	7.20	4.54	5.47	3.50	3.77	6.58	6.01	6.46	6.38
Hydrocarb.	%	0.39	0.15	0.08	1.72	0.16	7.30	2.81	0.89	0.70	0.65
Others	%	22.89	18.23	8.14	9.92	10.59	17.93	27.06	17.66	12.91	25.48
Total Phenols (mg/Kg oil)		137	219	240	260	336	450	616	706	826	992
Sensory Eval.		Sweet, Woody Fecal, Apple	Sharp,Lightly,Fruty,Varietal,Bitter	Very Sharp,Fruity, Varietal,Lightly bitter,Green leav.	Sharp,Lightly Fruity,Green Leaves, Ester Lightly bitt.	Bitter,Sharp	Green leaves	Higly Sharp, Bitter Anomalus flavour	Nasty taste Vegetative	Nasty taste Vegetative	Nasty taste Vegetative

TABLE XIX - Phenolic compounds, progressive volatile costituents contest and sensory evaluation of different virgin olive oils.

	O			I		L	
	M	N	D	O	E	P	C
Total Phenol (mg/Kg oil)	333	304	260	320	336	318	240
Volatile Costituents							
Areas x 10^6	8.13	8.60	9.05	9.30	10.40	11.20	15.24
Aldehydes %	50.59	44.32	75.70	72.55	69.34	75.85	78.98
Alcohols %	24.98	14.65	7.19	12.11	16.41	7.32	8.26
Esters %	4.19	9.49	5.47	3.97	3.50	5.50	4.54
Hydrocarbons %	0.29	0.13	1.72	5.13	0.16	1.07	0.08
Others %	19.95	31.41	9.92	6.24	10.59	10.26	8.14
Sensory Eval.	Lightles green leaves	Lightly green leaves Fruity , Ester	Sharp, Lightly frui-ty, Green leaves	as D Bitter	Bitter, Sharp Almoud Green Leaves, Fruity	Sharp, Bitter Green Leaves Fruity	Very sharp,Fruity Va-rietal Lightly, Bit-ter, Green Leaves

dity and the limited degree of oxydation of the tested oils,as is shown by the results reported in Table XX.

These last considerations lead to the conclusions that the harvesting period, which depends of the latitude, the condition of preservation of the fruit and the technology used for the extraction of the oil are very important factors.

TABLE XX - Analytical characteristics of virgin olive oils.

Oils samples	Free acids g/100 g oil	Perokide value O_2meq/100 g oil	U.V. $R = \frac{K\ 232}{K\ 270}$
A	0.56	10	10.00
B	0.25	18	12.44
C	0.38	19	11.57
D	0.38	25	13.84
E	0.28	13	10.52
F	0.24	20	11.86
G	0.49	27	12.11
H	0.48	21	12.38
I	0.24	14	10.81
L	0.28	22	11.43
M	0.54	17	11.11
N	0.49	17	10.36
O	0.48	29	12.35
P	0.29	16	11.14

Finally we think is our duty to emphazise that our studies are only a part of what has to be done later. In particular it is important to examine thoroughly the different phenolic co-stituents and the organolectic characteristics so that we may be able give both a regional or national characterization of the olive oils, which cannot be neither based upon the distribution of the fatty acids nor limited to some minor constituents, and an objiective index of the quality of these oils.

References

1. Montedoro,G.,Cantarelli,C., Riv. Ital. Sost. Gras. 46:115 (1969)
2. Montedoro,G., Sci. Tecnol. Alim. 2:177(1972)

3. Montedoro,G., Annali Facoltà Agraria,Perugia, 28:1(1973)
4. Ragazzi,E.,Veronese,G., Riv. Ital. Sost. Gras. 50:443(1973)
5. Vasquez Roncero,A.,Janer Del Valle,C., Janer Del Valle,L.A. Grasas y Aceites, 24:350(1973)
6. Vasquez Roncero,A., Janer Del Valle, L.,Grasas y Aceites 27:185(1976)
7. Vasquez Roncero,A., Revue Française des Corps Gras, 25:21 (1978)
8. Solinas,M., Di Giovacchino,L., Cucurachi,A.,Annali Ist. Elaiotec. 5(1975)
9. Flath,A.R., Forrey,R.R., Guadagni, D.G., J. Agric. Food Chem., 21:948(1973)
10. Olias, J.M., Del Barrio,A., Gutierrez, R., Grasas y Aceites, 28:107 (1977)
11. Olias Jimenez, J.M., Cabrera Martin,J., Gutierrez Gonzales Quijano,R., Grasas y Aceites, 25:34 (1974)
12. Gutierrez Gonzales Quijano,R., M-sti Vega,M., Colakoglu,M., Cabrera Martin,J., Grasas y Aceites, 23:351 (1972)
13. Nawar,W.W., 44th Nat. Meeting Am. Oil Chem. Soc., Chicago III (1970)
14. Fedeli,E., Camurati,F., Cortesi,N., Favini,G., Cirio,U., Vita,G., Lipids,vol. 2,p. 385,Raven Press,New York, 1976
15. Fedeli,E., Riv. Ital. Sost. Gras.,54:202(1977)
16. Gutierrez GonzalesQuijano,R., Olias Jimenez,J.M., Gutierrez Rosalez,F., Cabrera Martin,J., Barris Perez, Cerezol, A., Grasas y Aceites 26:21 (1975)
17. Lerker,G., Capella,P., Deserti,P.L., Sci. Tecnol. Alim. (S.T.A.), 299:(1973)
18. Bertuccioli,M., in press in Food Chemistry'
19. Bertuccioli,M., Montedoro,G., manuscript in preparation
20. Bertuccioli,M., Anichini,F., Montedoro,G., Tecnol. Alim. 2 (1978)
21. Bertuccioli,M.,Anichini,F., Montedoro,G., in press Tecnol. Alim., 3 (1978)
22. Anichini,F., Bertuccioli,M., Montedoro,G., in press Riv. Ital. Sost. Grasse 5 (1978)
23. Anichini,F., Montedoro,G., in press in Tecnol. Alim. 3(1978)
24. Anichini,F., Montedoro,G., in press in Techol. Alim. 4(1978)
25. Anichini,F., Montedoro,G., in press in Tecnol. Alim. 5(1978)
26. Dupuy,H.P., Rayner,E.I.,Wadswort,J.I., J.A.O.C.S. 53:628 (1968)

27. Blumenthal,M.M.,Trovit,J.R., Chang,S.S., J.Am. Oil Chem. Soc. 53:496 (1976)
28. Williams,J.L.,Applewhite,T.H., J.Am. Oil Chem. Soc. 54:461 (1977)
29. Jakson,H.W., Gicherio,D.J., J.Am. Oil Chem. Soc. 54:458 (1977)
30. Waltking,A.E.,Zmachinski,EL, J.Am. Oil Chem. Soc. 54:454 (1977)
31. Vasquez Roncero,A., Graciani Constante,E., Maestro Duran, R., Grasas y Aceites, 5:269 (1974)
32. Bertuccioli,M., Montedoro,G., J. Sci. Food Agric. 25:675 (1974)
33. Bertuccioli,M., Montedoro,G., Riv. Sci. Tecn. Alim. Nutr. Um., 1:39 (1975)
34. Jennings,W.G., Symposium On. Analysis of Food and Beverages Chicago, 8/29-9/2 (1977)
35. Montedoro,G., Riv. Ital. Sost. Gras., 54:441 (1977)
36. Montedoro,G., Petruccioli,G., Riv. Ital. Sost. Gras.,55:305 (1978)
37. Martinez Moreno,J.M.,Gomez Herrera,C., Janer Del Valle, C., Grasas y Aceites, 8:112 (1975)
38. Marinez Moreno,J.M., Gomez Herrera,C., Janer Del Valle, C., Grasas y Aceites, 12:166 (1961)
39. Martinez Moreno,J.M., Gomez Herrera,C., Janer Del Valle,C., Grasas y Aceites, 12:118 (1961)
40. Singleton, V.L., Sullivan,A.R., Kramer,C., An. J. Enol. Vit. 22:161 (1971)
41. Aninchini,F., Fantozzi,P., Petruccioli,G., Montedoro,G., manuscript in preparation
42. Drawert,F., Tressl,R., Heimann,W., Emberger,R., Speck,M., K. Mikrobiol. Tecnol. Lebens. 2:10 (1973)
43. Drawert,F., Heimann,W., Emberger,R., Tressl,R., Naturwissen schaften, 52:304 (1975)
44. Drawert,F., Heimann, W., Emberger,R., Tressl,R., Ann. 694:200 (1966)
45. Sieso,V., Crouzet,J., Food Chem., 2:241 (1977)
46. Stone,E.J., Hall,R.M., Kazeniac, S.J., J. Food Sci., 40:1138
47. Panizzi,L., Scarpati,M.L., Oriente,G., Gazzetta Chim. Itals 90:1449 (1960)
48. Ragazzi,E., Veronese,G., Guiotto,A., Ann. Ch. 63:13 (1973)
49. De Angelis,N., Doctoral thesis, Fac. Agr. Univ. Perugia, A/A 1976/77

50. Cantarelli,C., Atti I Conv. Naz. Oliv. Spoleto, 1/3 giugno (1962)
51. Solinas,M., Di Giovacchino,L., Mascolo,A.,Riv. Ital. Sost. Grasse 1:19 (1978)
52. Mazliak,P., in "The Biochem. of fruits and their products" (A.C. Hulme ed.) p. 229. Academic Press, New York, 1970.

THE VOLATILE FRACTION OF ORANGE JUICE. METHODS FOR EXTRACTION AND STUDY OF COMPOSITION.

José Alberola
Luis J. Izquierdo

Instituto de Agroquímica y Tecnología de Alimentos
C.S.I.C.
Valencia, España

I. INTRODUCTION

Flavour of a given food is considered as a combination of stimuli of our taste and odour receptor. Although taste is also important it is generally recognized that odour or aroma is mainly responsible for the characteristic flavour of foods.

The name of aroma is given either to the olfactory sensation produced by a food or to the volatile fraction of the food which causes this sensation (1) (2). Only aroma as volatile fraction of food is considered in this paper.

Many authors have paid attention to the volatile fraction or food aroma since several years ago. These studies have allowed the identification of aroma compounds and their proportions what, in addition to a better knowledge of food composition, leads to the preparation of artificial mixtures improving food quality (3).

Concerning to citrus fruits it has been proved that all varieties contain approximately the same aroma components but odour sensations are different, what means that not only the chemical nature of the volatile components but also their proportions in the mixture are responsible for the specific aroma of a variety (4).

ISBN 0-12-169060-1

In industrial processing of fruit juices the volatile components of peel oil are incorporated to the juice contributing to juice flavour (5). Citrus juices have also specific components not found in peel oil and all together form the characteristic juice aroma (6).

The complexity of the volatile fraction of orange juice made its study difficult until new modern techniques for separation and identification of components were developed. The study of orange juice volatile fraction usually requires four definite steps (7):

1. Separation of volatile fraction from non volatile matter.
2. Concentration of volatile fraction.
3. Breaking down of the aroma into their constituents.
4. Identification of constituents.

Many methods for separation and concentration of volatile fraction (steps 1 and 2) have been tried with more or less success but nearly all of them fall into two categories

a) Headspace analysis. Or the analysis of the vapors over the juice.
b) Total volatile analysis. Either by distillation of juice and extraction of the distillate with organic solvents or by direct extraction of juice with organic solvents.

Breaking down of the aroma into their constituents (step 3) is accomplished by gas-liquid chromatography. Identification of components (step 4) is carried out either by relative retention times or by using instrumental techniques.

This review is devoted to the discussion of the more important separation and concentration procedures. In the last part of this work volatile components which are identified in orange juice are listed and a brief commentary on off-flavour of orange juice is given.

II. HEASPACE ANALYSIS

The simplest, fastest and most precise technique of sampling the orange juice volatiles for analysis is to take a measured volume of air from the free space in a vessel above the juice and inject it directly into a gas chromatograph (8) (9).

This technique has been used by several authors in order to obtain chromatograms of the volatiles emanating from the products (10-17), and was recommended by the Working Group on Methods of Analysis of IOFI (International Organization of the Flavor Industry) to their members, since direct injection method is very suitable for routine control of the quality of orange juice volatile fraction (18) (19).

The chromatogram thus obtained has been named "aromagram" (20). The advantages of the aromagram technique are numerous: sample handling is minimized and no artifacts are introduced due to solvent use; sample preparation requiring time reduced, etc.

Moreover, the vapor over the food contains the olfactory stimuli responsible for the subjectively perceived odour. Composition of this mixture can be related to sensory response if those components which influence odour are detectable and measurable. Available instruments limite the minimum quantity which can be detected by direct vapor analysis (16).

This limitation has been the main disadvantage of headspace analysis. Volatile fraction components are present in very variable concentrations; some of them are in lower concentration than that necessary to be detected by gas chromatographic or mass spectrometric detector but not by the nose. This fact is not important when the method is used as a routine control in which only major components are required.

Nevertheless, in order to detect those minor components, often critically important to aroma, large volumes of headspace vapour would be required. Direct injection of a large volume gas sample, however, is not consistent with sharp peaks and high resolutions. Efforts have been made to concentrating the volatiles from moderately large volumes of headspace vapour.

The most commonly used approach has involved the use of cryogenic trapping in which not more than a few hundred milliliters of headspace vapour is swept with an inert gas into a cooled trap which may be part of the chromatographic system (21-23). This technique, however, presents the disadvantage that water is the main component recovered (24) (25).

For this reason, adsorbent materials have been utilized to minimize water. Activated charcoal has been the first adsorbent used (26-28). Other commercial products such as Porapak Q (ethylvinyl--benzene-divinylbenzene copolymer), Chromosorb 101 and Chromosorb 102 (styrene-divinylbenzene copolymer) have been succesfully used in a number of studies in recent years (17) (23) (29) (30). More recently a newer adsorbent material, Tenax GC (a porous polymer of 2,6 diphenyl-p-phenylene oxide) with greater temperature stability than Porapak Q has been used for analysis of trace volatiles and is likely to be useful in aroma studies (28). Some authors evaluated the efficiency of a number of porous polymers and concluded that not a single one was universally suitable (17) (31).

Recently an ample monography on headspace analysis of food volatiles has been published (32).

III. TOTAL VOLATILE ANALYSIS

Distillation is the most widely used technique to isolate volatiles from orange juice. This is obvious since volatility is a common property to all interesting compounds, so distillation sets them apart from the rest of the product (8). It is also the technique which better allows a later concentration of the distilled volatiles, usually by solvent extraction and further evaporation of the solvent leading to a better response of detectors to minor components.

However, this method has some disadvantages. The composition of volatile fraction in the distillate is less closely related to aroma than the composition of volatile fraction in the vapor over the food (headspace analysis) (33), and there is a higher probability of artifacts formation, mainly due to sample handling and extracting solvent used after distillation (16) (34), as it was mentioned above.

Both single plate and multiplate distillation have been used with orange juice. Single plate distillation has been frequently used for isolation of orange juice volatiles in the laboratory, usually at low pressure, to minimize heat damage of sample. For this purpose rotary vacuum evaporator may be used with good results (4). In our laboratory it has been developed a method for single plate distillation based on flash juice evaporation. The distillate is condensed with chilled water and liquid nitrogen traps and extracted with methylene chloride. The method shows a very good reproducibility (35).

However, single plate distillation produces very dilute aqueous solutions of volatiles which must be extracted several times with high amounts of solvent To avoid this problem the Likens and Nikerson apparatus (36) or anyone of its modifications described by different authors may be used (37)(38). This apparatus makes distillation and extraction simultaneously. Sample and solvent (less dense than water) boil in different flasks but vapors condense in the same condenser. The two liquid phases of the condensate continually return to their respective flask via two tubes at different level being the solvent constantly enriched in volatiles.

One of the modified apparatus was tested with model solutions with the most usual volatile components of fruit juices (38). Good extraction and reproducibility were obtained but differences were found among optimal conditions for extraction of components. At atmospheric pressure both citronellal and linalool showed instability and at reduced pressure (vapor temperature 52°C) the recovery of ethyl-3-hydroxy-hexanoate was only 6% after an hour of extraction. It can be concluded that no single set of operating conditions is suitable for all applications. When ethanol or other very soluble, low molecular weight compounds are of no interest, hexane may be the most convenient solvent.

A comparison of several techniques (distillation -extraction, direct headspace analysis, and porous polymer adsorption) carried out with model solutions showed that distillation gave results which more nearly agreed with those for direct injection of the neat model solution; headspace analysis presented higher peaks for more volatile, higher partial pressure compounds (heptane and ethanol) and porous polymer adsorption chromatograms were dominated by limonene, esters, and n-octanone (17).

Multiplate distillation technique, producing a concentration of the volatiles from the water by reflux stripping, has been widely used at laboratory or pilot plant scale (35)(39)(40) and, over all, it is a common practice in industrial concentrate orange juice manufacture (WURVAC process, (Western Utilization Research Vacuum Aroma Column)) (41). Aroma recovery process is based on vaporization of a part of the water present in the juice and on the tendency of this vapor to entrain the aroma components due to their high volatility in dilute aqueous solutions (42). This volatility increases with molecular weight as it has been proved for alcohols from methanol to octanol and for aldehydes, ketones, and esters up to C-9 (43) (44). The vapors obtained from freshly extracted juice, evaporator condensate or from some stages of the TASTE (Temperature Accelerated Short Time Evaporator) are concentrated by reflux (42). A concentrated aqueous solution (aqueous essence) about 100-150 fold is obtained and used to improve the flavour of concentrated orange juice.

A small oil fraction (essence oil) mainly constituted by limonene is also obtained in the process. The essence oil, normally considered as a by-product of aqueous essence production is also a useful flavoring agent due to its essence-like odour quality, its colorless appearance and its increasing availability (45).

In preparation of aqueous essence for gas chromatography many techniques have been tested depending on particular purposes. Solvent extraction porous polymer adsorption and direct injection are discussed here.

Solvent extraction is the most common method for studying the aqueous essences and a technique using methylene chloride has been extensively used (5)(14) (19)(35)(39)(46-49). The methylene chloride technique consists basically in several manual extractions of the essence saturated with sodium sulphate or sodium chloride in separatory funnels. The extracts are combined and concentrated using a rotary vacuum evaporator at 35-40°C up to almost complete removal of solvent. The concentrate may be directly analyzed by gas chromatography. Attempts made to substitute methylene chloride by methyl chloride, a lower boiling solvent, were not successful in retaining a greater percentage of the more volatile components (49). However, a good practice may be the use of diethyl ether to extract methanol and a few minor organic components from the residual aqueous solution after methylene chloride extraction (47-49). The use of diethyl ether as a unique solvent gave also good results (40)(51-53).

More sophisticated techniques to carry out the solvent extraction as continuous liquid-liquid extraction are not usual with aqueous essence because of manual extraction simplicity. In addition equilibria are seldom reached with other techniques as fully as they are possible with vigorous shaking in separatory funnel (8).

Attemps have been made in order to avoid the difficulties of solvent extraction methods as incomplete extraction, loss of volatiles during concentration, and transferring small quantities of sample. Adsorption on porous polymer has been one of these attempts (29). Aqueous essence was injected into a little column packed with Porapak Q, which practically does not retain water, methanol, acetal--dehyde and ethanol, the major components of the essence. The column was rapidly heated after those compounds have passed through, and the rest of the essence components were eluted being retained in a liquid nitrogen trap. The concentrated sample was ready to injection after heating the trap. The sharpness of the chromatographic peaks and the resolution of the chromatogram obtained with this method are similar to those obtained with conventional solvent extraction, but there are some qualitative differences which can be explained either by a loss of material during the solvent evaporation or by incomplete extraction of certain components.

All techniques involving concentration of samples are adequate for qualitative analysis because of the enrichment in compound concentration which makes the identification easier after gas chromatographic separation. However, these techniques are not suitable for quantitative analysis due to extraction and concentration problems mentioned above.

The interest in the quantitative evaluation of essence strength and quality is obvious from a commercial point of view. For this reason several analytical methods have been developed to determine concentrations in total aldehyde, oxygenated terpenes saturated aliphatic aldehydes and esters, and chemical oxigen demand (54-57). However, all these methods give global evaluations without any information about particular compounds. This information can only be obtained by separating procedures as gas chromatography. In this sense only direct injection of samples, which avoids the possibility of losses is completely reliable.

Direct injection of orange aqueous essences gives chromatograms with smaller peaks than those obtained by concentration procedures as methylene chloride extraction (19), but the former method is easier, faster and more suitable for identification and quantification of the most abundant aroma components of the essence. In addition, major compounds of very low boiling point as methanol and ethanol which are hardly extracted with methylene chloride can be easily quantified allowing an objective evaluation of essence composition. A problem may be the injection of large amounts of water which can reduce column life and produce distorsions on detector response.

As it has already been said, a small oil fraction (essence oil) mainly constituted by limonene, is also obtained in aroma recovery process. This fraction has been extensively analized because its potential as flavoring agent (45)(58-60).

Methods used by different authors for qualitative analysis of essence oils are very similar. They basically consist in a distillation of the sample at reduced pressure in a rotary evaporator with liquid nitrogen traps between the condenser (refrigerated with chilled water) and the vacuum pump. Thus three fractions are obtained: trap condensate, chilled water condensate and flask residue.

Occasionally, flask residue and chilled water condensate are fractioned on an adsoption chromatographic column using Florisil or neutral alumina and eluting with hexane, ethyl ether and ethanol or mixtures of them. Eluates are concentrated at reduced pressure and the residue is injected in the gas chromatograph.

Results obtained with this procedure showed that liquid nitrogen trap condensate had a full essence-like odour, while flask residue and chilled water condensate had less essence-like odour than the original oil (59). So, the most volatile components in essence oil are the most important contributors to essence-like odour of the oil.

Accurate quantitative measures of essence oil components are feasible by direct injection of the oil because its high concentration. Some papers about quantitative evaluations of the main component concentrations of orange oil have been published (45)(59)(60).

Although less used than distillation procedures, direct solvent extraction of juice may be a suitable technique for volatile studies of fruits because these materials contain only small amounts of non volatile material soluble in organic components (8). The operation may be discontinuous when using separatory funnels for shaking juice and sòlvent (usually methylene chloride)(61).or continuous when using liquid-liquid extractors which avoid emulsion problems of shaking step. In our laboratory it has been used a liquid-liquid extractor with methylene chloride as solvent to extract volatiles from fresh orange juice, pasterized juice and reconstituted concentrated juice as well as for orange comminuted and orange beverages. After solvent evaporation, the extracts are directly injected in the gas chromatograph.

With carefully controlled conditions reproducibility is very good and extraction of volatiles tested with model solutions, is between 80 and 90%.

IV. VOLATILE COMPOUNDS IDENTIFIED IN ORANGE JUICE

Table I listes 217 components of the volatile fraction of orange juice identified by different authors. Numbers after the name of every compound indicate the authors who reported it. These numbers are the same as those indicate in the references.

With the exception of those components reported in two papers (6)(62), the components present in the volatile fraction of orange juice have been broken down by using gas chromatography, with packed or capillary column, and identified either by its retention time or by instrumental techniques as infrared spectrophotometry nuclear magnetic resonance or mass spectrometry.

Most of the compounds listed have been identified more than once as orange juice volatile constituent.

Alcohols appear as the most numerous compounds identified in orange juice volatiles (forty nine compounds). All straight-chain saturated primary alcohols from methanol to dodecanol are present. Linalol is the most frequently cited alcohol.

TABLE I. Volatile compounds identified in orange juice

ACIDS

Acetic acid (4)(5)(63-65)
Butyric acid (4)(5)(66)
Decanoic acid (4)(5)
Formic acid (4-6)(66)
Hexanoic acid (4)(5)(66)
Isohexanoic acid (4)(5)(66)
Isopentanoic acid (4)
3-Methylbutanoic acid (5)
4-Methylpentanoic acid (5)
Octanoic acid (4)(5)(66)
Pentanoic acid (4)(5)
Propionic acid (4)(5)(66)

ALCOHOLS

Borneol (4)
1-Butanol (4)(5)(23)
cis-Carveol (4)(50)(59)(60)(72)(73)
trans-Carveol (4)(5)(14)(19)(40)(46)(50)(58)(59)(65)(69)(72)(73)
Citronellol (4-6)(14)(39)(46)(58)(60)(65)(68)(69)(71)(73)
1-Decanol (4)(5)(46)(58)(65)(71)
n-Dodecanol (58)
Ethanol (4-6)(14)(19)(23)(40)(50)(51)(59)(60)(62)(65)(68-73)
Farnesol (4)
Geraniol (4-6)(14)(39)(46)(65)(68)(69)(71)
Heptanol (50)(70)(72)
3-Hepten-1-ol (4)(5)(68)(71)
1-Hexanol (4)(5)(14)(19)(23)(39)(40)(46)(51)(65)(69-72)(74)
2-Hexanol (68)
2-Hexen-1-ol (4)
cis-2-Hexen-1-ol (4)(5)(14)(39)(40)(46)50)(51)(65)(70-72)
trans-2-Hexen-1-ol (40)(51)
3-Hexen-1-ol (4)
cis-3-Hexen-1-ol (4)(19)(74)
Isobutanol (4)(23)(40)(50)(65)(69-72)
Isohexanol (4)(65)
Isopentanol (4)(6)(23)(39)(40)(50)(51)(65)(69-72)(74)
Linalool (4)(5)(14)(19)(39)(40)(46)(50)(51)(58-60)(63)(64)(68-74)

cis-2,8-p-Menthadien-1-ol (50)(59)(72)(73)
trans-2,8-p-Menthadien-1-ol (19)(39)(50)(59)(69)(72)(73)
1,8-p-Menthadien-9-ol (39)(50)(58)(72)(73)
8-p-Menthen-1,2-diol (4)(39)(58)(73)
trans-p-Menthen-9-ol (69)
Methanol (4)(5)(14)(19)(40)(46)(50)(62)(65)(68)(71)
2-Methyl-1-butanol (23)
2-Methyl-3-buten-2-ol (4)(23)(40)(50)(51)(72)
3-Methyl-2-buten-1-ol (69)
Methyl-Heptenol (4)(5)(14)(65)
Neo-isopulegol (4)
Nerol (4)(5)(14)(46)(50)(65)(68)(69)(71)(72)
Nerolidol (69)
1-Nonanol (4)(5)(14)(46)(58)(65)(71)
2-Nonanol (4)(14)(46)(65)(73)
1-Octanol (4)(5)(14)(19)(39)(40)(46)(50)(51)(58)(64)(69) (71-74)
1-Pentanol (4)(5)(14)(19)(46)(65)(71)
2-Pentanol (4)(5)(69)
1-Penten-3-ol (19)(40)(50)(52)(70)
2-Phenylethanol (4)(6)(69)
1-Propanol (4)(5)(14)(23)(40)(51)(65)(71)
Isopulegol (4)
Terpinen-4-ol (4)(5)(14)(19)(39)(40)(50)(51)(60)(65)(68)(69)(71)(72)(74)
α-Terpineol (4)(5)(6)(14)(19)(39)(40)(46)(50) (51)(58)(60)(65)(68)(69)(71-74)
Thymol (60)
n-Undecanol (58)

ALDEHYDES

Acetaldehyde (4)(5)(6)(14)(19)(23)(40)(46)(50)(51)(60)(62)(64)(68)(70)(72)
Benzaldehyde (23)(40)
Butyraldehyde (51)(67)
Citral (4)(6)
Citronellal (4)(5)(40)(46)(60)(63)(68)(71)
Decanal (4)(5)(14)(40)(46)(51)(63)(68-71)(73)(74)
2-Decenal (4)
α- *β*-Diehptylacrolein (4)
α- *β*-Dioctylacrolein (4)
Dodecanal (73)
2-Dodecenal (4)
3-Ethoxyhexanal (70)
α-Ethylbutyraldehyde (4)(5)(14)(46)(68)
3-Ethylhexanal (68)
Furfural (4)(14)(62)(64)(68)

Geranial (4)(5)(14)(39)(40)(46)(51)(64)(68)(69)(72)(73)
Heptanal (4)(73)
α-Heptyl-β-nonylacrolein (4)
Hexanal (4)(5)(14)(19)(23)(39)(46)(50)(51)(59)(63)(64)(68-74)
2-Hexenal (4)(5)(14)(19)(23)(39)(40)(50)(52)(63)(64)(68)(71)
trans-2-Hexenal (23)
α-Hexyl-β-heptylacrolein (4)
α-Hexyl-β-nonylacrolein (4)
α-Hexyl-β-octylacrolein (4)
Malonaldehyde (66)
2-Methyl-1-butanal (23)
Neral (4)(5)(14)(39)(40)(46)(50)(51)(64)(68)(69)(71-73)
Nonanal (4)(5)(14)(46)(60)(63)(68)(70-72)
Octanal (4)(5)(14)(19)(23)(39)(40)(46)(50)(59)(60)(63)(64)(68)(70-74)
2-Octenal (4)(5)(14)(46)(64)
α-Octyl-β-heptylacrolein (4)
Pentanal (4)
trans-2-pentenal (50)(72)
Perillaldehyde (4)(40)(50)(51)(58)(60)(72)(73)
Propionaldehyde (70)
α-Sinensal (4)(60)
β-Sinensal (4)
Undecanal (4)(5)(14)(46)(68)(71)

ESTERS

Bornyl acetate (4)
Butyl butyrate (60)
Butyl octanoate (69)
Citronellyl acetate (4)(68)(69)
Citronellyl butyrate (4)(5)
Decyl acetate (4)
Ethyl acetate (4)(14)(23)(40)(46)(50)(51)(59)(60)(69)(70)(72)(73)
Ethyl benzoate (4)
Ethyl butyrate (4)(5)(14)(19)(23)(39)(40)(50)(51)(58)(59)(68-74)
Ethyl decanoate (5)(69)
Ethyl formate (4)(6)(14)(51)(68)(70)
Ethyl heptanoate (4)
Ethyl hexanoate (4)(40)(51)(69)
Ethyl 3-hydroxybutyrate (69)
Ethyl 3-hydroxyhexanoate (4)(19)(39)(40)(51)(69)(74)
Ethyl 3-hydroxyoctanoate (69)

Ethyl isobutyrate (23)
Ethyl isopentanoate (4)
Ethyl 2-methylbutyrate (4)(23)(69)
Ethyl octanoate (4)(5)(6)(14)(40)(46)(51)(68)(71)
Ethyl pentanoate (4)
Ethyl propionate (4)(5)(23)(50)(59)(72)(73)
Geranyl acetate (4)(60)
Geranyl butyrate (4)
Geranyl formate (4)
Linalyl acetate (4)(5)
1,8-p-Menthadien-9-yl-acetate (73)
Methyl acetate (60)
methyl anthranilate (4)
Methyl benzoate (60)
Methyl butyrate (23)(39)(40)(50)(59)(70)(73)(74)
Methyl 2-ethylhexanoate (4)
Methyl hexanoate (23)(40)(69)
Methyl 3-hydroxyhexanoate (4)(51)(69)(74)
Methyl isopentanoate (4)
Methyl N-methylanthranilate (4)(14)(46)(68)
Methyl 3-methylbutyrate (14)
Methyl octanoate (63)(68)(69)
Neryl acetate (4)(69)
Octyl acetate (4)(40)(51)(69)
Octyl butyrate (4)(14)(68)(72)
Octyl 3-methylbutyrate (4)(69)
Terpinyl acetate (4)(5)(69)
Terpinyl formate (4)(5)

HYDROCARBONS

Benzene* (4)(51)(70)
Biphenyl* (40)(51)
δ-Cadinene (4)(74)
Camphene (4)(52)
Δ-3-Carene (4)(5)(14)(46)(71)
Caryophyllene (4)(40)(69)(74)
α-Copaene (4)(58)
β-Copaene (4)(73)
β-Cubenene (73)
Cyclohexane* (4)(51)
p-Cymene (4)(5)(14)(46)(58)(68)(69)
β-Elemene (4)(58)(60)(73)(74)
Ethylbenzene* (51)
Farnesene (4)(52)
Heptane* (72)(73)
Hexane (4)
α-Humulene (4)(72)
β-Humulene (4)

Isoprene (4)(73)
p-Isopropenyltoluene (4)(14)
Limonene (4)(5)(14)(19)(23)(39)(40)(46)(50)(51)(58-
-60)(63)(68-74)
2,4-p-Menthadiene (4)
Methane* (4)(51)
Methylciclohexane (58)
Methylciclopentane* (51)(58)(73)
2-Methylpentane* (51)
3-Methylpentane* (4)(51)
Myrcene (4)(5)(14)(23)(40)(46)(51)(58)(60)(63)
(68-71)(73)(74)
Nonane (4)(73)
Allo-Ocimene (69)
Octane (4)(73)
α-Phellandrene (4)(5)
β-Phellandrene (4)(5)
α-Pinene (4)(5)(14)(23)(46)(58)(59)(60)(63)(68-71)
(73)(74)
β-Pinene (4)(5)(60)(68)(70)(71)
Sabinene (4)(58)(59)(73)(74)
Selinadiene (69)(74)
α-Terpinene (4)(5)(14)(46)(69)
β-Terpinene (4)
γ-Terpinene (4)(5)(14)(23)(46)(60)(71)(74)
Terpinolene (4)(5)(14)(23)(46)(60)
α-Thujene (4)(23)
Toluene* (4)(51)
Calencene (4)(40)(58)(63)(68)(69)(70)(73)(74)
m-Xylene* (4)(51)(60)
O-Xylene* (4)(51)(60)
p-Xylene* (4)(51)(60)
Ylangene (4)(52)

* Possibly an artifact

KETONES

Acetone (4)(5)(14)(39)(46)(50)(58)(59)(60)(62)(67)
(68)(70-73)
Butanone (51)(67)
2-Butanone (4)(51)(72)
Carvone (4)(5)(14)(39)(40)(46)(50)(51)(58)(59)(63)
(64)(68)(71-73)
2-Decanone (4)
3-Hydroxy-2-butanone (69)
β-Ionone (69)
Methylheptenone (4)(5)(71)
4-Methyl-2-pentanone (4)(51)
Nootkatone (4)(63)

2-Octanone (46)
2-Pentanone (23)(67)
1-Penten-3-one (50)(59)(72)(73)
Piperitenone (4)(50)(58)(69)(72)(73)

MISCELLANEOUS
1,8-Cineole (4)
γ-Decanolactone (40)
1,1-Diethoxyethane (4)(39)(50)(51)(59)
Diethylcarbonate (23)
1,1-Ethoxymethoxyethane (50)(60)
Ethyl sec-butyl ether (4)(51)
Hydrogen sulfide (62)
cis-Limonene oxide (4)(5)(70)(71)
trans-Limonene oxide (4)(5)(70)(71)
cis-Linalool oxide (4)(5)(19)(39)(69)(71)
trans-Linalool oxide (5)(19)(39)(71)
Methyl chloride* (51)

* Possibly an artifact

Forty eight hydrocarbons have been identified. However, some authors establish that in certain cases (thirteen hydrocarbons), these compounds are present in orange juice not as components of the aromatic fraction, but as artifacts introduced by solvents used in the extraction processes. In the case of biphenyl, this compound is likely to come from the fungistatic bath used to avoid decay (51).

Among the thirty eight aldehydes identified all the straight-chain saturated compounds from acetal--dehyde to dodecanal, have been also found.

Only twelve acids have been identified. They have been only reported in seven studies. Eight of them are straight-chain acids, and four branch-chain acids.

Forty four esters have been reported as orange juice volatile fraction constituents. Ethyl butyrate ethyl acetate, ethyl octanoate, ethyl propionate and methyl butyrate are the most often found. Methanol and ethanol are the most cited alcohols as esters constituents.

Only fourteen ketones have been identified. Most of them are straight-chain ketones, and only three are branch-chain compounds.

A miscellaneous category of compounds is also compiled. The four identified oxides only correspond to two compounds as both isomers cis-trans have been detected.

Hydrogen sulfide, detected in one of the studies reviewed (62), has been the only sulfur--containing compound identified in the volatile fraction of orange juice.

V. OFF FLAVOUR OF ORANGE JUICE

During storage of single-strength or concentrated orange juices, mainly if storage is at warm temperature, the product develops detrimental off--flavours (61). Since there are a great lot of substances in orange juice, interactions among them may frequently occur, and some of the reaction products may be the origin of aroma and taste alterations. Many theories have been proposed to explain the formation of these detrimental products (4)(75-83).

In a recent study eleven degradation compounds have been isolated in canned orange juice stored at 35°C during twelve weeks (61). Ten of these compounds have been identified and are listed in table II.

Two compounds have been reported for the first time as degradation products of orange juice (4-vinyl guaiacol and 2,5-dimethyl-4-hydroxy-3(2H)--furanone).

Comparation of tables I and II shows that there are two compounds (furfural and -terpineol) that have been reported as components of the volatile fraction of orange juice and as degradation compounds in off flavoured orange juice (62)(68)(84)(85).

Although furfural doesn't contribute to off--flavour, its concentration in the juice increases as storage time increases, or with thermal abuse. For this reason furfural has been proposed as an index of storage time and/or temperature abuse (86-89).

α-Terpineol imparted a stale, musty or piney odour to the juice (61). As the content in this product increases linearly with storage time, it was suggested as an indicator for predicting storage time of orange juice (63).

TABLE II. Degradation compounds identified in off--flavoured orange juice

Benzoic acid
2,5-Dimethyl-4-hydroxy-3(2H)furanone
Furfural
2-Hydroxyacetyl furan
5-Hydroxyacetyl furfural
3-Hydroxy-2-pyrone
cis-1,8-p-menthanediol
trans-1,8-p-menthanediol
α-Terpineol
4-Vinyl guaiacol

2,5-dimethyl-4-hydroxy-3(2H)-furanone also imparts an unnatural flavour and aroma to orange juice like pineapple, since this compound is the main compound in pinneaples (90). Nevertheless, this flavour was not objectionable, although it maskes the natural orange flavour if the concentration exceeded its threshold level (0.05 ppm)(61).

The most isolated and identified detrimental compound that affects the flavour of orange juice appears to be 4-vinyl guaiacol. When added to a fresh juice, most taste panelists remarked that this compound imparted an "old-fruit" or "rotten" flavour to the juice. In conclusion, qualitative and quantitative determination of these three last mentioned compounds in orange juice might provide the basis for a definitive test of orange juice quality as affected by storage time and/or heat-abused industrialization process (61).

Not only the formation of degradation products but also big changes in proportions of compounds may contribute to undesirable flavours. In a comparative study between orange aqueous essences with bad and good sensorial characteristics was observed that the former had lower contents in acetaldehyde, ethyl butyrate and octanal, all of them good flavour contributors, and higher concentrations of hexanal, trans-2-hexenal, α-terpineol, 1-hexanol, trans-2,8-p-menthadien-1-ol, and trans-carveol, which may produce off-flavour (19).

Finally, an alteration characterized by a vinegary to buttermilk off-flavour may be occasionally detected in orange juice. This alteration is originated by the growth of certain acid tolerant bacteria of the genera Lactobacillus and Leuconostoc (91)(92). These bacteria produce acetyl--methylcarbinol which originated either diacetyl or 2,3-butyleneglycol by further bacterial action or autooxidation. If diacetyl is produced in orange juice or concentrate during processing or storage, the aroma and flavour imparted to the product is quite similar to that of butter-milk. Determination of diacetyl has been applied as an indication of the growth of bacteria that produce butter-milk off-flavours (94).

REFERENCES

1. Maier, H.G., Angew. Chem., 82, 965 (1970).
2. Kovats, E. sz., Trav. Chim. alim. Hyg., 64, 39 (1973).
3. Chang, S.S., Vallese, F.M., Hwang, L.S., Hsieh, O.A.L., and Min, D.B.S. J. Agr. Food Chem., 25, 450 (1977).
4. Bielig, H.J., Askar, A., and Treptow, H. "Aromaveranderanderungen von Orangesaft". Fortschritte in der Lebensmittelwisenshaft, Nr. 1. Techn. Univ. Berlin (1974).
5. Wolford, R.W., and Attaway, J.A. J. Agr. Food Chem., 15, 360 (1967).
6. Hall, J.A., and Wilson, C.P., J. Am. Chem. Soc., 47, 2575 (1925).
7. Mehlitz, A., and Gierschner, K., Intern. Fed. Fruit Juice Producers, Symp. Volatile Fruit Flavours. Bern. pag. 25 (1962).
8. Weurman, C., J. Agr. Food Chem. 17, 370 (1969).
9. Nawar, W.W., Food Technol. 20/2, 115 (1966).
10. Nawar, W.W., Sawyer, F.M., Beltran, E.G., and Fagerson, I.S. Anal Chem. 32, 1534 (1960).
11. Buttery, R.G., and Teranishi, R., Anal. Chem. 33, 1439 (1961).
12. Mackay, D.A.M., Lang, D.A., and Berdick, M., Anal. Chem. 33, 1369 (1961).
13. Weurman, C., Food Technol., 15, 531 (1961).
14. Wolford, R.W., Attaway, J.A., Alberding, G.E., and Atkins, C.D. J. Food Sci., 28, 320 (1963).
15. Paillard, N., Pitoulis, S., and Mattei, A., Lebensm. Wiss. u. Technol., 3, 107 (1970).
16. Issemberg, P., J. Agr. Food Chem., 19, 1045 (1971).

17. Jennings, W., and Filsoof, M., J. Agr. Food Chem. 25, 440 (1977).
18. Anonym. Flavours, 6, 128 (1975).
19. Lund, E.D., and Bryan, W.L., J. Food Sci., 42, 385 (1977).
20. Teranishi, R., and Buttery, R.G., Intern. Fed. Fruit Juice Producers Symp. Volatile Fruit Flavours. Bern. päg. 257 (1962).
21. Nawar, W.W., and Fagerson, I.S., Food Technol. 16/11, 107 (1962).
22. Flath, R.A., Forrey, R.R., and Teranishi, R., J. Food Sci., 34, 382 (1969).
23. Schultz, T.H., Flath, R.A., and Mon, T.R., J. Agr. Food Chem., 19, 1060 (1971).
24. Jennings, W.G., Wohleb, R., and Lewis, M.J., J. Food Sci., 37, 69 (1972).
25. Tressl, R., and Jennings, W.G., J. Agr. Food Chem., 20, 189 (1972).
26. Carson, J.F., and Wong, F.F., J. Agr. Food Chem. 9, 140 (1961).
27. Jennings, W.G., and Nursten, H.E., Anal. Chem., 39, 521 (1967).
28. Clark, R.G., and Cronin, D.A., J. Sci. Fd. Agric 26, 1615 (1975).
29. Moshonas, M.G., and Lund, M.G., J. Food Sci., 36, 105 (1971).
30. Bertuccioli, M., and Montedoro, G., J. Sci. Fd. Agric. 25, 675 (1974).
31. Butler, L.D., and Burke, M.F., J. Chromatog. Sci., 14, 117 (1976).
32. Charalambous, G., "Analysis of food and beverages: Headspace techniques". Academic Press, New York, 1978.
33. Nawar, W., J. Agr. Food Chem., 19, 1045 (1971).
34. Jeon, I., Reineccius, G., and Thomas, E., J. Food Sci., 24, 433 (1976).
35. Lafuente, B., Benedito, J., Mansanet, A., and Nadal, M.I., Agroq. Tecnol. Alim., 16, 89 (1976)
36. Nickerson, G.B., Likens, S.T., J. Chromatogr. 21, 1 (1966).
37. Maarse, H., and Kepner, R., J. Agr. Food Chem. 18, 1095 (1970).
38. Schultz, T., Flath, R., Mon, T., Eggling, S., and Teranishi, R. J. Agr. Food Chem., 25, 446 (1977).
39. Moshonas, M.G., Lund, E., Berry, R., and Veldhuis, M., J. Agr. Food Chem. 20, 688 (1972).

40. Schultz, T.H., Black, D., Bomben, J., Mon, T.R. and Teranishi, R. J. Food Sci. 32, 698 (1967).
41. Bomben, J.L., Kitson, J.A., and Morgan, Jr.A.I. Food Technol., 20, 1219 (1966).
42. Wolford, R., Dougherty, M., and Petrus, D. Intern. Fed. Fruit Juice Prod. pag. 151 (1968).
43. Butler, J., Ramchandini, C., and Thompson, D., J. Chem. Soc. pag. 280 (1935).
44. Buttery, R., Ling, L., and Guadagni, D., J. Agri. Food Chem. 17, 385 (1969).
45. Coleman, R., and Shaw, P., J. Agr. Food Chem. 19, 520 (1971).
46. Wolford, R.W., Alberding, G.E., and Attaway,J.A. J. Agr. Food Chem., 10, 297 (1962).
47. Moshonas, M.G. and Shaw, P., J. Agr. Food Chem. 20, 70 (1972).
48. Moshonas, M.G. and Shaw, P., J. Agr. Food Chem. 20, 1029 (1972).
49. Moshonas, M.G. and Shaw, P., J. Agr. Food Chem. 19, 119 (1971).
50. Moshonas, M.G. and Shaw, P., J. Food Sci., 38, 360 (1973).
51. Schultz, T.H., Teranishi, R., McFadden, W.H., Kilpatrick, P., and Corse, J. J. Food Sci. 29, 290 (1964).
52. Teranishi, R., Schultz, T.H., McFadden, W.H., Lundin, R.E., and Black, D.R., J. Food Sci. 28, 541 (1963).
53. Teranishi, R., Lundin, R., McFadden, W.H., Mon, T.R., Schultz, T.H., Stevens, K.L., and Wasserman, J., J. Agr. Food Chem., 14, 447 (1966).
54. Ismail, M., and Wolford, R., J. Food Sci., 35, 300 (1970).
55. Attaway, J., Wolford, R., Dougherty, M., and Edwards, G., J. Agr. Food Chem. 15, 688 (1967).
56. McNary, R., Dougherty, M., and Wolford, R., Sewage and Ind. Wastes, 29, 894 (1957).
57. Dougherty, M., Food Technol., 22, 1145 (1964).
58. Coleman, R.L., Lund, E.D., and Moshonas, M.G., J. Food Sci., 34, 610 (1969).
59. Shaw, P.E., and Coleman, L.R., J. Agr. Food Chem. 19, 1276 (1971).
60. Coleman, R.L., and Shaw, P.E., J. Agr. Food Chem. 20, 1290 (1972).
61. Tatum, J., Nagy, S., and Berry, R., J. Food Sci. 40, 707 (1975).
62. Kirchner, J.G., and Miller, J.M. J. Agr. Food

Chem. 5, 283 (1957).
63. Askar, A., Bielig, H.J., and Treptow, H., Dts. Lebensmittel-Rdsch., 69, 162 (1973).
64. Attaway, J.A., Wolford, R.W., and Edwards, G.J. J. Agr. Food Chem., 10, 102 (1962).
65. Attaway, J.A., Wolford, R.W., Alberding, G.E., and Edwards, G.J., J. Agr. Food Chem., 12, 118 (1964),
66. Braddock, R.J., and Petrus, D.R., J. Food Sci., 36, 1095 (1971).
67. Dinsmore, H.L., and Nagy, S. J.Agri. Food Chem. 19, 517 (1971).
68. Rymal, K.S., Wolford, R.W., Ahmed, E.M., and Dennison, R.A., Food Technol., 22, 1592 (1968).
69. Schreier, P., Drawert, F., Junker, A., and Mick, W., Z. Lebensm. Unters.-Forsch. 164, 188 (1977).
70. Shaw, P.E., and Moshonas, M.G., Proc. Fla. Sta. Hort. Soc., 87, 305 (1974).
71. Wolford, R.W., Attaway, J.A., and Barabas, L.J. Proc. Fla. Sta. Hort. Soc., 78, 268 (1965).
72. Moshonas, M.G., and Shaw, P.E., Flavours, 6, 133 (1975).
73. Coleman, R.L., and Shaw, P.E., J. Agr. Food Chem., 19, 520 (1971).
74. Radford, T., Kawashima, K., Friedel, P.K., Pope, L.E., and Gianturco, M.A., J. Agr. Food Chem. 22, 1066 (1974).
75. Boyd, J.M. and Peterson, G.T., Jud. Eng. Chem. 37, 370 (1945).
76. Nagy, S. and Nordby, H.E. J. Agr. Food Chem. 18 593 (1970).
77. Senn, V.J. J. Food Sci., 28, 531 (1963).
78. Blair, J.S., Godar, E.M., Masters, J.E. and Riester, D.W. Food Res. 17, 235 (1952).
79. Huskins, C.W. and Swift, L.S. Food Res. 18, 305 (1953).
80. Kefford, J.F., McKenzie, H.A., and Thompson, P.C.O. Food Preserv. Quart. 10, 44 (1955).
81. Blair, J.S., Godar, E.M., Reinke, H.G., and Marshall, J.R. Food Technol. 11, 61 (1957).
82. Swift, L.J. Proc. Fla. Sta. Hort. Soc. 64, 181 (1951).
83. Huskins, C.W., and Swift, L.J. Food Res. 18, 305 (1953).
84. Tatum, J.H., Shaw, P.E., and Berry, R.E. J. Agr. Food Chem., 15, 773 (1967).

85. Tatum, J.H., Shaw, P.E., and Berry, R.E. J. Agr. Food Chem. 17, 38 (1969).
86. Dinsmore, H.L., Nagy, S. J. Food Sci. 37, 768 (1972).
87. Nagy, S. and Randall, V. J. Agr. Food Chem. 21, 272 (1973).
88. Nagy, S., and Dinsmore, H.L. J. Food Sci., 39, 1116 (1974).
89. Dinsmore, H.L., and Nagy, S. J.A.O.A.C., 57, 332 (1974).
90. Rodin, J.O., Himel, C.M., and Silverstein, R.M. J. Food Sci., 30, 280 (1965).
91. Hays, G.L., and Riester, D.W. Food Technol. 6, 386 (1952).
92. Murdock, D.I., Troy, V.S., and Folinazzo, J.F. Food Technol. 6, 127 (1952).
93. Hill, C.E., Wenzel, F.W., and Barreto, A. Food Technol. 8, 168 (1954).
94. Hill, C.E., and Wenzel, F.W. Food Technol. 11, 240 (1957).

THE AROMA OF VARIOUS TEAS

Tei Yamanishi

Department of Food & Nutrition
Ochanomizu University
Bunkyo-ku, Tokyo, Japan

I. INTRODUCTION

Tea is one of the most popular beverages and there are many types and grades of teas available in the market of the world.

The starting material for tea manufacture is the tender, rapidly growing shoot tips of tea plant (tea flush), or tender young leaves.

Various teas are broadly classified by the manufacturing process, i.e. fermented tea (black tea), partially fermented tea (oolong tea and pouchong tea) and non-fermented tea (green tea).

The high acceptability of tea may be due to many factors, but one of the most contributory factors seems to be its aroma.

Tea aroma is determined by the quality of tea flush as well as the conditions of manufacturing process. The quality of tea flush is largely influenced by the season, the geographical location and the variety of tea plant, *Camellia sinensis*.

An amazing complexity of tea aroma components has been shown first by M.A. Gianturco and coworkers at the Coca-Cola Co., Atlanta, Georgia, U.S.A.(64). They analyzed an aroma concentrate from black tea by capillary column gas chromatography and found that black tea aroma contains about three hundred different compounds.

In order to find out the most important constituents which are responsible for the tea aroma characteristics, a number of research has been carried on, and the investigations are gradually focused on a miner component. Thus, the total

ISBN 0-12-169060-1

number of compounds identified related with tea aroma arised to almost 300 up to date (1-2), as shown in Table I. Here, it may be said that the compound which first reported year is later, the amount supposed to be smaller.

Although, such a numerous compounds have been identified from tea aroma, real contributory components which determine the aroma character of various teas, seemed to be not so large number.

This paper deals with the aroma characteristics of several distinctive teas, such as Sri Lankan (Ceylon) flavory black tea, Keemun black tea, pouchong tea, lotus tea and Japanese green tea (Sen-cha) and roasted green tea.

TABLE I. Aroma Constituents of Tea

Aroma Constituent	(Reference)	The first reported year	Aroma Constituent	(Reference)	The first reported year
I. Hydrocarbons (26)					
1. Benzene	(f)	'62	14. trans-β-Ocimene	(j)	'67
2. Toluene	(d,f,h)	'62	15. β-Myrcene	(j)	'67
3. Vinylbenzene	(f)	'62	16. Terpinolene	(d)	'72
4. Dimethylbenzene	(f)	'62	17. Undecane	(f)	'62
5. Ethylbenzene	(f,h)	'62	18. α-Cubebene	(d)	'71
6. Trimethylbenzene	(f)	'62	19. α-Copaene	(d)	'71
7. Ethyl-methyl benzene	(f)	'62	20. Caryophyllene	(d)	'71
8. Propylbenzene	(f)	'62	21. γ-Muurolene	(d)	'71
9. Isopropylbenzene	(h)	'74	22. α-Muurolene	(d)	'70
10. Diethylbenzene	(f)	'62	23. δ-Cadinene	(d)	'70
11. Methyl-propylbenzene	(f)	'62	24. Calamenene	(d)	'71
12. Limonene	(d,i)	'67	25. α-Humulene	(d)	'71
13. cis-β-Ocimene	(i,j)	'67	26. β-Sesquiphellandrene	(d)	'71
II. Alcohols (42)					
1. Methanol	(g,h,i)	'16	8. 2-Methyl-3-buten-2-ol	(i)	'67
2. Ethanol	(e,g,i)	'62	9. 1-Penten-3-ol	(d,h,i)	'66
3. 2-Propanol	(i)	'67	10. cis-2-Penten-1-ol	(d,h,i)	'65
4. 2-Butanol	(i)	'67	11. trans-2-Penten-1-ol	(h)	'67
5. 2-Methylpropanol	(b,d,i)	'34	12. 2-Methylbutanol	(b)	'34
6. Butanol	(b,d,i)	'34	13. 3-Methylbutanol	(b,d,h,i)	'34
7. Furfurylalcohol	(d)	'70	14. Pentanol	(d,h,i)	'66

TABLE I. (continued)

Aroma Constituent (Reference)	The first reported year
15. cis-3-Hexen-1-ol (a,b,c,d,h,i)	'20
16. trans-3-Hexen-1-ol (h,i)	'67
17. trans-2-Hexen-1-ol (d,h,i)	'66
18. Hexanol (b,c,d,h,i)	'35
19. Benzylalcohol (b,c,d,h,i)	'35
20. Heptanol (d)	'71
21. 2-Phenylethanol (b,c,d,h,i)	'35
22. 1-phenylethanol (d)	'70
23. 1-Octen-3-ol (d,h,i)	'67
24. Octanol (b,c,d,i)	'36
25. 2-Ethyl-1-hexanol (n)	'74
26. 3-phenylpropan-1-ol (c)	'36
27. Nonanol (d,i)	'71
28. 3,7-Dimethyl-1, 5,7-octatrien-3-ol (d)	'69
29. Linalool (b,c,d,h,i)	'36
30. Nerol (d,i)	'65
31. Geraniol (b,c,d,h,i)	'36
32. α-Terpineol (d,h,i)	'66
33. 4-Terpineol (n)	'74
34. Linalooloxide I (trans, furanoid) (d,h,i)	'64
35. Linalooloxide II (cis, furanoid) (d,h,i)	'64
36. Linalooloxide III (trans, pyranoid) (d,h,i)	'66
37. Linalooloxide IV (cis, pyranoid) (d,h,i)	'64
38. Nerolidol (d,i)	'67
39. α-Cadinol (d)	'71
40. Cubenol (d)	'71
41. epi-Cubenol (d)	'70
42. Phytol (g)	'65

<u>III. Aldehydes (49)</u>

Aroma Constituent (Reference)	The first reported year
1. Acetaldehyde (d,e,g,h,i)	'57
2. Acrolein (g)	'65
3. Propanal (g,h,i)	'65
4. 2-Methylpropanal (b,c,d,e,g,h,i)	'34
5. Butanal (b,c,d,h,i)	'34
6. cis-3-Pentenal (n)	'74
7. trans-2-Pentenal (d,h)	'72
8. 3-Methylbutanal (b,c,d,g,h,i)	'34
9. 2-Methylbutanal (b,c,h)	'35
10. Pentanal (d,h,i)	'66
11. Furfural (e,h,i)	'57
12. trans, trans-2,4-Hexadienal (h,i)	'67
13. cis-4 isomer of 12 (n)	'74
14. trans-2-Hexenal (b,c,d,h,i)	'34
15. cis-3-Hexenal (m)	'73
16. Hexanal (c,d,h,i)	'37
17. 5-Methylfurfural (d,i)	'67
18. Benzaldehyde (b,c,d,h,i)	'35
19. trans,trans-2, 4-Heptadienal (h,i)	'67
20. cis-4 isomer of 19 (h)	'67
21. trans-2-Heptenal (n)	'74
22. Heptanal (d,h,i)	'66
23. Phenylacetaldehyde (d,h,i)	'65
24. 2-Methylbenzaldehyde (n)	'74
25. trans-2,trans-4-Octadienal (n)	'74
26. cis-4 isomer of 26 (n)	'74

TABLE I. (continued)

Aroma Constituent (Reference)	The first reported year
27. trans-2-Octenal (d,i)	'66
28. Octanal (i)	'67
29. 4-Methoxybenzaldehyde (n)	'74
30. 4-Ethylbenzaldehyde (h)	'74
31. trans-2,trans-4-Nonadienal (n)	'74
32. cis-4 isomer of 31 (n)	'74
33. trans-2, cis-6-Nonadienal (n)	'74
34. trans-2-Nonenal (n)	'74
35. Nonanal (d)	'71
36. 2-Phenyl-2-butenal (d,h)	'67
37. 2,4,6-Decatrienal (h)	'74
38. Safranal (n)	'74
39. β-Cyclocitral (n)	'74
40. trans-2, trans-4-Decadienal (d,i)	'67
41. cis-4 isomer of 40 (n)	'74
42. Geranial (d)	'70
43. Neral (n)	'74
44. trans-2-Decenal (n)	'74
45. trans-2-Undecenal (n)	'74
46. 4-Methyl-2-phenyl-2-pentenal (n)	'74
47. 5-Methyl-2-phenyl-2-hexenal (n)	'74
48. 4-Ethyl-7,11-dimethyl-trans-2, trans-6, 10-dodecatrienal (n)	'74
49. cis-6 isomer of 48 (n)	'74
IV. Ketones (41)	
1. Acetone (g,h,i)	'65
2. 2-Butanone (b,d,i)	'37
3. Diacetyl (e,h,i)	'57
4. trans-3-Penten-2-one (i)	'67
5. 2,3-pentandione (i)	'67
6. 4-Methyl-4-penten-2-one (i)	'67
7. 4-Methyl-3-penten-2-one (i)	'67
8. trans,trans-3,5-Heptadien-2-one (h)	'74
9. 2-Heptanone (n)	'74
10. Acetophenone (b,d,e)	'37
11. trans-3,trans-5-Octadien-2-one (d,h,i)	'67
12. cis-5 isomer of 11 (h,n)	'74
13. cis-3 isomer of 11 (h)	'74
14. 6-Methyl-trans,trans-3,5-heptadien-2-one (d,h)	'71
15. 2-Methyl-2-hepten-6-one (h,i)	'67
16. trans-3-Octen-2-one (d)	'72
17. 2-Octanone (n)	'74
18. 3-Octanone (n)	'74
19. 2,6,6-Trimethyl-cyclohex-2-en-1-one (n)	'74
20. 2,2,6-Trimethyl-cyclohexanone (i,n)	'67
21. 2-Nonanone (n)	'74
22. 2,6,6-Trimethyl-cyclohex-2-en-1,4-dione (n)	'74
23. 2,2,6-Trimethyl-6-hydroxycyclohexanone (h)	'69
24. Benzylethylketone (n)	'74
25. 2-Decanone (h)	'74
26. 5-Ethyl-6-methyl-2-heptanone (n)	'74
27. cis-Jasmone (d,h,i)	'65
28. β-Damascenone (n)	'74
29. α-Damascone (n)	'74
30. β-Damascone (n)	'74

TABLE I. (Continued)

Aroma Constituent (Reference)	The first reported year	Aroma Constituent (Reference)	The first reported year
31. 1,5,5,9-Tetramethyl-bicyclo [4,3,0]-non-8-ene-7-one (n)	'74	37. 6,10-Dimethyl-2-undecanone (n)	'74
32. cis-Theaspirone (d,e)	'69	38. 3-Oxo-β-ionone(k)	'71
33. α-Ionone (d,h,i,k)	'67	39. 5,6-Epoxy-β-ionone (h,l)	'69
34. β-Ionone (d,h,i,k)	'66	40. 2,3-Epoxy-α-ionone (h)	'69
35. Geranylacetone (h)	'74	41. 6,10,14-Trimethyl-2-pentadecanone (d,n)	'71
36. 2',2"-Dihydro-α-ionone (d,n)	'71		

V. Esters (38)

Aroma Constituent (Reference)	The first reported year	Aroma Constituent (Reference)	The first reported year
1. Ethyl formate (i)	'67	22. Hexyl butyrate (n)	'74
2. Ethyl acetate (i,j)	'67	23. Ethyl octanoate(n)	'74
3. Methyl succinate (h)	'74	24. Methyl 4-oxononanoate (k)	'72
4. trans-2-Hexenyl formate (n)	'74	25. Benzyl butyrate(n)	'74
5. cis-3-Hexenyl formate (n)	'74	26. Geranyl formate(i)	'67
6. Isomyl acetate (d)	'57	27. cis-3-Hexenyl 2-methylbutyrate (n)	'74
7. Hexyl formate (n)	'74	28. cis-3-Hexenyl trans-2-hexenoate(n)	'74
8. Methyl benzoate(h)	'67	29. trans-3-Hexenyl cis-3-hexenoate (n)	'74
9. Benzyl formate (d)	'70	30. Neryl acetate (d)	'72
10. cis-3-Hexenyl acetate (d,i)	'67	31. Geranyl acetate (d)	'72
11. trans-2-Hexenyl acetate (n)	'74	32. α-Terpinyl acetate(d)	'71
12. Benzyl acetate (c,d)	'37	33. cis-3-Hexenyl hexanoate (d)	'71
13. Phenylethyl formate(d)	'70	34. trans-2-Hexenyl hexanoate (n)	'74
14. trans-2-Hexenyl propionate (n)	'74	35. cis-3-Hexenyl benzoate (d)	'70
15. trans-3-Hexenyl propionate (n)	'74	36. Methyl jasmonate(d,n)	'73
16. cis-3-Hexenyl propionate (n)	'74	37. Methyl trans-dihydro-jasmonate (n)	'74
17. Methyl octanoate (n)	'74	38. Hexyl phenylacetate(n)	'74
18. Ethyl phenylacetate(n)	'74		
19. trans-2-Hexenyl butyrate (n)	'74		
20. trans-3-Hexenyl butyrate (n)	'74		
21. cis-3-Hexenyl butyrate (d)	'72		

TABLE I. (continued)

Aroma Constituent (Reference)	The first reported year	Aroma Constituent (Reference)	The first reported year
VI. Lactones (15)			
1. 4-Butanolide (i)	'67	9. 4-Nonanolide (d)	'73
2. 4-Pentanolide (h)	'74	10. Jasmine lactone (d,h)	'73
3. 4-Hexanolide (h,i)	'67	11. 5-Decanolide (d,h)	'73
4. 4-Methyl-5-hexen-4-olide (n)	'74	12. 4-Decanolide (h)	'74
		13. Dihydroactinidiolide (d,h,k,l)	'67
5. 4-Heptanolide (h)	'74		
6. 5-Heptanolide (h)	'74	14. 2,3-Dimethyl-2-nonen-4-olide (d)	'73
7. 4-Octanolide (d)	'73		
8. Coumarin (d)	'71	15. 4-Tetradecanolide (h)	'74
VII. Acids (25)			
1. Formic (d,f)	'57	14. 5-Methylhexanoic (f)	'62
2. Acetic (d,f,i)	'34	15. Heptanoic (f)	'62
3. Propionic (c,d,f,i)	'34	16. Phenylacetic (n)	'74
4. Isobutyric (d,f,i)	'34	17. trans-2-Octenoic (n)	'74
5. Butyric (b,d,f,i)	'34	18. 6-Methylheptanoic (f)	'62
6. 3-Methylbutyric (b,c,f,i)	'34	19. Octanoic (f)	'62
		20. 7-Methyloctanoic (f)	'62
7. 2-Methylbutyric (i)	'67	21. Nonanoic (f)	'62
8. Pentanoic (d,f,i)	'65	22. trans-Geranic (n)	'74
9. cis-3-Hexenoic (d,i)	'65	23. 8-Methylnonanoic (f)	'62
10. trans-2-Hexenoic (d,i)	'65	24. Decanoic (f)	'62
11. 4-Methylpentanoic (f,i)	'62	25. Dodecanoic (f)	'62
12. Hexanoic (b,c,d,f,i)	'34		
13. Benzoic (d)	'57		
VIII. Phenols (13)			
1. Phenol (b,d)	'35	8. 4-Vinylphenol (d)	'71
2. o-Cresol (b,d)	'35	9. Vanillin (d)	'78
3. m-& p-Cresol (d)	'66	10. 4-Ethylguaiacol (d)	'70
4. Guaiacol (d,o)	'72	11. Thymol (n)	'74
5. Salicylic acid (b,d)	'35	12. Carvacrol (n)	'74
6. 2,3-Dimethylphenol (o)	'72	13. Isoeugenol (d)	'78
7. Methylsalicylate (a,b,c,d,h,i)	'16		
IX. Miscellaneous oxygenated compounds (13)			
1. 1-1-Dimethoxyethane (i)	'67	4. Propionoin (i)	'67
2. 2-Ethylfuran (i)	'67	5. 1,4-Dimethoxybenzene (d)	'75
3. 2-Acetylfuran (d,n)	'73		
		6. 4-Ethylanisole (h)	'74

TABLE I. (continued)

Aroma Constituent	(Reference)	The first reported year	Aroma Constituent	(Reference)	The first reported year
7. Isoamylfuran	(h)	'74	11. Theaspirane	(d,n)	'74
8. Amylfuran	(i)	'67	12. 6,7-Epoxy-dihydro-theaspirane	(n)	'74
9. Safrole	(n)	'74	13. 6-Hydroxy-dihydro-theaspirane	(n)	'74
10. Anethol	(d)	'75			
X. Sulfur compounds (4)					
1. Hydrogensulfide	(d)	'55	3. Dimethylsulfide	(d,h,i)	'63
2. Methanthiol	(c)	'40	4. Thiophene	(h)	'74
XI. Nitrogenous compounds (32)					
1. 2-Formylpyrrole	(d)	'70	20. 2-Ethyl-3-methyl-		'73
2. 2-Methylbutanenitrile	(i)	'67	21. Trimethylpyrazine	(d)	'73
3. 3-Methylbutanenitrile	(i)	'67	22-23. 2,5 & 2,6-Diethylpyrazines	(d)	'73
4. 2-Acetylpyrrole	(c,d)	'40	24. 3-Ethyl-2,5-dimethylpyrazine	(d)	'73
5. N-Ethylsuccinimide	(h)	'74	25. Tetramethylpyrazine	(d)	'73
6. 1-Ethyl-2-formylpyrrole	(d,h)	'67	26. 2,5-Diethyl-3-methylpyrazine	(d)	'73
7. Indole	(d)		27. 2,6-Diethyl-3-methylpyrazine	(d)	'73
8. Phenylacetonitrile	(d,h)	'67	28. 6,7-Dihydro-5H-cyclopentapyrazine	(d)	'73
9. 1-Ethyl-2-acetylpyrrole	(d)	'73	29. 2-Methyl-6,7-dihydro-5H-cyclopentapyrazine	(d)	'73
10. Methyl anthranilate	(h)	'74	30. 5-Methyl-6,7-dihydro-5H-cyclopentapyrazine	(d)	'73
11. Quinoline	(c)	'37	31. 2-(2'-Furyl)-pyrazine	(d)	'73
12. Diphenylamine	(d)	'71	32. 2-(2'-Furyl)-5(or 6)-methylpyrazine	(d)	'73
13. Methylpyrazine	(d)	'73			
14-16. 2,3-,2,5- & 2,6-Dimethylpyrazines	(d)	'73			
17. Ethylpyrazine	(d)	'73			
18. 2-Ethyl-5-methylpyrazine	(d)	'73			
19. 2-Ethyl-6-methylpyrazine	(d)	'73			

REFERENCES

(a) : (3)	(e) : (49-50)	(i) : (64-65)	(m) : (73-74)
(b) : (4-14)	(f) : (51)	(j) : (66-68)	(n) : (75)
(c) : (15-21)	(g) : (52-56)	(k) : (69)	(o) : (76-77)
(d) : (22-48)	(h) : (57-63)	(l) : (70-72)	

Constituents are listed in order of Carbon number in each group.

II. TEA MANUFACTURING PROCESS

A. Black Tea

Manufacturing process of black tea : withering ⟶ rolling ⟶ fermentation ⟶ drying.

Withering is the preliminary step to rolling, resulting in a desired loss of water, a flaccid condition of the leaves which they can stand rolling without breaking. The wither is judged by loss weight (35～40%), softness and an agreeable fruity aroma developed during withering.

The object of the rolling is to impart the characteristic twist, break the leaf cells and expose the juice to the air, starting the oxidation of the cell sap. Generally, the withered leaves are rolled twice, the first rolling requires about 50 minutes after that, the leaves are sifted and the finer leaves are brought into fermenting room, while the coarser rolled again, continuing for 40 minutes with increasing pressure.After the second rolling, the leaves are separated by screen and brought in fermentation process.

The leaves brought into the fermenting room are spread evenly in shallow trays in 5 cm deep and the room temperature should be kept around 25°C and the relative humidity over 95%.During fermentation, oxidation of the components in tea leaves bring about chemical changes which largely determine the flavor, strength, body and color of its infusion.

Drying process has two purposes, the one is stopping the fermentation and the other is the reduction of water content to 4～5% at which level the tea can be stored safely.

B. Green Tea (Sen-cha)

Manufacturing process of Sen-cha : steaming ⟶ primary heating and rolling ⟶ rolling ⟶ secondary drying ⟶ final rolling ⟶ final drying.

The first step of the Sen-cha manufacturing is steaming by which the oxidizing enzyme in the leaf is inactivated and the green color of leaf can be maintained. In the "primary heating tea roller", tea leaves are agitated and dehydrated by hot air, introduced by fan from a furnance, and taking out when the loss of weight reached 48%.Then the leaves are rolled under pressure for 10 minutes without heating, and after taken out, the lumps should be broken. The object of this rolling is to break the leaf cells, and enable to dissolve the content of the leaves easily when brewed.

By "secondary drying" in the tea roller (a trough, placed over a heater), the leaves are reduced 68% of their weight, for about 20 minutes.

"Final rolling" is performed in the final tea roller, here the tea leaves are twisted under pressure and dried with the help of the heater beneath the trough. After about 35 minutes the tea leaves is taken out from the final tea roller, the tea leaves are dried again with hot air at about 65°C, for reducing the moisture to 3-4%.

C. Pouchong Tea

The highest quality of pouchong tea is made only from the variety Chin-shin-oolong. The tea flush is subjected first to withering by sun-light for 5-20 minutes, then withering indoors for 2-4 hours, before the usual process of manufacturing of pan fired green tea (Chinese green tea). The first step of pan fired green tea manufacture is parching in a pan at higher temperature about 230°C, for 8-10 minutes. Then, continue heating in the pan such a way that 150°C, 10 min., 110°C, 15 min., 100°C, 30 min., 80°C, 60 min., with repeating turn over the leaves. By this process the tea has a curly form.

III. AROMA COMPOSITION OF VARIOUS TEAS

A. Flavory Ceylon Black Tea

In Sri Lanka (Ceylon), there are two well-defined seasons during which Ceylon tea flavor is outstanding. These are Jan./Feb. in the Dimbula District, and July/Aug./Sept. in the Uva District (56).

1. Composition of The Top Note Aroma. Generally, tea is evaluated its quality first by topnote of aroma which arises from tea infusion.

Fresh unblended black tea produced during the flavory season in Uva District, was used for our investigation.

The Uva tea was kindly provided by Dr. R.L. Wickremasinghe of Tea Research Institute of Sri Lanka.

Aroma concentrate which had a typical topnote or first perceptible characteristic note of aroma from the tea infusion was prepared from the tea by relatively short time steam distillation (15 min.), followed by isopentane extraction (40). The aroma constituents were identified by combined gas chromatography—mass spectrometry (GC-MS) and IR.

A total of 57 compounds were identified and approximate composition of the topnote aroma was determined as shown in Table II.

Besides the constituents listed in Table II, twelve miner components were recognized but not identified.

Although there are still unidentified important components which will contribute to topnote aroma of flavory Ceylon tea, composite mixture of identified components prepared based on the quantitative data in Table II, seemed to improve the flavor of commercial instant black tea to some extent.

TABLE II. Approximate Composition of The Topnote Aroma

Aroma constituent	Relative quantity %	Aroma constituent	Relative quantity %
Linalool	53.6	Pentanal	0.1
trans-Linalooloxide (furanoid)	16.7	trans-2-Octen-1-al	0.1
		α-Terpineol	0.1
cis-Linalooloxide (furanoid)	4.8	Neryl acetate	0.1
		Nerol	0.1
trans-2-Hexen-1-al	4.7	Benzylalcohol	0.1
Hexanol	1.7	3-Methylbutanal	+
trans-3-Octen-2-one	1.6	trans-2-Pentenal	+
Benzaldehyde	1.2	Butanol	+
Linalooloxides (pyranoid)	0.9	3-Methylbutanol	+
1-Penten-3-ol	0.8	Myrcene	+
cis-3-Hexen-1-ol	0.6	d-Limonene	+
Furfurylalcohol	0.6	Pentanol	+
cis-3-Hexenyl acetate	0.5	Octanal	+
trans,trans-2,4-Heptadienal	0.5	Terpinolene	+
		Nonanal	+
cis-2-Penten-1-ol	0.4	trans,trans-2,4-Hexadienal	+
1-Octen-3-ol	0.4	trans-2-Hexen-1-ol	+
α-Terpinyl acetate	0.3	cis-3-Hexenyl butyrate	+
Geranyl acetate	0.3	trans-2,cis-4-Heptadienal	+
Hexanal	0.3	trans,trans-2,4-Heptadienal	+
Octanol	0.3	trans,trans-3,5-Octadienone	+
cis-3-Hexenyl hexanoate	0.3	6-Methyl-3,5-heptadienone	+
Heptanal	0.2	Nonanal	+
3,7-Dimethyl-1,5,7-octatrien-3-ol	0.2	Phenylacetaldehyde	+
		1-Ethylformylpyrrole	+
Acetophenone	0.2	γ-Caprolactone	+
Methyl salicylate	0.2	α-Ionone	+
Geraniol	0.2	β-Ionone	+
		2-Phenylethanol	+

Mark + means less than 0.1%

2. Aroma Characteristic of Ceylon Flavory Tea. It has been found that flavory Sri Lankan black tea appear to contain more linalool, linalool oxides, geraniol and cis-jasmone than non-flavory black tea (off season or low grown), and also, found that non-flavory black tea appear to have lower boiling volatiles than flavory black tea (54). From organoletic stand point, the most attractive character of Ceylon flavory tea

TABLE III. Fractionation of Aroma Concentrate by Silica-gel Column Chromatography

Fraction No.	Solvent Ether % in Hexane	Relative[a] quantity of concentrate	Aroma of Each Fraction
1	0	0.4 %	Citrus-like
2	12	6.0	Unpleasant, Rubber-like
3	2	10.8	Methyl salicylate-like(strong)
4	3	14.1	Castor oil-like (strong)
5	4	12.5	Cucumber and Mint-like
6	6	13.2	Linalool-like, slightly acidic
7	8	21.5	Cucumber and Linalool-like
8	10	4.3	Oily and Japanese parsley-like
9	20	7.1	One of characteristic odor
10	Methanol 100%	0.4	Flowery aroma
[11]	"	2.7	Most significant sweet tea aroma (very strong)
12	"	7.0	Sweet and greenish aroma

a) The concentrate was obtained by distillation of solvent, which could not be removed perfectly.

TABLE IV. Gas Chromatographic Separation of Fraction [11]

Trap No.	Rt (min.)	Relative quantity	Aroma of each trap	Main components in each group
1	7-32	12.8%	Greenish odor	Pentanol cis-3-Hexen-1-ol
2	32-35	43.3	Flowery, Pleasant	Linalool oxide I and II 2-Phenylethanol
3	35-37	28.3	Woody	Linalool oxide III and IV
4	37-75	15.6	Peach, Apricot-like, Sweet floral	Compounds shown in Table V

seemed to depend on its fruity and floral (particularly, of jasmine flower) sweet flavor. The investigation was focused on an elucidation of the constituents responsible to this flavor (41).

The best quality of Dimbula Tea produced during flavory season, February, 1971, was used for preparation of the aroma concentrate (Yield 0.02%). The aroma concentrate was separated into thirteen fractions by silica-gel column chromatography as shown in Table III.

The most significant fraction No.[11], which containing the characteristic aroma of Ceylon flavory tea was further separated into four traps by gas chromatography, as explained in Table IV. A typical fruity and jasmine flower-like aroma was concentrated in trap [IV]. By GC-MS analysis using three different columns, the constituents were identified and their approximate relative quantities were obtained from the gas chromatograms. The results are shown in Table V.

Methyl jasmonate possessed a sweet-floral, tenacious jasmine-like aroma and jasmine lactone had a powerful fruity aroma, resembles to peach or apricot, with floral note.

The presence of methyl jasmonate in black tea was reported one year later by Renold et al.(75). They found methyl trans-dihydrojasmonate besides methyl jasmonate.

TABLE V. Composition of Trap [IV]

Constituent (listed in order of Rt on column Carbowax 20M)	Semi quantitative[a)] value %
2-Methyl-hept-2-en-6-one	1.8
retro-Ionone b)	2.5
Linalool oxide III and IV	11.9
Geraniol	6.4
4-Octanolide	2.1
cis-3-Hexenyl cis-3-hexenoate	8.2
4-Nonanolide	4.8
Theaspirone	3.0
2,3-Dimethyl-2-nonen-4-olide	1.1
5-Decanolide	5.3
Jasmine lactone	11.9
Methyl jasmonate	4.6
Dihydroactinidiolide	15.5

a) obtained from peak area percentages.

b) Not published yet.

Jasmine lactone and methyl jasmonate had been identified from jasmine absolute as the characteristic components resemble to fragrance of jasmine flower (78-79). Dihydroactinidiolide (2,6,6-trimethyl-2-hydroxycyclohexylidene-1-acetic acid lactone) was found in black tea aroma first by Bricont et al. (60), then by Ina et al.(70). Theaspirone (2,6,10,10-Tetramethyl-1-oxa-spiro-[4,S]-dec-6-ene) was first found by Ina et al. along with dihydroactinidiolide, and reported to have a distinctly black tea like flavor when enhanced by small amounts of dihydroactinidiolide (70). 2,3-Dimethyl-2-nonen-4-olide had a celery-like pleasant aroma and had been isolated from cigar tobacco leaves (80). 4-Octanolide, 4-nonanolide and 5-decanolide had similarity in tenacious sweet creamy, nut-like odor with a heavy fruity undertone.

In conclusion, aroma of flavory Sri Lankan black tea seemed to be characterized by the constituents listed in Table V, especially methyl jasmonate and jasmine lactone, together with high level contents of linalool and its oxides.

B. Keemun Black tea

Keemun black tea is one of the highest quality of Chinese black tea and is known for its characteristic strong heavy flavor including a rose-like-aroma.

To investigate the aroma characteristics of Keemun black tea, the aroma composition was compared with that of flavory Sri Lankan black tea.

Keemun tea, Uva and Dimbula teas were generously given by China National Native Product & Animal By-Products Import & Export Corporation Shanghai Tea Branch, Grenanore Estate Haptale Sri Lanka and Tea Research Institute of Sri Lanka, respectively. All these three tea samples were produced in 1976.

The aroma concentrates from the three tea samples were analyzed by GC-MS.

The most remarkable differences between Keemun tea and Sri Lankan tea (both Uva and Dimbula)were found in the amount of geraniol, geranic acid, benzylalcohol, 2-phenylethanol, linalool and its oxides. The former four compounds were much larger quantities in Keemun tea than in Sri Lanka tea, while the total of linalool and its oxides was much smaller amount in Keemun tea, as shown in Table VI. This seemed us to explain the aroma characteristics of Keemun tea which has a rosy and heavy thick flavor (46).

TABLE VI. Compositional Differences of Main Aroma Compounds between Keemun and Sri Lanka Teas

Main Compound	Peak area percentages		
	Keemun	Uva	Dimbula
Hexanal	2.1	2.2	5.4
trans-2-Hexenal	3.1	5.4	8.5
cis-2-Penten-1-ol	0.1	2.0	1.2
cis-3-Hexen-1-ol	1.3	5.9	4.0
trans-2-Hexen-1-ol	0.3	1.3	1.3
Linalool oxide I	2.8	3.8	3.2
Linalool oxide II	7.2	19.7	16.3
Linalool	6.8	33.1	27.9
Phenylacetaldehyde	4.6	1.2	1.6
Linalool oxide III and IV	6.4	2.4	3.7
Methyl salicylate	1.5	9.1	3.9
Geraniol	29.6	3.1	10.5
Benzylalcohol	11.0	1.8	1.7
2-Phenylethanol	7.1	0.4	0.9
β-Ionone	5.2	1.9	2.0
trans-Geranic acid	6.0	1.2	1.4

C. Green Tea

Organoleptically, the aroma of green tea is considerably different from that of black tea.

In the manufacture of black tea, the tea fermentation process is allowed to proceed to near completion. Therefore, remarkable chemical changes take place during fermentation, and various aroma compounds are produced as described by Sanderson, G.W. (1,81).

As mentioned earlier, in the manufacture of green tea, the enzyme in the leaf is inactivated at the first step of the manufacturing processes. Therefore, the most of aroma constituents of green tea are those compounds that originally contained in fresh leaves, and only a few is formed in the course of manufacturing.

As for the major difference in aroma components between green tea and black tea, Hara, T. et al.(82) reported that the yield of aroma concentrate from black tea was 4-5 times as much as that from green tea and compositional differences were observed in the amounts of indole, benzylalcohol, nerolidol, B-ionone, linalool, linalool oxides, cis-3-hexen-1-ol, methylsalicylate and hexanoic acid. The first four compounds were found in larger amounts in green tea, while the rest compounds in larger quantities in black tea.

1. <u>Aroma Composition of Green Tea.</u> Aroma constituents were identified from the contributory fractions to the green tea flavor (i.e. intermediate and high boiling fraction) by silica gel column chromatographic fractionation and gas chromatographic isolation followed by IR and GC-MS analysis. Relative quantities of these compounds were roughly obtained by calculation based on the yield of each fraction separated by silica gel column chromatography mentioned above and peak area percentage on gas chromatogram of each fraction(38). The results are shown in Table VII.

TABLE VII. Composition of The Contributory Fraction to The Distinctive Green Tea Flavor

Linalool	19.9	Nerol	0.6
δ-Cadinene	9.4	cis-3-Hexenyl benzoate	0.6
Geraniol	5.5	Furfurylalcohol	0.5
Nerolidol	4.0	α-Ionone	0.5
Indole	3.9	Phenol	0.5
Benzylalcohol	3.7	Nonanal	0.5
α-Muurolene	3.2	Limonen	0.5
α-Terpineol	3.0	cis-3-Hexen-1-ol	0.5
cis-Jasmone	3.0	α-Copaene	0.5
Linalool oxide I	2.6	α-Cubebene	0.4
β-Ionene	2.6	2',2"-Dihydro- -ionone	0.4
Octanol	1.6	Acetophenone	0.4
Linalool oxide II	1.4	α-Humulene	0.4
epi-Cubenol	1.2	Linalool oxides IV	0.4
Dihydroactinidiolide	1.2	6,10,14-Trimethyl-2-pentadecanone	0.4
Caryophyllene	1.1		
Cubenol	1.1	m-,& p-Cresol	0.3
Calamenene	1.0	β-Sesquiphellandrene	0.3
cis-3-Hexenyl hexanoate	1.0	4-Ethylguaiacol	0.3
2-Acetylpyrrole	0.9	trans,trans-3,5-octadien-2-one	0.2
1-Ethylformylpyrrole	0.9		
-Muurolene	0.8	5-Methylfurfural	0.2
2-Phenylethanol	0.7	α-Terpinyl acetate	0.1
Coumarin	0.7	6-Methyl-trans,trans-3,5-heptadien-2-one	0.1
α-Cadinol	0.6		
3,7-Dimethyl-1,5,7-octatrien-3-ol	0.6	Heptanol	0.1
4-Vinylphenol	0.6	Unidentified compounds	ca. 14%
Diphenylamine	0.6		

2. Aroma Characteristics of Spring Green Tea. Aroma of spring green tea which made from the tea flush harvested in early Spring is especially favorite in Japan, because of its briskness. The components which responsible for the briskness were found to be cis-3-hexen-1-ol and its esters, i.e. hexanonate and trans-2-hexenoate(45). Also an adequate concentration of indole and dimethylsulfide seems to make the aroma of spring green tea more attractive.

3. Roasted Green Tea (Hoji-cha). Hoji-cha is usually made from lower grade of green tea, i.e. Ban-cha (a coarse tea, made from coarser leaves, or by separating from raw Sen-cha in the refining process), by roasting around 200°C for a few minutes. Roast flavor makes Ban-cha much more acceptable.

Aroma concentrate from Hoji-cha had a strong typical roast flavor and its yield was much larger than that of original Sen-cha, i.e. 45mg% (about three times as much as Sen-cha).

TABLE VIII. Peak Identification of Fig. I.

[A] Neutral Fraction

Peak	Compound
a.	same as a'
b.	same as b'
c.	same as c'
d.	same as d' small amount of trans-2-Hexen-1-ol
e.	same as e'
f.	same as f_1'
g.	Linalool oxide II
h.	Linalool
i.	Phenylacetaldehyde Furfurylalcohol
j.	α-Terpineol
k.	Linalool oxide III, IV
l.	Nerol, same as l'
m.	Geraniol 3,4-Dihydro-β-ionone
o.	β-Ionone
p.	2-Acetylpyrrole 2-Furylpyrazine 2-Formylpyrrole Nerolidol
Q.	Theaspirone
R.	Dihydroactinidiolide

[B] Basic Fraction

Peak	Compound
a'.	Methylpyrazine
b'.	2,5- & 2,6-Dimethylpyrazine
c'.	2-Ethylpyrazine 2,3-Dimethylpyrazine
d'.	2-Methyl-5-ethylpyrazine 2-Methyl-6-ethylpyrazine
e'.	2,3,5-Trimethylpyrazine 2-Methyl-3-ethylpyrazine
f_1'.	2,5 & 2,6-Dimethylpyrazine
f_2'.	2,5-Dimethyl-3-ethyl-pyrazine
g'.	Tetramethylpyrazine
h'.	2,5-Diethyl-3-methyl pyrazine
i'.	5-Methyl-6, 7-dihydro-5H-cyclopentapyrazine
j'.	6,7-Dihydro-5H-cyclopenta-pyrazine
k'.	same as k
l'.	2-Methyl-6, 7-dihydro-5H-cyclopentapyrazine
m'.	2-Ethyl-6, 7-dihydro-5H-cyclopentapyrazine

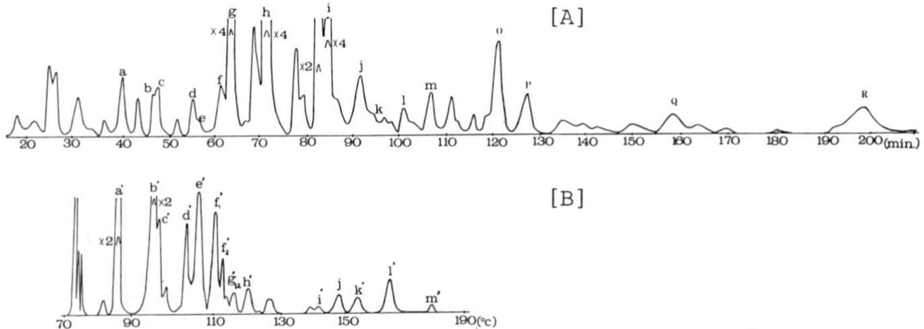

Fig. I. Gas Chromatograms of Neutral and Basic Fractions
[A] Neutral fraction; 23% Carbowax 6000 on Shimalite W 60/80 mesh, 3mm (i.d.) x 3m, TCD.
[B] Basic fraction; Capillary column (0.5mm x 50m) coated with Carbowax 20m, TIC monitor

Since it is well known that the main components of roast flavor in various foods are pyrazines, the aroma concentrate from Hoji-cha was separated into basic, acidic and neutral fractions. The proportion of three fractions was roughly 29% (basic), 24%(acidic) and 47%(neutral), while in the case of original Sen-cha, the basic fraction was only a trace amount(42).

The gas chromatograms of the neutral and basic fractions and their peak identification are shown in Fig. 1 and Table VIII.

TABLE IX. Furans and Pyrroles in The Aroma Concentrates from Hoji-cha and its Original Sen-cha

	Hoji-cha	Original Sen-cha
	% *	% *
Furans		
Furfurylalcohol	13.7	0.5
Furfural	9.3	-
2-Acetylfuran	4.7	-
5-Methylfurfural	2.9	0.2
Pyrroles		
2-Acetylpyrrole	5.4	0.9
2-Formylpyrrole	2.6	-
1-Ethyl-2-formylpyrrole	4.7	0.9
a Pyrrole compound	9.1	0.9

* The values were obtained by measuring the area of each peak as a percentages of total area of all the peaks on the gas chromatograms of the aroma concentrates from Hoji-cha and original Sen-cha.

Besides pyrazines, furans and pyrrols increased remarkably as shown in Table IX.

D. Pouchong Tea and Jasmine Tea

Pouchong tea is a kind of partially fermented tea. Fermentation degree of pouchong tea is presented as one-third of black tea, when oolong tea is expressed as half-fermented tea.

The highest quality of pouchong tea which has a superior floral elegant flavor, is made only from the tea flush of var. Chin-shin-oolong plucked in the two flavory seasons (early April and end of September or beginning of October in north Taiwan).

Common grade of pouchong tea is often scented by blending with jasmine flower to enhance the flavor.

Pouchong tea used in our investigation was the highest-grade and non-blended sample, manufactured from both autumn leaves (1976) and spring leaves (1977) of high grown in the north Taiwan.

Jasmine tea was the popular pan fired Chinese tea scented with jasmine flower, manufactured in 1976.

Both tea samples were generously provided by Mr. Wan-Kan Sie of the Agriculture Products Laboratory, Bureau of Commodity Inspection & Quarantine, Taiwan.

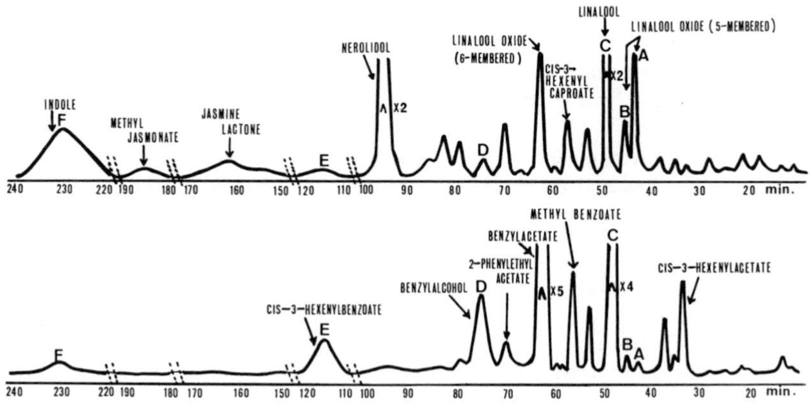

Column : Carbowax 20M. 23% on Shimalite W 60/80
3mm(i.d.) x 3m
Temp.: 70 - 200°C (2°C/min), TCD.

Fig. 2. Gas Chromatograms of the Aroma Concentrates from Pouchong Tea and Jasmine Tea

The aroma concentrates from teas were prepared by atmospheric steam distillation and analyzed by GC-MS using three columns of different polarities (i.e. Carbowax 20M, SE30 and SP 1000).

Fig. 2 shows the comparison of pouchong tea with jasmine tea. As can be seen, the aroma pattern of jasmine tea is much different than that of pouchong tea.

It is noteworthy that jasmine lactone and methyl jasmonate, which are known as the most contributory constituents to jasmine flower (79), were much larger amount in pouchong tea than in jasmine tea, and benzyl alcohol and benzyl acetate were predominant in jasmine tea.

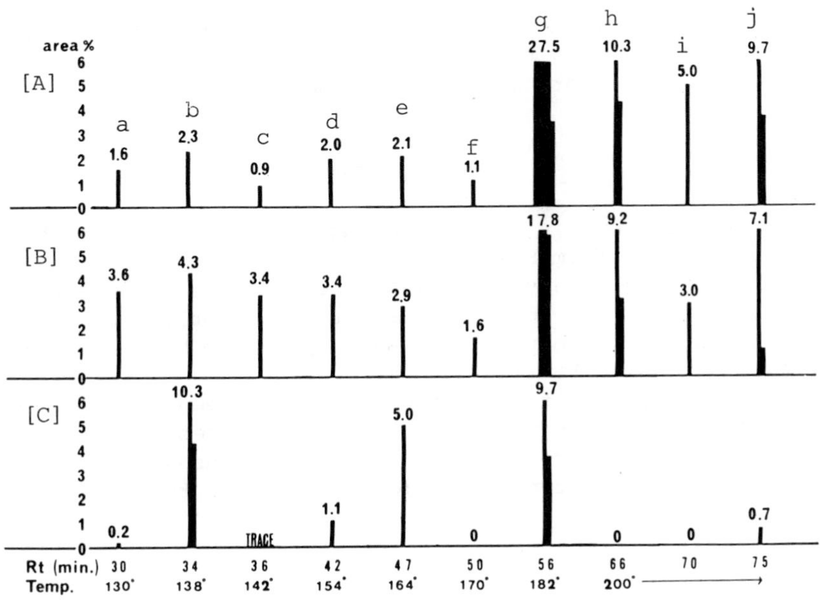

[A]; Pouchong tea, Spring, 1977
[B]; Pouchong tea, Autumn, 1976
[C]; Japanese Spring Green tea, 1976

Fig. 3. Comparison of The Peak Area Percentages of Pouchong tea with Japanese Green Tea

Peak	Constituent	Peak	Constituent
a.	Linalool oxide I & II	f.	Benzylcyanide
b.	Linalool	g.	Nerolidol
c.	3,7-Dimethyl-1,5,7-octatrien-3-ol	h.	cis-Jasmine lactone
d.	Linalool oxide III & IV	i.	cis-Methyl jasmonate
e.	Benzylalcohol	j.	Indole

From organoleptic standpoint, pouchong tea has much attractive flowery and elegant flavor, extremely different from that of jasmine tea.

To make clear the aroma characteristics of the highest quality of pouchong tea, the comparison of GC peak area percentages of main components in spring and autumn pouchong teas with that of Japanese spring green tea was performed. As can be seen from Fig. 3, nerolidol, jasmine lactone, methyl jasmonate, indole and benzylcyanide were found at significant amounts in pouchong tea (both spring and autumn). In addition to these compounds, linalool oxides, 3,7-dimethyl-1,5,7-octatrien-3-ol (i.e. oxidized product of linalool) were much more in quantity than Japanese green tea.

Therefore, this may enable us to explain the aroma characteristics of pouchong tea which is heavier and has a superior floral, sweet and very attractive aroma.

E. Lotus Tea

Tea plant grown in Vietnam belongs to the family of theacease originated from Shan District.

Lotus tea is purely Vietnamese special tea, scented with natural lotus pollen (600 flowers were used for 1Kg of pan fired green tea). Lotus tea is said to be the best liked and the most expensive tea in Vietnam.

The aroma composition of the pan fired green tea used for lotus tea is shown in Table X. The large contents of pyrazines and pyrroles in the volatiles seemed to be caused by pan firing process of which heating condition is stronger than that of Japanese green tea.

TABLE X. Aroma Composition of Pan Fired Vietnamese Green Tea (peak area percentages)

Compound	Composition	Compound	Composition
	%		%
Linalool	20.7	trans,trans-3,5-Octadien-2-one	5.3
1-Ethyl-2-formylpyrrole	13.6	5-Methylfurfural	
3,7-Dimethyl-1,5,7-Octatrien-3-ol	11.0	α-Terpineol	4.2
2,5-(or 2,6-)Dimethylpyrazine	10.9	2-Acetylpyrrole	3.2
Linalool oxide I	5.9	Pentanol	2.7
Linalool oxide II	6.5	2-Acetylfuran	2.6
2-Methyl-5-ethylpyrazine	5.6	cis-2-Penten-1-ol	1.6
		Anethole	1.3
		cis-3-Hexen-1-ol	0.6
		Limonene	0.3

The aroma concentrate from lotus tea contained a large amount of 1,4-dimethoxybenzene which was oftained as a crystalline form when the aroma concentrate was kept in the refrigerator. 1,4-Dimethoxybenzene was found in amount to 93% of aroma concentrate prepared from dry lotus pollen.

Organoleptically, it was clear that 1,4-dimethoxybenzene was the key substance of the characteristic aroma of lotus tea.

CONCLUSIONS

The character and quality of various teas are determined, largely by their aroma constituents and compositions.

The author mentioned the aroma characteristics of the most distinctive several teas in relation with chemical components.

As a whole, the aroma characteristics of various teas seemed to be determined by the balance of several aroma constituents except scented teas.

Black tea is the most widely consumed tea (world consumption of black tea is about 80% of the total tea consumption). Among black teas, there are numerous grades and type of odor notes, and if the aroma character could catch by objective method, the evaluation of black tea and blending proportion will be more easily decided.

Dr. M.A. Gianturco et al.(83) have reported a numerical approach to the correlation between composition and quality of tea aroma. They computerized some selected parameters derived from gas chromatographic data, and got a pretty good correlation with sensory scores.

When the contributory constituents to the distinctive favorite flavor of each kind of tea such as Sri Lanka, Dahjeeling, and Keemun black teas, green tea, pouchong tea, oolong tea, etc., are clarified computerizing method will be employed as the most efficient method for the evaluation of their qualities.

ACKNOWLEDGEMENTS

The author is grateful to Dr. R.L.Wickremasinghe of Tea Research Institute of Sri Lanka for his kindly providing flavory Sri Lanka black tea.

The author wishes to express her indebtedness to her coworkers appeared in the references No.22-48, especially Dr. A. Kobayashi and Dr. Y. Nakatani.

The author is also grateful to Dr. M.A. Gianturco and Dr. M. Koshika of The Coca-Cola Company for their great support and valuable suggestions for describing this paper.

REFERENCES

1. Sanderson, G.W. "Geruch-und Geschmack-Stoffe" ed. Drawert F., Page 65, Verlag Hans Carl Nürnberg (1975)
2. Yamanishi, T., J.Agr.Chem. Soc. Japan, 49, No.9, P.R-1-R9 (1975)
3. Van Romburgh, P., Proc. Acad. Sci. Amsterdam, 22, 758(1920) (Proc. Kon. Ned. Akad. Wetensch, 22, 758 (1920)
4. Takei, S. and Sakato, Y., Bull. The Institute of Physical and Chemical Research(Riken Iho), 12, 13 (1932)
5. Takei, S., Sakato, Y. and Ohno, M., ibid., 13, 116 (1934)
6. Takei, S., Sakato, Y. and Ohno, M., ibid., 13, 1561 (1934)
7. Takei, S., Sakato, Y. and Ohno, M., ibid., 14, 303 (1935)
8. Takei, S., Sakato, Y. and Ohno, M., ibid., 14, 507 (1935)
9. Takei, S., Sakato, Y. and Ohno, M., ibid., 14, 1262 (1935)
10. Takei, S., Sakato, Y. and Ohno, M., ibid., 15, 626 (1936)
11. Takei, S., Sakato, Y. and Ohno, M., ibid., 16, 7 (1937)
12. Takei, S., Sakato, Y. and Ohno, M., ibid., 16, 773 (1937)
13. Takei, S., Sakato, Y. and Ohno, M., ibid., 17, 216 (1938)
14. Takei, S., Sakato, Y. and Ohno, M., ibid., 17, 871 (1938)
15. Yamamoto, R. and Kato, Y., J. Agr. Chem. Soc. of Japan (Nippon Nogei Kagaku Kaishi) 10, 661 (1934)
16. Yamamoto, R. and Kato, Y., ibid. 11, 639 (1935)
17. Yamamoto, R. and Kato, Y., Sci. Papers Inst. Phys. Chem. Research (Tokyo), 21, 122 (1935)
18. Yamamoto, R. and Itoh, K., J.Agr.Chem.Soc.Japan, 13, 736 (1937)
19. Yamamoto, R., Itoh, K. and Chin, H., ibid., 16, 781 (1940)
20. Yamamoto, R., Itoh, K. and Chin, H., ibid., 16, 791 (1940)
21. Yamamoto, R., Itoh, K. and Chin, H., ibid., 16, 800 (1940)
22. Tsujimura, M., Yamanishi, T., Akiyama, R. and Tanaka, S., J. Agr. Chem. Soc. of Japan, 29, 145 (1955)
23. Yamanishi, T., Takagaki, J. and Tsujimura, M., Bull. Agr. Chem. Soc. of Japan, 20, 127 (1956)
24. Yamanishi, T., Takagaki, J., Kurita, H. and Tsujimura, M., ibid., 21, 55 (1957)
25. Kiribuchi, T. and Yamanishi, T., Agr. Biol. Chem., 27, 56 (1963)
26. Yamanishi, T., Kiribuchi, T., Sakai, M., Fujita, N., Ikeda, Y. and Sasa, K., ibid., 27, 193 (1963)
27. Yamanishi, T., Sato, H. and Ohmura, A., ibid., 28, 653(1964)
28. Yamanishi, T., Kiribuchi, T., Mikumo, Y., Sato, H., Ohmura, A., Mine, A. and Kurata, T., ibid., 29, 300 (1965)
29. Kobayashi, A., Sato, H. and Yamanishi, T., ibid., 29, 488 (1965)
30. Kobayashi, A., Sato, H., Arikawa, R. and Yamanishi, T., ibid., 29, 902 (1965)

31. Yamanishi, T., Kobayashi, A., Sato, H., Ohmura, A. and Nakamura, H., ibid. 29, 1016 (1965).
32. Kobayashi, A., Sato, H., Nakamura, H. and Yamanishi T., ibid., 30, 779 (1966)
33. Yamanishi, T., Kobayashi, A., Sato, H., Nakamura, H., Osawa, K., Uchida, A., Mori, S. and Saijo, R., ibid., 30, 784(1966)
34. Yamanishi, T., Kobayashi, A., Uchida, A and Kawashima, Y., ibid., 30, 1102 (1966)
35. Yamanishi, T., Kobayashi, A., Nakamura, H., Uchida A., Mori, S., Osawa, K. and Sasakura, S., ibid., 32, 379 (1968)
36. Nakatani, Y., Sato, S. and Yamanishi, T., ibid., 33, 967 (1969)
37. Yamanishi, T., Nose, M. and Nakatani, Y., ibid., 34, 599 (1970)
38. Sato, S., Kobayashi, A., Nakatani, Y. and Yamanishi, T., ibid. 34, 1355 (1970)
39. Nose, M., Nakatani, Y. and Yamanishi, T., ibid., 35, 261 (1971)
40. Yamanishi, T., Kita, Y., Watanabe, K. and Nakatani Y.,ibid., 36, 1153 (1972)
41. Yamanishi, T., Kawatsu, M., Yokoyama, T. and Nakatani, Y., ibid., 37, 1075 (1973)
42. Yamanishi, T., Shimojo, S., Ukita, M., Kawashima, K. and Nakatani, Y., ibid. 37, 2147 (1973)
43. Yamanishi, T., Uchida, A., Kawashima, Y., Fujinami, Y. and Miyamoto, M., Tea Research Journal, 41, 48 (1974)
44. Nguyen, T.T. and Yamanishi, T., Agr. Biol. Chem., 39, 1263 (1975)
45. Takei, Y., Ishiwata, K. and Yamanishi, T., ibid., 40, 2151 (1976)
46. Aisaka, H., Kosuge, M. and Yamanishi, T., ibid., (in press)
47. Yamanishi, T., Maeda, R. and Kosuge, M., Abstracts of the Scientific Program, VII International Congress of Essential Oils, (October 7-11, 1976) page 89
48. Yamanishi, T. and Tokitomo, Y., Agr. Biol. Chem. (in press)
49. Bokuchava, M.A. and Skobeleva, N.I., Ber. Akad. Wiss,USSR 112, 896 (1957)
50. Skolbeleva, N.I., Petrova, T.A. and Bokuchava, M.A. Soobshch. Akad. Nauk Gruz. SSR 85 (2), 437 (1977)
51. Brandenberger, H. and Müller, S., J. Chromatog., 7,137(1962)
52. Wickremasinghe, R.L. and Swain, T., J. Sci. Food. Agr., 16 57 (1965)
53. Tirimanna, A.S.L. and Wickremasinghe, R.L., Tea Quarterly, 37, 134 (1966)
54. Yamanishi, T., Wickremasinghe, R.L. and Perera, C., ibid., 39, 75 (1968)
55. Wickremasinghe, R.L., Wick, E. and Yamanishi, T., J. Chromatog. 79, 75 (1973)
56. Wickremasinghe, R.L., Phytochemistry, 13, 2057 (1974)

57. Müggler-Chavan, F., Viani, R., Bricout, J., Reymond, D. and Egli, R.H., Helv. Chim. Acta 49, 1763 (1966)
58. Reymond, D., Müggler-Chavan. F., Viani, R., Vuataz, L. and Egli, R.H., J. Chromatog. 4, 28 (1966)
59. Muggler-Chavan, F., Viani, R., Bricout, J., Reymond, D. and Egli, R.H., Chimia, 20, 28 (1966)
60. Bricout, J., Viani, R., Müggler-Chavan, F., Marion, J. P., Reymond, D., and Egli, R.H., Helv. Chim. Acta, 50, 1517(1967)
61. Müggler-Chavan, F., Viani, R., Bricout, J., Marion, J.P., Mechtler, H., Reymond, D. and Egli, R.H., ibid,52,549(1969)
62. Cazenave, P., Horman, I., Müggler-Chavan, F. and Viani, R., ibid., 57, 206 (1974)
63. Cazenave, P. and Horman, I., ibid. 57, 209 (1974)
64. Bondarovich, H.A., Giammarino, A.S., Renner, J.A., Shephard, F.W., Shingler, A.J. and Gianturco, M.A., J. Agr. Food. Chem., 15, 36 (1967)
65. Gianturco, M.A., Biggers, R.E. and Ridling, B.H., ibid., 22, 758 (1974)
66. Saijo, R., Agr. Biol. Chem., 31, 389, 1265 (1967)
67. Saijo, R. and Takeo, T., ibid., 34, 227 (1970)
68. Saijo, R. and Takeo, T.,Plant and Cell Physiol.,13,991(1972)
69. Fukushima, S., Akahori, Y. and Tsuneya, T., Yakugaku Zasshi, 88, 646 (1968)
70. Ina, K. and Sakato, Y., Tetrahedron Letters, 23, 2777(1969)
71. Ina, K. and Eto, H., Agr. Biol. Chem., 35, 962 (1971)
72. Ina, K. and Eto, H., ibid., 36, 1027 (1972)
73. Hatanaka, A. and Harada, T., Phytochem., 12, 234 (1973)
74. Hatanaka, A. and Harada, T., Agr. Biol.Chem., 39, 243(1975)
75. Renold, W., Näf-Müller, R., Keller, U., Willhalm, B. and Ohloff, G., Helv. Chim. Acta., 57, 1301 (1974)
76. Gogiya, V.T. and Tkeshelashvili, C.F., Prikl, Biokhim. Mikrobiol., 8, 600(1972)
77. Tkeshelashvili, C.K. and Gogiya, V.T., ibid., 8, 965(1972)
78. Demole, E., Lederer, E. and Mercier, D., Helv. Chim. Acta., 45, 675 (1962)
79. Winter, M., Malet, G., Pfteiffer, M. and Demole, E., ibid., 45, 1250 (1962)
80. Kaneko, H. and Mita, M., Agr. Biol. Chem., 33, 1525 (1969)
81. Sanderson, G.W. and Graham, H.N., Agr. Food Chem. 21, 576 (1973)
82. Hara, T. and Kubota E., Chagyo Gijutsu Kenkyu (Study of Tea ed. National Research Institute of Tea, Shizuoka, Japan) 50, 68 (1976)
83. Gianturco, M.A., Biggers, R.E. and Ridling, B.H., Agr. Food Chem., 22, 758 (1974)

RELATIONSHIP BETWEEN PHYSICAL AND CHEMICAL ANALYSIS AND TASTE TESTING RESULTS WITH BEERS

Manfred Moll
That Vinh
Roland Flayeux

Centre de Recherche et Développement
TEPRAL
(Branche Alimentaire de BSN)
Champigneulles, France

I. INTRODUCTION

For various reasons the sensory analysis of beer was largely neglected until 1970 because of :

- the difficulty of comparison of tasting results between centres ;
- insufficient definition of flavour terms ;
- lack of liaison with other food industries in regard to flavour testing ;
- lack of training of tasters.

Several publications during recent years have allowed a degree of credibility to be attached to flavour testing in breweries (1-7). Statistical interpretation presents some problems and a few contributions (8-14) mention applications of statistical analysis of factors such as analysis by principal components or factorial discriminant analysis.

In our study, we consider the relationships between physical and chemical results and those of tasting utilizing the above statistical methods in :

- a 5 litres microbrewery ;
- a 10 hl pilot brewery ;
- a commercial brewery producting batches of volume more than 300 hl.

ISBN 0-12-169060-1

II. RELATIONSHIPS BETWEEN PHYSICAL/CHEMICAL ANALYSES OF BEERS FROM MICROBREWING AND TASTING RESULTS

The use of micro-brewing has the advantage of costing little in raw material and energy. However, it should be emphasised that the validity of an extrapolation of the results obtained on this scale to the pilot or industrial scale is limited. This technique serves, above all, to test feasibility.

In earlier studies (12,13) it was shown over four years, using pure varieties of barley cultivated in different regions of France, malted and brewed in identical fashion, that it would be possible by statistical treatment of the analytical data obtained from the beers to distinguish the varieties of barley used (figure 1) and the general areas in which they were cultivated (figure 2).

The physical and chemical factors which permit this classification by analysis of principal components are :

- nitrogeneous compounds (total nitrogen, free amino-nitrogen, amino-acids)
- volatile compounds of carbon disulphide extracts (alcohols, esters, fatty acids)
- mineral elements (K, Mg, Ca, Na)
- various classical factors : residual reducing sugars, viscosity, bitterness, colour.

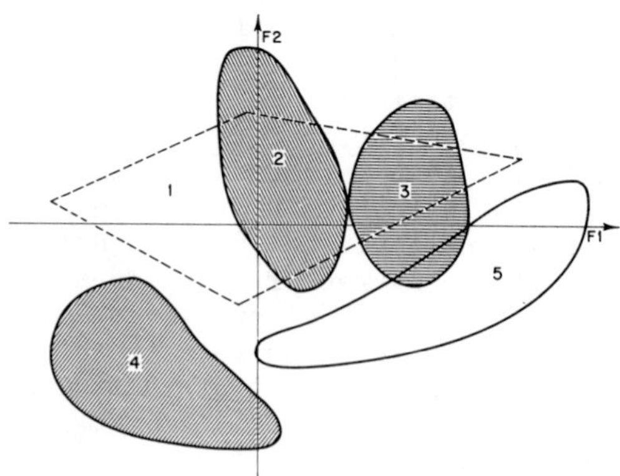

Fig. 1. Use of analysis by principal components to distinguish barley variety : Carina (1), Berenice (2), Berac (3), Julia (4), six row barley (5).

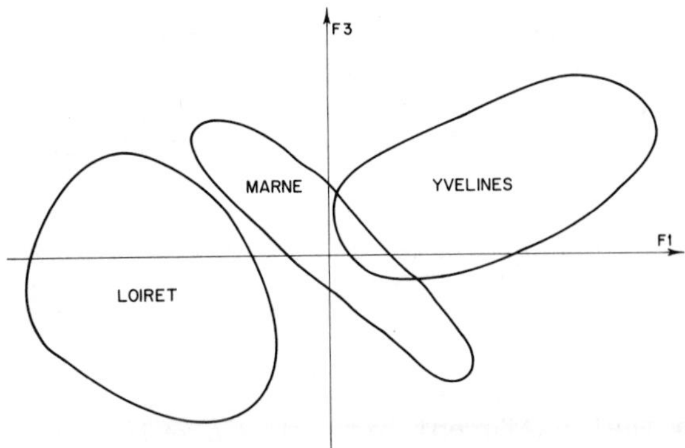

Fig. 2. Use of analysis by principal components to distinguish places of growth of the barley : Loiret, Marne, Yvelines.

Figure 3 shows that there are three main groups of variables near the circumference of the correlation circle. The extract and the sugars are oppositely situated to the volatiles, while the group based on nitrogenous substances is practically independent of the other groups.

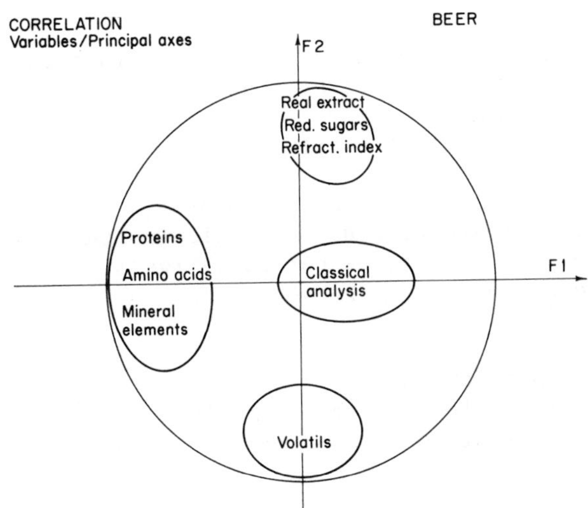

Fig. 3. Correlation circle for beer analyses (analysis by principal components).

For classification of beers as good or bad with the help of factorial discriminant analysis, the same families of factors are concerned. With twelve physical/chemical characteristics of beer, 90 % of beers are correctly classified.

III. RELATIONSHIP BETWEEN ANALYSES OF BEERS FROM THE PILOT PLANT AND TASTING RESULTS

A study has been made in the 10 hl pilot brewery which was designed to test different bittering agents : resin extracts and various isomerized extracts. This pilot installation of 10 hl capacity represents the smallest unit allowing a beer to be made of quality identical with that produced industrially and thus permitting extrapolation of results to the industrial scale. Forty brews were made and the addition of bittering agents was done according to the recommendations of the suppliers. The beers were analysed and tasted at bottling and then after six months to estimate flavour stability.

Multidiscriminant analysis applied to the forty-four factors measured, in order to separate the two types of production, shows that the factors important for discrimination are :

- bitterness
- tasting results after six months
- concentrations of volatiles at bottling and after six months
- mineral salts

On the other hand, colloidal stability factors intervene very little and other classical analytical factors such as proteins not at all.

Figure 4 represents the relative importance of these factors.

In order better to define the important factors, a step-by-step factorial discriminant analysis was carried out, which allowed 39 of 40 beers to be correctly identified according to their method of hopping by :

- bitterness (EBU)
- colour (EBC)
- ethyl caprylate content
- caprylic acid content after six months
- beta-phenylethanol content after six months
- pH
- viscosity
- overall flavour impression after six months

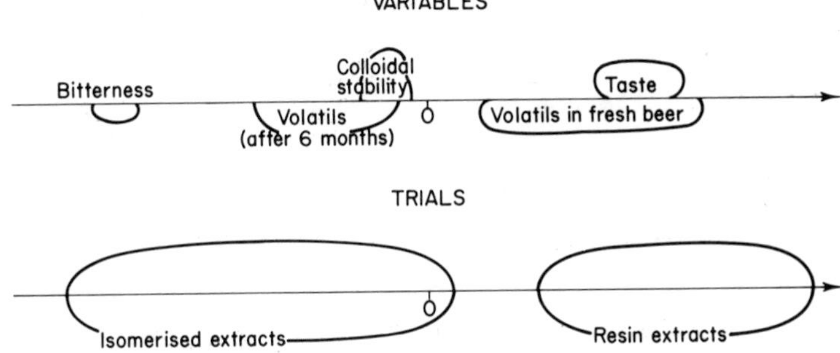

Fig. 4. Use of factorial discriminant analysis to distinguish beers made with resin extracts of hops from those made with pre-isomerised extracts.

This analysis allowed, with the help of the linear discrimination function found, to place the results of each trial or control and to relate to a group any trial on which eight analyses had been carried out ; in figure 4, we show a representation of the two groups on the discrimination axis.

By analysis by principal components, there was no grouping as a function of the type of hop extract used, i.e., the effect of type of extract is less than the effect of the dispersion between the repetitions of brewings with a given extract.

IV. RELATIONSHIPS BETWEEN PHYSICAL/CHEMICAL PARAMETERS OF INDUSTRIAL BEERS AND TASTING RESULTS

In each of eight breweries, six brews in succession were made from each of four varieties of barley malted in six malthouses. Twenty tastings were carried out on these beers.

A. Classification of Beers by Tasting

Analysis by principal components allows correct visualization of the variables and the samples using three principal axes.

The discriminating power (or inertia absorbed) by the different axes are :

axis 1	58.1 %
axis 2	17.8 %
axis 3	9.1 %
	85.0 %

To put this result otherwise, 85 % of the phenomenon is explained by these three axes.

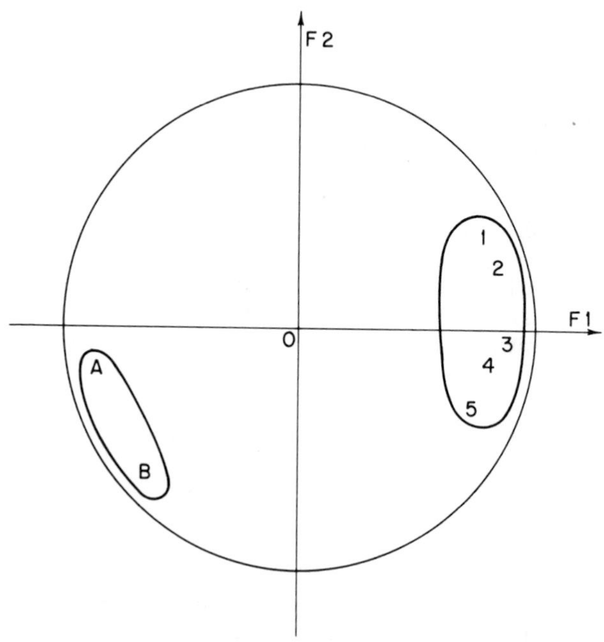

Fig. 5. Correlation circle for tasting test results (analysis by principal components) A : defect ; B : off flavor 1 : Overall impression ; 2 : Odor ; 3 : Bitterness ; 4 : Taste 5 : Mouthfeel.

Figure 5 shows the correlations of tasting variables using axes 1 and 2 (the correlation circle is shown). We have a group of variables representing the quality of beer (e.g. bitterness, taste, odour, general impression) strongly correlated with axis 1 and in opposition on this axis with off-flavours and flavour defects. This confirms previous results (11).

The projection of the results on the first principal plane (constituted by axes 1 and 2) allows the beers to be correctly separated into three groups : good, average and poor (figure 6) It should be noted that in this classification there is no relationship to individual malthouse, brewery or control. However, on axis 1 there is separation of the varieties Carina and Berac to the right and Malta and Sonia to the left.

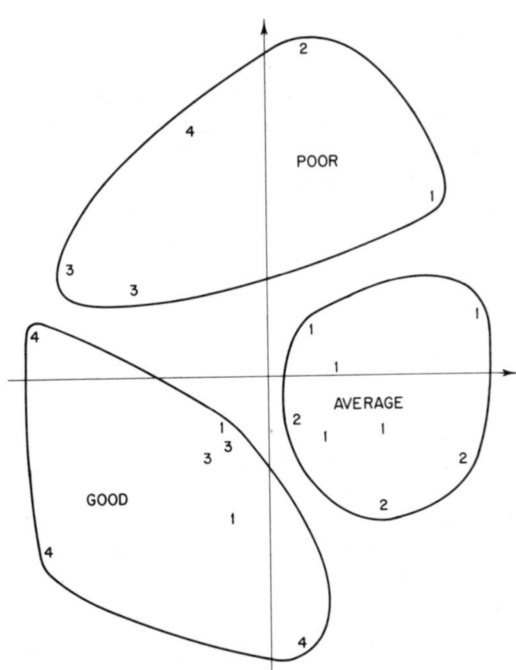

Fig. 6. Use of analysis by principal components to distinguish beers having poor, average and good overall flavours. Barley varieties : Carina (1), Berac (2), Sonia (3), Malta (4).

B. Comparison Between Beer Properties and Tastings

As a result of factorial discriminant analysis, it was possible to classify beers correctly as good, average or poor with only nine physical and chemical factors of 22 examined, as follows :

- colloidal stability (7 days at 40°C/1 day at 0°C)
- cold sensitivity (24 hours at 0°C)
- brightness at 12°C
- six months test
- beta-phenylethanol
- ethyl caprylate
- isoamyl acetate
- isobutanol
- foam stability

V. CONCLUSION

Statistical analysis of data, especially by analysis by principal components and factorial discriminant analysis permits it to be shown that there are links between certain physical/chemical factors and the results of tastings tests applied to beers. These results have been confirmed over several years and the same groups of factors appear both in the examination of beers from microbrewing on the one hand and pilot and industrial-scale brewing on the other.

REFERENCES

1. Brown, D.G.W., Clapperton, J.F., Meilgaard, M.C. and Moll, M., J. Amer. Soc. Brew. Chem. 36:73 (1978).
2. Clapperton, J.F., Dalgliesh, C.E. and Meilgaard, M.C. Master Brewers Assoc. Amer. Tech. Quart. 12:273 (1975).
3. Clapperton, J.F., Dalgliesh, C.E. and Meilgaard, M.C., in "The practical brewer", 2nd ed. (H.M. Broderick ed.) p. 433, Master Brewers Assoc. Amer., Madison, 1977.
4. Dalgliesh, C.E. Proc. Eur. Brew. Conv., Amsterdam, 623, 1977.
5. Ferdinandus, A., Oosterrom-Kleijngeld, I. and Runneboom, A.J.M. Master Brewers Assoc. Amer. Tech. Quart., 7:210 (1970).

6. Lindsay, R.C., in "Objective Measurements of flavor quality of beer" (R.A. Scanlan ed.), p. 89, Amer. Chem. Soc., Washington, Series 51, 1977.
7. Meilgaard, M.C. in "Geruch und Geschmackstoffe" (F. Drawert ed.) p. 211, Verl. Hans Carl, Nürnberg, 1975.
8. Brown, D.G.W., Clapperton, J.F. and Dalgliesh, C.E. Proc. Amer. Soc. Brew. Chem., 32:1 (1974).
9. Helbert, J.R. and Hoff, J.T. Proc. Amer. Soc. Brew. Chem. 32 ; 43 (1974).
10. Hoff, J.T. and Herwig, W.C.J., J. Amer. Soc. Brew. Chem. 34:1 (1976).
11. Moll, M., Flayeux, R., Vinh, T. and Noel, J.P., Bios, 5:328 (1974).
12. Moll, M., Flayeux, R., Vinh, T. and Bastin, M., Proc. Eur. Brew. Conv., Nice, 39 (1975).
13. Moll, M., Flayeux, R., Vinh, T., Annales de nutrition in press.
14. Noel, J.P. and Bauer, G., Bios, 5:355 (1974).

THE AROMA COMPOSITION OF DISTILLED BEVERAGES AND THE PERCEIVED AROMA OF WHISKY

Paula Jounela-Eriksson

Research Laboratories of the State Alcohol Monopoly (Alko), Helsinki, Finland

The more than 400 compounds identified in the aroma of alcoholic beverages include alcohols, fatty acids and esters, lactones and carbonyl, phenolic, sulphur and nitrogen compounds. Although yeast produces largely the same aroma compounds irrespective of sugar source, some characteristic components are produced during fermentation, distillation and maturing. E.g. 2-ethyl-3-methylbutyric acid seems indicative of rum and β-methyl-γ-octalactone and some phenols of oak casks, while the composition of sulphur compounds in neutral grain spirits reflects the effectiveness of the distillation. The increasing number of aroma components makes it difficult to determine a characteristic whisky aroma, a perceived aroma cannot be described in terms of quantitative chemical composition. Sensory assessment combined with quantitative gas chromatography has been used to describe the strength and quality of whisky aroma. The core of the aroma was delineated with a whisky model and using odor thresholds and GLC results. The dependence of whisky quality on the quantities of known important aroma compounds was also studied in different brands of genuine whisky.

I. INTRODUCTION

Alcoholic beverages are stimulants, and therefore the flavor transmitted by our senses has traditionally been one of the most significant qualitative factors in their use. Since the flavor of distilled beverages in particular is mainly

ISBN 0-12-169060-1

based on odor sensations caused by volatile compounds, this paper focuses only on aroma, the most general concept used even in commonplace discussion. Intensive aroma research recently carried out with increasingly refined methods continuously expands the number of identified aroma components. On the other hand, each of us can easily distinguish between cognac and whisky and even various types of whisky. In spite of this the gap between the abundance of chemically analyzed components and the perceived aroma impression is still wide. Specification of the aroma of a beverage does not succeed even by means of the quantitative composition of known aroma compounds but only becomes more difficult as the aroma composition becomes more precisely known. Knowledge of aroma composition with all its related analytical and methodological problems is, however, the necessary basis in endeavouring to analyze how the sensory experience of a human being, meaning aroma, is formed by the results of chemical analysis, to mean aroma components.

II. THE CHEMICAL COMPOSITION OF ALCOHOLIC BEVERAGES

The aroma of the most commonly consumed alcoholic beverages already consist of several hundreds of chemical compounds (1). Fusel alcohols, acids and fatty acid esters quantitatively form the greatest part of aroma, but also carbonyl, phenolic, sulphur and nitrogen compounds as well as lactones, acetals, hydrocarbons, sugars and some other unclassified compounds rank among aroma components. Yeast and fermentation conditions determine the most important aroma components in beverages produced by fermentation. Approximately the same compounds appear in the aroma of distilled beverages (Fig. 1) as well as in beer and wines irrespective of the raw materials used (3,4). The choice of distillation methods allows, however, a high degree of regulation of the quantitative concentrations of aroma components in particular. Changes also occur during the phases of maturation, when new compounds may be formed and others consumed in chemical reactions.

A. Fusel Alcohols

The fusel alcohols form quantitatively the largest group of aroma components. The average concentration of fusel alcohols is 1.5 g/l in cognacs, about 1 g/l in whiskies and approximately 0.6 g/l in rums (5). Table 1 presents the pro-

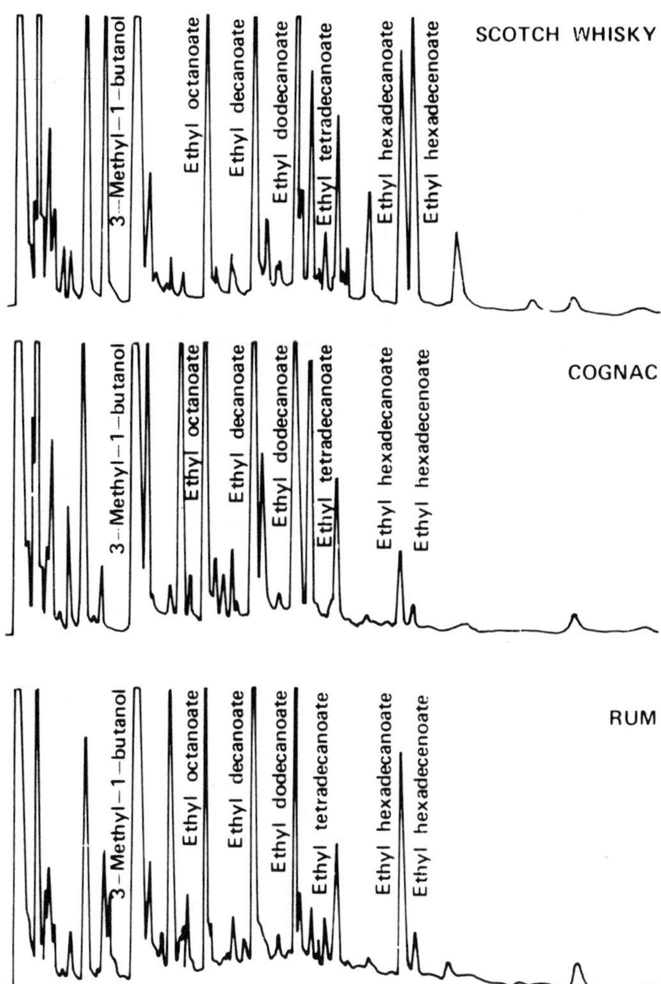

FIGURE 1. Gas chromatograms of the major aroma compounds isolated from Scotch whisky, cognac and rum (2).

portions of the most important fusel alcohols in various beverages. The formation of fusel alcohols depends on fermentation conditions, such as yeast nutrients, temperature and the type of the yeast used, but very little on the raw materials of a beverage (8). 2-Butanol, most probably produced from 2,3 butanediol through bacterial fermentation, is a fusel alcohol typical of rum only. Jamaican rum in turn contains more 1-propanol than other distilled beverages (9).

TABLE 1. Proportions of the most important fusel alcohols in some beverages (6,7)

Beverage	3-Methyl-1-butanol %	2-Methyl-1-butanol %	2-Methyl-1-propanol %	1-Butanol %	2-Butanol %	1-Propanol %
Jamaican rum	54	10	15	1	5	15
Martinique rum	67	12	15	traces	traces	6
Scotch whisky	42	15	37	traces	0	6
French red wine	64	15	16	1	0	4
French white wine	58	15	23	1	0	3
Finnish beer	49	19	13	-	0	19

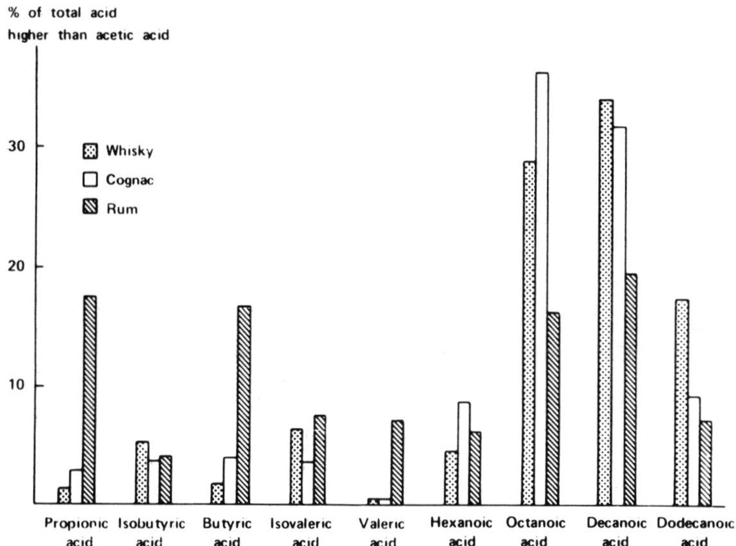

FIGURE 2. The relative proportions of acids higher than acetic acid in whisky, cognac and rum (9)

B. Fatty Acids

Yeast also synthesizes the fatty acids present in alcoholic beverages. The concentration of acids is a good 100 mg/l in Scotch whisky, while Bourbon whisky contains at least four times this level. In cognacs it is approximately 200 mg/l and can even attain 600 mg/l in rum with a strong aroma (10). Acetic acid is quantitatively the largest component of the steam-volatile acids in alcoholic beverages. Its proportion of the quantity of titratable acids varies from 50 percent in Scotch whisky and 80 percent in rum up to 90 percent in Bourbon whisky. The proportions of the other volatile acids presented in Figure 2 indicate that rum has relatively more propionic, butyric and valeric acid than whisky and cognac, but on the other hand less octanoic acids and especially higher fatty acids. The acids in alcoholic beverages are for the most part saturated, straight chain acids. Palmitoleic acid is the only significant unsaturated acid and is nearly solely found in Scotch whisky. 2-Ethyl-3-methylbutyric acid, encountered up till now in no other alcoholic beverages, has been recently identified in rum at the Research Laboratory of the Finnish State Alcohol Monopoly (Alko) (11).

C. Fatty Acid Esters

Fatty acid esters form an essential group of aroma components causing aroma intensity of a beverage to vary according to their concentration. For instance light-flavored Puerto Rico rum contains only an approximate 40 mg/l concentration of esters, whereas the concentration of esters in heavy-flavored Martinique rum can be as high as 600 mg/l (9). The perceived aroma intensity would seem to correlate to the ester concentration while the qualitative aroma composition is highly similar.

The main ester component in alcoholic beverages is ethyl acetate, although ethyl, iso-butyl and 3-methylbutyl esters of short-chain fatty acids, called "fruit esters" due to their pleasing aromas, also appear. Quantitatively significant components are ethyl esters of octanoic, decanoic and dodecanoic acids. Of the higher esters ethyl hexadecenoate is interesting in that significant amounts of it have been found mainly in Scotch whisky (Fig. 1). The yeast used has a great influence on the production of esters in the fermentation process. The ester content of distilled beverages also depends on whether or not the yeast is present at the time of distillation. If distillation occurs in the presence of yeast, the ethyl ester concentrations of at least decanoic, dodecanoic and hexadecenoic acids increase (12).

D. Carbonyl Compounds

The aldehyde content of different beverages can vary considerably. Aldehydes are intermediate products in the formation of fusel alcohols, and thus the aldehyde content usually is at a maximum in fermentation medium. The mash for Scotch whisky contains some 160 mg/l aldehydes after fermentation (13), whereas the total aldehyde content of Scotch grain whisky is approximately 10 mg/l and that of malt whisky 30-40 mg/l after distillation. Aldehyde content increases during ageing, and is about 40-50 mg/l in unblended, mature malt whisky, as an example (2). Brandies contain some 150 mg/l of aldehydes (13). During the ageing of distillates with an alcohol content of over 50 % (v/v), the aldehydes present react with ethanol to form acetales, thus softening the pungent odor of the aldehydes. Table 2 presents the proportions of aldehydes determined by gas chromatography in Scotch whisky as well as the concentration of the vicinal diketones also produced in fermentation. Their content decreases during the maturation of the distillates. Also other aldehydes, such

TABLE 2. Proportions of the most volatile aldehydes and average concentrations of the vicinal diketones in the Scotch malt and grain whisky (14,19)

	Malt whisky %	Grain whisky %
Acetaldehyde	63	58
Propionaldehyde	1.0	3
iso-Butyraldehyde	20	26
Valeraldehyde	0.3	0
iso-Valeraldehyde	15	13
Hexanal	0.2	0
	mg/1	mg/1
Diacetyl	1.6	0.35
2,3 Pentadion	0.2	0.05

as acrolein, furfural and several ketones have additionally been identified, for instance in whiskies (14).

E. Phenolic compounds and lactones

The phenolic compounds appearing in alcoholic beverages can be formed in two different ways, either during fermentation or by alcoholic extraction from the oak casks during maturing. In addition to phenol and guaiacol, cresols, eugenol and vanillin have been found in the ethanol extract of American white oak chips, in which eugenol was observed to be the main component (15). Table 3 presents concentrations of phenol compounds in some different types of whiskies. Phenol and cresols seem to be typical of Scotch whisky, whereas 4-ethylphenol and 4-ethylguaiacol was observed to be the main phenolic compounds in Bourbon and Canadian whisky.

Of the lactones found in alcoholic beverages, at least β-methyl-γ-octalactone has its origin in wood. Its *trans*-diastereomer has been discovered both in distilled beverages and wines matured in barrels, and its *cis*-diastereomer at least in whiskies and rums (15).

F. Sulphur Compounds

Nearly all the alcoholic beverages contain some sulphur compounds. In distilled beverages the most common compounds

TABLE 3. Concentrations of phenols in some brands of whisky(2)

	Scotch, malt µg/l	Scotch, blended µg/l	Bourbon whisky µg/l	Canadian whisky µg/l	Irish whisky µg/l
Phenol	110	190	traces	-	traces
o-Cresol	75	75	-	-	-
m-Cresol	30	35	-	-	-
p-Cresol	50	50	-	-	-
Guaiacol	90	120	90	-	-
4-Ethylphenol	70	40	390	100	35
Unknown	30	20	-	20	-
4-Ethylguaiacol	100	30	360	80	60
Eugenol	50	100	195	50	55
	mg/l	mg/l	mg/l	mg/l	mg/l
Total phenols (analyzed spectrophotometrically)	100	95	210	110	80

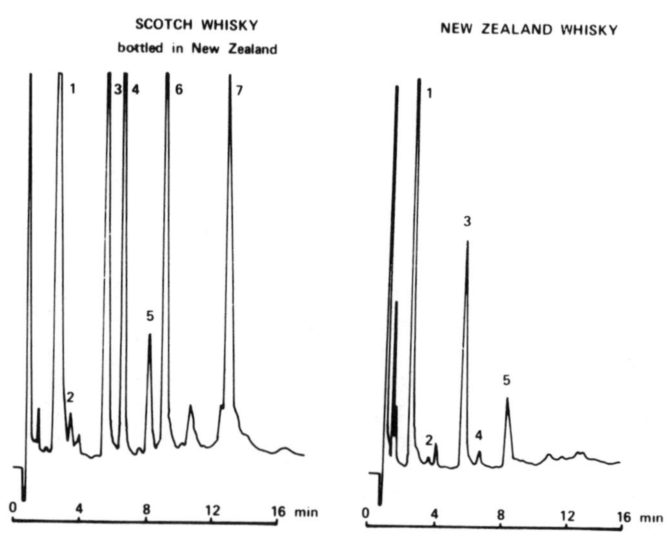

FIGURE 3. Gas chromatograms of sulphur compounds in Scotch and New Zealand whisky. 1 Dimethyl disulphide, 2 methylethyl disulphide, 3 diethyl sulphite, 4 dimethyl trisulphide, 5 internal standard (di-isobutyl disulphide), 6 and 7 unknown (17,18).

are ethanethiol, diethyl sulphide and dimethyl and diethyl disulphides (16). Approximately 100 μg/l of dimethyl disulphide has been found in blended Scotch whisky using a flame photometric detector at Alko's Research Laboratory (17,18). Additionally small quantities of methylethyl disulphide, dimethyl trisulphide and diethyl sulphite have been detected (Fig. 3). On the other hand, New Zealand Bourbon type whisky contains fewer sulphur compounds.

Sulphur compounds play a more significant role in pure spirits since they often have an offensive odor and low threshold values. Most of the sulphur compounds in raw grain spirit are removed with the predistillate during rectification, but some can still be found in pure spirits. Sulphur compounds primarily of a sulphide type in spirits originate in pyrolytic reactions occurring during distillation from sulphur-containing amino acids and other proteins. Table 4 presents concentrations of sulphur compounds in vodkas from different countries (19). Since sulphides have an unpleasant odor, their concentration in a beverage also reveals the effectiveness of the distillation.

TABLE 4. Concentrations (μg/l) of sulphur compounds analysed in some brands of vodka (19)

	Finnish vodka	Russian vodka	Polish vodka	American vodka	Swedish vodka
Dimethyl disulphide	-	0.31	1.10	0.45	1.10
Methylethyl disulphide	0.11	0.11	0.49	0.06	0.18
Diethyl disulphide	-	-	0.06	-	0.15
Dimethyl trisulphide	-	-	0.07	-	0.05
Total sulphides	0.11	0.42	1.72	0.51	1.48

III. THE PERCEIVED AROMA OF WHISKY

A. The Core of Whisky Aroma

Although only a few examples of volatile aroma components in alcoholic beverages are presented above, they already indicate that the aroma of an alcoholic beverage is always a mixture of several aroma components. The greater the number of identified compounds, the more difficult it is to determine what components are essential from the standpoint of the perceived aroma. Aroma compounds and their concentrations are known, but concentrations do not determine the perceived aroma intensity. Threshold values are useful basic units for measuring sensation intensity if they are based on iterative and statistically controlled methods (20). Compounds essential to the perceived aroma can be isolated by comparing the concentrations and threshold values. In the total aroma however, interactions of the aroma components might exert a greater influence on the significance of some components than the mere relation between the concentration and threshold value would indicate. By using a whisky imitation of Scotch whisky made from 9 carbonyls, 13 alcohols, 21 acids, 24 esters and pure grain spirit, it was not only endeavoured to chart possible interactions but also to discover the compounds and compound mixtures most essential to whisky aroma (21,22). Table 5 lists compounds which these tests revealed to be most important in the whisky imitation. The contribution of each compound mixture to the total aroma and correspondingly the contribution of a single compound in a particular mixture or fraction have been measured with relative values based on the concentration and threshold. According to these results, the most significant compounds for perceived aroma intensity would be 3-methyl-1-butanol, ethyl acetate and the ethyl esters of C_6-C_{12} acids as well as some aldehydes and diacetyl. The whisky imitation in fact lacked some compounds, such as all the phenols, and, by using these relative values, the contribution of carbonyl compounds turned out to be obviously too great due to their extremely low threshold levels.

The concentrations of many aroma components in different production batches of the same whisky vary nearly to the same extent as do concentrations in different brands of whisky. This posed the question of what magnitude of an increase in the concentration of aroma components can be discerned in the perceived aroma intensity. Tests with the whisky imitation revealed that approximately a threefold concentration of fusel alcohols and esters as well as about a fourfold concentration of acids and carbonyl compounds was distinguished from the

TABLE 5. Contribution of aroma compounds to the strength of the total aroma in the whisky imitation and in the fractions of compound, respectively (20,21).

	Contribution to strength of aroma in whisky imitation %	Contribution to strength of aroma in fraction %
Alcoholc	4.4	
3-Methyl-1-butanol		82
Other alcohols		18
Acids	3.8	
Esters	28.5	
Ester mixture of		88
Ethyl acetate		
Ethyl hexanoate		
Ethyl octanoate		
Ethyl decanoate		
Ethyl dodecanoate		
3-Methylbutylacetate		
Other esters		12
Carbonyl compounds	63.3	
Carbonyl mixture of		97
iso-Butyraldehyde		
Butyraldehyde		
iso-Valeraldehyde		
Valeraldehyde		
Diacetyl		
Other carbonyl compounds		3

original whisky imitation in a triangular test with statistical significance (23). On the other hand, preliminary tests with a few genuine Scotch whiskies indicated that, when added to whisky, certain esters serving an important contribution in the aroma of the whisky imitation, can be distinguished from genuine whiskies already at 1.25-1.5 times higher concentrations.

B. The Perceived Strength and Quality of Whisky Aroma

The quality of aroma is as important as its strength for a positive sensory response. In the quantitative sensory assessment the so-called magnitude estimation method can be used. In this method panelists are instructed to assign numbers to a sequence of stimuli so that the ratios of the numbers reflect the ratios of the perceived sensory intensities or relative preferences (24,25). In psychophysical experiments in which ratio scales have their origin, one attempts to assess a graded series of stimuli, which vary in one or more physically measurable attributes. The physical measurements are related to the perceptual assessments by a series of equations. The ratio scales usually yield a curvilinear function which becomes linear in log-log coordinates

$$\log S = n \log C + \log k \quad \text{or} \quad S = kC^n$$

where S = sensory response
C = concentration

The power function form for the magnitude estimation function has been shown empirically in psychophysical experiments.

When applying the magnitude estimation method to aroma reseach, endeavours are made to follow the dependency of a perceived aroma intensity or a particular qualitative property on the concentrations of components producing a sensory response. In this study the method was used to analyze how for example the components forming the core of aroma of the artificial whisky, contribute to the aroma of genuine whiskies. Table 6 indicates the concentrations of some fusel alcohols, fatty acid esters and carbonyl compounds in three different types of whisky. Figure 4 depicts the relationship of the total aroma intensity and a typical whisky aroma to concentrations of the components mentioned above, beginning with genuine whisky and increasing the concentration of components at a ratio of 1:1.5.

When lines are fitted to the functions using regression analysis, their slopes as estimates for the exponents of the power function indicate that the components selected for the graphs in Figure 4 also have an importance in the aromas of genuine whiskies. Of fusel alcohols, the contribution of 3-methyl-1-butanol to aroma intensity is similar in all three whisky types. The slope is on the average 0.6, so a fivefold increase in concentration results in a 2.6 times increase in perceived aroma intensity. The effect of 3-methyl-1-butanol is more pronounced than that of 2-methyl-1-butanol and 2-methyl-1-propanol combined. 3-Methyl-1-butanol has an effect on the typical whisky aroma with a contribution of the same

TABLE 6. Concentrations of the aroma compounds studied with magnitude estimation method in three types of whisky (2).

Compound	Canadian whisky mg/1	Bourbon whisky mg/1	Scotch whisky mg/1
3-Methyl-1-butanol	150	1050	230
2-Methyl-1-propanol	65	330	315
2-Methyl-1-butanol	62	360	80
Ethyl acetate	66	230	86
Ethyl hexanoate	0.5	3.6	0.8
Ethyl octanoate	0.6	16.2	2.8
Ethyl decanoate	0.9	28.5	8.8
Ethyl dodecanoate	0.5	12.4	8.0
3-Methylbutylacetate	0.3	5.1	5.6
Total	2.8	65.8	26.0
Acetaldehyde	13.2	33.0	18.7
iso-Butyraldehyde	5.5	13.8	7.8
iso-Valeraldehyde	3.3	8.3	4.5
Total	22.0	55.1	31.0
Diacetyl	0.5	1.0	0.8

magnitude as it has on aroma intensity, except for Bourbon whisky where its contribution is lower.

The contribution of esters and especially ethyl esters of C_6-C_{12} acids to the aroma intensity of whisky is the highest. Considerable differences exist between the different whisky types so that a slope of 1.2 for the ester mixture in Bourbon whisky is nearly twice that of the slopes for Canadian and Scotch whiskies. The slopes of these whiskies do not greatly differ from each other, regardless of a tenfold difference in the ester concentrations. Ethyl acetate alone seems to have a part in the aroma intensity primarily in Bourbon and Canadian whiskies. Conversely, typical whisky aroma seems to disappear only in Scotch whisky as rapidly as ethyl esters of C_6-C_{12} acids affect aroma intensity. The diagrams of the ester mixture additionally indicate that an 1.5-fold increase in the concentration of esters in Bourbon and Scotch whiskies causes a change in both perceived aroma intensity and quality. Canadian whisky requires a 2.3-fold increase of esters.

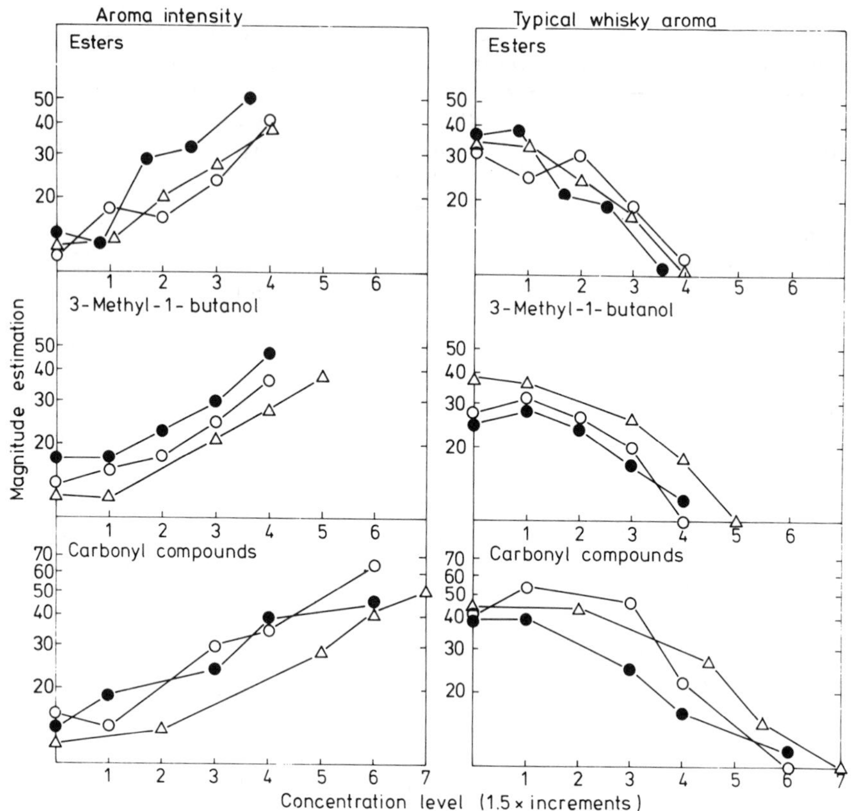

FIGURE 4. Relation between concentration levels of some aroma compounds in Bourbon (●-●), Canadian (○-○) and Scotch whisky (△-△) and perceived aroma intensity and typical whisky aroma estimated by magnitude estimation. Composition of ester and carbonyl mixtures are given in Table 6 (25).

The carbonyl compounds seem in general to have the highest contribution in Canadian and the lowest one in Scotch whisky, both to the aroma intensity and to the typical whisky aroma, the values of the slopes varying 0.60 - 0.45. Nearly 3-fold increment of the carbonyls is needed before a change is observed in the typical aroma of Scotch whisky, but in Canadian whisky their influence is nearly the same as that which 3-methyl-1-butanol has in this whisky. The typical aroma of Bourbon whisky also disappear more rapidly when adding the carbonyls than do the increments of 3-methyl-1-butanol in this whisky.

The conclusions presented above, concerning the dependence of the typical aroma and its intensity in different types of whiskies upon certain aroma compounds, have been drawn from preliminary results. The utility of the method - for instance how well panelists can actually utilise ratio scales - is to be tested and more complete results are to be published at a later date. These results relate the contribution of various aroma components in perceived whisky aroma and also aim to obtain data important in the technology of whisky on the inter-relations of various component groups such as fusel alcohols and esters in different types of whisky distillates.

REFERENCES

1. Kahn, J. H., *J. Ass. Offic. Anal. Chem. 52*: 1166 (1969).
2. Lehtonen, M. unpublished.
3. Suomalainen, H., Nykänen, L. and Eriksson, K., *Amer. J. Enol. Viticult. 25*: 179 (1974).
4. Suomalainen, H., *Ann. Technol. Agr. 24*: 453 (1975).
5. Suomalainen, H. and Hykänen, L., *Naeringsmiddelindustrien 23*: 15 (1970).
6. Sihto, E., Nykänen, L. and Suomalainen, H., *Tek. Kem. Aikak. 19*: 753 (1962).
7. Arkima, V., *Monatsschr. Brau. 21*: 25 (1968).
8. Suomalainen, H. and Nykänen, L., *Wallerstein Lab. Commun. 35*: 185 (1972).
9. Lehtonen, M. and Suomalainen, H., in "Economic Microbiology, Vol. 1. Alcoholic Beverages" (A. H. Rose, ed.) p. 595. Academic Press, London 1977.
10. Nykänen, L. Puputti, E. and Suomalainen, H., *J. Food. Sci. 33*: 88 (1968).
11. Lehtonen, M. J., Gref, B. K., Puputti, E. V. and Suomalainen, H., *J. Agr. Food Chem. 25*: 953 (1977).
12. Suomalainen, H. and Nykänen, L., *J. Inst. Brew., London 72*: 469 (1966).
13. Suomalainen, H. and Nykänen, L., XXXVI Congress International de Chimie Industrielle, Bruxelles 1966, *Compt. Rend. 3*: 807 (1967).
14. Suomalainen, H. and Nykänen, L., *Process Biochem. 5* (7): 13 (1970).
15. Suomalainen, H. and Lehtonen, M., *Kemia-Kemi 3*: 69 (1976).
16. Dellweg, H., Miglio, G. and Niefind, H-J., *Branntweinwirtschaft 109*: 445 (1969).
17. Ronkainen, P., Denslow, J. and Leppänen, O., *J. Chromatogr. Sci. 11*: 384 (1973).
18. Leppänen, O. and Forss, D. A., unpublished.

19. Leppänen, O., Ronkainen, P. and Denslow, J., unpublished.
20. Salo, P., *J. Food Sci.* *35*: 95 (1970).
21. Salo, P., Nykänen, L. and Suomalainen, H., *J. Food Sci.* *37*: 394 (1972).
22. Salo, P., *Lebensm. Wiss. Technol.* *6* (2): 52 (1973).
23. Jounela-Eriksson, P., in "Olfaction and Taste VI, Paris 1977", Proc. (J. Le Magnen and P. MacLeod eds.). Information Retrieval, London 1977.
24. Moskowitz, H. R., *Food Technol.* *28* (11): 16 (1974).
25. Moskowitz, H. R., *Tech. Quart., Master Brew. Ass. Amer.* *14*: 111 (1977).
26. Jounela-Eriksson, P., unpublished.

WINE FLAVOR

Pascal RIBEREAU-GAYON

Institut d'OEnologie
Université de Bordeaux II
33405 Talence
France

I - INTRODUCTION

Wine is the result of the transformation of a natural product, the grape, by alcoholic fermentation. It is quite remarkable to note that with this one term we group a variety of products whose organoleptic characteristics, or flavor, are extremely different; for instance there is little resemblance between a Champagne and a fine Bordeaux red wine, or between a Sherry, a Port wine and a Moselle wine.

Another characteristic of vinicultural production, unlike that of any other food or drink, is the vast hierarchy of quality and price that exists, differentiating the various types of production on one hand, and wines of the same type on the other.

This diversity of course expresses differences in chemical composition responsible for the differences in odor and taste; but they are not always easily distinguished by classical analyses. PEYNAUD (1970) points out the example of two wines having pratically the same basic analysis (alcohol, acidity, extract...) although the price of one (Medoc grand cru classé) was twenty times higher than the other (ordinary table wine); an

ISBN 0-12-169060-1

even larger scale of prices can exist.

In order to interpret the differences in odor and in taste which justify such a large price range, it is necessary to explain the composition of wine and the causes of its modification in more detail. Numerous natural factors are involved in the constitution of wine : a) particular properties of the variety; b) the state of maturity; c) eventual intervention of parasites which develop on the grape, particularly *Botrytis cinerea*, the fungus which causes rot; d) conditions of the alcoholic fermentation by yeast; e) in certain cases, the intervention of malic acid fermentation by lactic acid bacteria; f) and interventions, always to be feared, of chemical or bacteriological spoilage.

Thus we understand the complexity of the chemical composition of wine, a product which has probably given rise to a greater amount of research and analytical study than any other food or drink. Although our knowledge is still very fragmentary, we will try to indicate the current ideas concerning the taste and aroma of wine.

II - THE TASTE OF WINE

A - The equilibrium of tastes

The four basic tastes are found in wine (PEYNAUD, 1970; RIBEREAU-GAYON et al., 1975 b) : a) the acid taste comes from the numerous organic acids whose free functions correspond to 100 meq/l (N/10); b) the sweet taste which is found even in wines which contain no sugar is due to the alcohol and glycerol (8 to 10 g/l); c) the salty taste is due to 2 to 4 g/l of mineral salts; d) the bitter taste comes primarily from the phenolic compounds and tannins (2 to 4 g/l in red wine).

The quality of wine depends on the harmony of these different tastes; one should not dominate the others. In particular, the acid and bitter tastes should be balanced by the sweet taste, the only one which is pleasant by itself. In wines which contain no sugar, the alcohol plays an important role; a solution of 30 g/l of ethanol has a sweet taste much like a solution of 20 g/l of saccharose; a slightly alcoholic solution of saccharose is sweeter than a similar solution in water. A harmonious balance must therefore exist :

$$\text{Sweet taste} \rightleftharpoons \text{acid taste} + \text{bitter taste}$$

With red wine, a demonstration can be given by separating the alcohol through vacuum distillation or by steam distillation; the distillate has a mellow taste, truly sweet, which couples with the vinosity of the alcohol; the stongly acidic and bitter residue is entirely undrinkable.

This equilibrium helps explain why red wine, rich in tannins and thus in bitterness, cannot tolerate as high a level of acidity as can white wine, whose level of tannins is low.

Wine also has a relatively high level of acidity; with a pH between 2.8 and 3.8, it is the most acidic of all fermented drinks; this acidity is linked to a relatively strong acid found in the grape, tartaric acid. Such acidity is tolerable only because it is counterbalanced by the sweet taste of the alcohol, since wine is also the most alcoholic of all fermented drinks; reciprocally, the alcohol is tolerable because of wine's high acidity. Moreover, wine owes its microbiological stability to its acidity and alcohol content, which permit a certain conservation without the use of highly specific techniques.

B - The Suppleness Index

It follows from the preceeding ideas that wines, especially red wines must be supple, that is they should not have an excessively bitter or acid taste. It was thus found necessary to define a suppleness index (PEYNAUD, 1970; RIBEREAU-GAYON et al., 1975 b) which is explained with examples in table I; an increase in the value of the index corresponds to an increase in the sensations of volume, softness and fullness of body, characteristics highly desirxable in red wines.

This suppleness index constitutes a relatively unique attempt to translate organoleptic qualities by a relation which combines several elements of the chemical composition. The index could be improved upon as it remains relatively imprecise, yet little work is being done in this field. Of course it is not true that an increase of alcohol or a decrease of acidity would improve quality in every case; these variations rapidly become excessive and the suppleness index is valid only within a certain compositional range. In addition, the units employed are certainly not the best possible; an increase in acidity of 1 g/l probably has more effect on the decrease of suppleness than a decrease in alcohol of 1° GL. Also the total quantity of acid is not the only factor involved in wine flavor; the pH is involved as well as the nature of acids, for instance succinic acid has a characteristic vinous taste.

The hardness of red wines, which is the opposite of suppleness, depends upon additional compositional elements, particularly the level of acetic acid andethyl acetate. When present in excessive quantities, these compounds are an indication of bacterial development and give wines particular organoleptic characteristics which reduce their quality. Even below the organoleptic perception threshold (on the order of 1 g/l for acetic acid and 150 mg/l for ethyl acetate), these compounds, especially the latter, intervene in sensory evaluation, particualrly on the aftertaste; they reinforce the impressions of hardness and burning; the suppleness index should take these compounds into account as well.

C - Influence of phenolic compound structure (tannins)

TABLE I

THE SUPPLENESS INDEX AND EXAMPLES OF ITS APPLICATION

Suppleness index = Alcohol (degree G.L) — [Acidity (g/l, H_2SO_4) + tannin (g/l)]

Examples	Wine n° 1	Wine n° 2	Wine n° 3
Alcohol (°GL)	11°0	11°5	12°0
Acidity (g/l, H_2SO_4)	4,0	4,0	3,5
Tanin (g/l)	3,0	2,5	2,5
Suppleness index	4	5	6
Wine characteristic	firm, thin	balanced	soft, full-bodied

Another important element in the suppleness of red wines, as well as in their general composition, is the presence of phenolic compounds, or more specifically tannins; these compounds are involved in the taste, not only by their quantity, but also by their nature which is not taken into account by the suppleness index.

Tannins are formed by the condensation of 2 to 10 elementary flavan molecules (catechins). The polymerization level affects the tannins capacity to combine with the proteins, this capacity governs all of their properties; in particular, astringency results from a loss of the saliva's lubricating effect by denatuation of its proteins. Thus, as an example, a tannin composed of 2 flavan dimers will not have the same properties, organoleptic in particular, as a tannin composed of 1 tetramer, even though the total weight remains the same.

The polymerization of flavans can set off several mechanisms which lead to tannins with different properties. Other molecules as well (anthocyanins, mineral salts, polysaccharides) can intervene in the structure of these tannins; table II gives a recent classification developed by GLORIES (1978). Each class of phenolic compounds has its own organoleptic properties; thus explaining the enologist's evaluation which differentiates the "good tannins" which give both body and a certain mellowness to the wine, and the "bad tannins" which give wine an agressive astringency. A precise chemical interpretation of these sensory differences remains to be done, they are related to the different tannin structures which cannot be distinguished by traditional analytical methods. High quality vine varieties, cultivated in reputed vineyards, are distinctive precisely because they produce grapes rich in "good tannins"; this characteristic dominates the tasting of fine red wines; the variations between vintage years as well concern a modification of tannins, linked to the climatic conditions during ripening.

TABLE II

DIFFERENT CLASSES OF CONDENSED PHENOLIC COMPOUNDS (TANNINS) IN RED WINE (GLORIES, 1978)

Constitution	Key	Molecular weight
Catechins, procyanidins (flavans of low polymerization	C-P	600
Tannins (flavans of average polymerization)	T	1000 à 2000
Tannin and anthocyanin complexes	TA	1000 à 2000
Condensed tannins (highly polymerized flavans also containing anthocyanins)	TC	2000 à 3000
Highly condensed tannins (very highly polymerized flavans also containing anthocyanins)	TtC	3000 à 5000
Complexes of tannins with polysaccharides and mineral salts	T-P T-S	5000

III - WINE AROMA

A - Aroma and bouquet

Aroma, in enology, generally designates the oderiferous constituents of young wines; the transformation of the aroma during aging produces bouquet. The chemical transformations which correspond to the appearance of bouquet are still poorly understood. Two types of bouquet are distinguished : a) the bouquet of oxidation (Rancio, Madeira) whose character is due to the presence of aldehydes and acetals; b) the bouquet of reduction which appears in fine table wines after long aging periods in bottles; ordinary wines under similar conditions do not improve, to the contrary they loose their freshness.

When dealing with aroma, a distinction is made between primary aroma which comes from the grape and secondary aroma produced by fermentation.

B - Primary aroma from the grape

Grape varieties are characterized, either by neutral aromas, or by distinct aromas (Muscat, Pinot, Sauvignon). It is only with Muscat varieties that there are some fairly precise chemical givens. This aroma results from a mixture of 8 to 10 terpene compounds, related to linalool; their total concentration is from 1 to 3 mg/l (table III). The threshold of olfactory perception for each substance is extremely varied, ranging from 100 to 7000 μg/l (table III); they have an additive effect, since their mixture is more aromatic (olfactory perception threshold 90 μg/l) than the most aromatic among them. The relative concentrations of the different substances, each having its own odor, account for the subtle differences between Muscat aromas.

These same terpenes intervene as well in the aroma of other aromatic vine varieties, such as

TABLE III

AVERAGE CONCENTRATION OF PRINCIPLE TERPENE COMPOUNDS IN MUSCAT GRAPES AND THEIR OLFACTORY PERCEPTION THRESHOLD. THERE IS NO DOSAGE FOR THE RELATIVELY ABUNDANT HOTRIENOL DUE TO ABSENCE OF REFERENCE PRODUCT (TERRIER, 1972; RIBEREAU-GAYON et al., 1975 c)

	Concentration (μg/l)	Approximate perception threshold (μg/l)
Linalool	400	100
Geraniol	300	130
Nerol	80	400
α-terpineol	60	450
Linalool oxide A	130	7000
Linalool oxide B	80	65000
Linalol oxide C	130	3000
Linalool oxide D	60	5000

Riesling, with a concentration from 0.1 to 0,3mg/l. On the other hand, no terpenes are found in the varieties known to produce grapes lacking characteristic aroma.

C - Secondary aroma from fermentation

The origin of secondary aroma is totally different than that of primary aroma; it is formed by alcohols and esters produced by yeast during alcoholic fermentation. Chemically, it is much better known, nevertheless olfactory interpretation is difficult since the overall aroma is the result of the superposition of numerous basic aromas which are present in relatively weak doses as compared to their perception thresholds.

Tables IV and V list the principle alcohols and esters identified in wine (BERTRAND, 1975). Methanol is the only alcohol not produced by fermentation. The high alcohols (table V) are present in a slightly higher concentration than their organoleptic perception threshold (table VI). Their odor is not particularly pleasant, but sufficiently diluted, it reinforces wine aroma. When their concentration becomes excessive (approximately above 400 mg/l), it is judged that their effect becomes negative.

Among esters (table IV), ethyl acetate plays an important role in wine aroma, apart from its previously mentioned influence on the actual taste. Its olfactory perception threshold in water is 25 mg/l, while in wine it is close to 160 mg/l; these figures show the importance the interaction between different constituants can have on the total taste of wine. Above 160 mg/l, the suffocationg odor, characteristic of ethyl acetate, can be perceived and distorts the aroma; below, this ester cannot be identified but contributes to the hard character of wine; at very weak doses (50 to 80 mg/l), it becomes part of the pleasant bouquet of red wines.

TABLE IV
PRINCIPLE ESTERS INVOLVED IN WINE AROMA (mg/l)
(BERTRAND, 1975)

	white wine		red wine	
	min.	max.	min.	max.
Ethyl formate	0,02	0,84	0,03	0,20
Methyl acetate	0	0,11	0,08	0,15
Ethyl acetate	4,50	180	22	190
Ethyl propionate	0	7,50	0,07	0,25
n-propyl acetate	0	0,04	0	0,08
Ethyl-2 methyl propionate	0	0,60	0,03	0,08
2-methylpropyl acetate	0,03	0,60	0,01	0,08
Ethyl n-butyrate	0,04	1,0	0,01	0,20
Ethyl 2-methyl butyrate	0	0,02	0	0,08
Ethyl 3-methyl butyrate	0	0,04	0	0,09
3-methyl butyl acetate	0,04	6,10	0,04	0,15
Ethyl n-hexanoate (caproate)	0,06	0,60	0,06	0,13
n-hexyl acetate	0	0,63	0	0,60
Ethyl lactate	3,80	15	9	17
Ethyl n-octanoate (caprylate)	1,10	5,10	1,0	6,0
Ethyl n-decanoate (caprate)	0,90	3,50	0,60	4,0
Diethyl succinate	0,01	0,80		
2-phenylethyl acetate	0,20	5,10		
Ethyl n-dodecanoate (laurate)	0,10	1,20		
Ethyl n-tetradecanoate	0,10	1,20		
Ethyl n-hexadecanoate	0,10	0,85		

TABLE V

PRINCIPLE HIGHER ALCOHOLS INVOLVED IN WINE AROMA (mg/l) (BERTRAND, 1975)

	white wine		red wine	
	min.	max.	min.	max.
Methanol	38	118	43	222
1-propanol	9	48	11	52
2-methyl 1-propanol	28	170	45	140
1-butanol	1,4	8,5	2,1	2,3
2-methyl 1-butanol	17	82	48	150
3-methyl 1-butanol	70	320	117	490
1-hexanol	3	10	3	10

Ethyl lactate and diethyl succinate are not true products of fermentation; taking into account their olfactory perception threshold on the order of 100 mg/l, their role in wine tasting is slight, indeed even nonexistent.

Fatty acid esters (ethyl caproate, caprylate, caprate, laurate) are important in the aroma of wines, especially in the case of white wines. Their concentration is low (5 to 10 mg/l); yet is approximately equal to 10 times their perception threshold. Table VI shows the influence of the medium in which the esters are found on their particulary low threshold. In addition, each ester contributes a particular note and influences the nature of the overall aroma; the odor ef ethyl caproate is dominated by a fruity quality; as the number of carbons increases, the odor of these esters tends to become soft, then soap-like, and finally stearic.

The content of higher alcohols and esters in wine is affected by the kind of yeast and by the fermentation conditions. Each yeast species has its own characteristics with respect to the formation of volatile aroma compounds (PARK, 1974; BERTRAND, 1975; SOUFLEROS, 1978). The figures in table VII, obtained in a synthetic medium under identical experimental conditions, are very significant. However the fact that _Sacch. cerevisiae_ is the dominant species in all cases of wine fermentation, suggests that the type of yeast has little effect on the organoleptic character of wine; elsewhere the numerous attempts to improve the quality of wines by yeasting have not had very significant results.

On the other hand, the conditions of the yeast's development, i.e. the fermentation conditions, are much more important than the type of yeast. Table VIII shows the effect of pH and temperature on the formation of higher alcohols and esters (SOUFLEROS, 1978). The large effect of low temperatures on the total quantity of higher esters explains why the vinification of white wines at low temperatures improves its aromatic characteristics.

TABLE VI

OLFACTORY PERCEPTION THRESHOLD OF ETHYL CAPROATE, ETHYL ACETATE, AND AMYL ALCOHOL (MIXTURE OF 2-METHYL 1-BUTANOL AND 3-METHYL 1-BUTANOL, 1/3) (μg/l) (BOIDRON, 1978)

	Water	Tartaric acid in hydro alcoholic solution 12° GL at pH 3	Same solution + 100 g/l sugar	Dry white wine
Ethyl caproate	36	37	56	850
Ethyl acetate	25.000	40.000		160.000
Amyl alcohol	1.900	12.500	60.000	180.000

TABLE VII

FORMATION OF HIGHER ALCOHOLS AND ESTERS BY DIFFERENT YEAST STOCKS UNDER IDENTICAL LABORATORY CONDITIONS (SOUFLEROS, 1978)

	Total higher alcohols (mg/l)	Total fatty acid esters (mg/l)
Sacch. cerevisiae *Sacch. bayanus* *Sacch. chevalieri* *Sacch. italicus* *Sacch. heterogenicus*	150	7
Sacch. capensis *Sacch. rosei* *Sacch. bailii* *Sacch. prostoserdovii* *Torulopsis stellata*	82	1
Hanseniaspora	32	1

TABLE VIII

FORMATION OF HIGHER ALCOHOLS AND ESTERS AS FUNCTION OF TEMPERATURE AND pH. THE NUMBERS ARE THE MEAN OF NUMEROUS DETERMINATIONS (SOUFLEROS, 1978)

Temperature (0°C)	pH	Total higher alcohols (mg/l)	Total fatty acid esters (mg/l)
20	3,4	201	10,8
20	2,9	180	9,9
30	3,4	188	7,8
30	2,9	148	5,4

TABLE IX

INFLUENCE OF JUICE CLARIFICATION ON THE FORMATION OF VOLATILE AROMA COMPOUNDS DURING FERMENTATION (RIBEREAU-GAYON et al., 1975 a)

	Total higher alcohols (mg/l)	Total fatty acid esters (mg/l)
Juice clarified by racking	225	11
Juice not clarified by racking	423	7

The aroma of such wines is marked by both a fruity and a soap-like odor typical of these esters.

In the production of red wines, the fermentation temperature should be considerably higher (30°C) in order to improve maceration of grape seeds and skins; fermentation aromas are less important in these wines whose richness in tannins is dominant in their character.

An analogy can be drawn wxith another vinification technique which has a large effect on the aromatic character of white wines; pre-fermentation clarification of musts allows fermentation to take place in clear juice (RIBEREAU-GAYON et al., 1975 a). It has been shown that this technique increases both the intensity and quality of the aroma. Laboratory experimentation has found that juice clarification by racking increases the total content of higher esters (table IX), a factor known

IV - FACTORS OF QUALITY

A - Natural factors

The production of quality wines implies the use of sound, ripe grapes; of course this can only be obtained by the choice of varieties adapted to specific soil and climatic conditions.

An overly severe ripening must be avoided, as it leads to a loss of aroma and a reinforcement of tannic characterisxtics. For this reason the finest wines are produced in areas situated at the limits of possible vine cultivation; however under such conditions, ripening is submitted to the hazards of climatic variations which are responsible for large differences in quality from one year to the next. An approximate relationship exists between the quality of vintage years as determined by sensory evaluation and the temperature and rainfall of that year (table X).

The influence of climatic conditions on the composition of grapes is shown in table XI; in

TABLE X

RELATION BETWEEN VINTAGE YEAR QUALITY AND BORDEAUX CLIMATIC CONDITIONS (APRIL TO SEPTEMBER INCLUDED)

The years are classed in order of the descending values of the difference : sum of average temperatures - rainfall. The result is an approximate classification in order of sensory wine quality.

Year	Sum of average temperatures (°C) a	Rainfall (mn) b	Difference a - b	Wine quality
1947	3478	259	3219	Exceptional
1949	3458	286	3172	Exceptional
1945	3361	253	3108	Exceptional
1961	3294	213	3081	Exceptional
1943	3308	282	3026	Very good
1964	3327	339	2988	Very good
1959	3310	331	2979	Very good
1952	3270	306	2964	Very good
1970	3184	232	2952	Very good
1973	3299	354	2945	Good
1944	3212	302	2910	Good
1950	3252	346	2906	Good
1962	3123	249	2874	Good
1955	3247	375	2872	Very good
1967	3120	278	2842	Good
1953	3198	367	2831	Very good
1942	3227	414	2813	Good
1948	3126	315	2811	Good
1966	3161	359	2802	Good
1946	3061	261	2800	Fairly good
1971	3269	496	2773	Good
1940	3094	339	2755	Fairly good
1957	3043	371	2672	Average
1960	3130	459	2671	Average
1969	3168	521	2647	Average
1958	3142	497	2645	Average
1956	3083	441	2642	Mediocre
1968	3102	458	2644	Mediocre
1963	3010	370	2640	Bad
1951	3052	443	2609	Mediocre
1954	2904	311	2593	Mediocre
1972	2900	331	2569	Mediocre
1965	3005	461	2544	Bad
1941	3009	536	2473	Mediocre

TABLE XI

CABERNET SAUVIGNON GRAPE COMPOSITION NEAR MATURITY, FROM TWO BORDEAUX REGION WINEYARDS OVER 5 CONSECUTIVE YEARS (RIBEREAU-GAYON et al., 1976)

Year	Date	Weight of 100 grapes	Sugar (g/l)	Acidity (méq/l)	Anthocyanins from skins (g in 200 grapes)	Tanins from skins
"Medoc grand cru" vineyard						
1969	29.IX	85	176	126	0,25	0,36
1970	28.IX	183	200	95	0,42	0,76
1971	27.IX	110	185	105	0,23	0,60
1972	9.X	101	180	150	0,22	0,41
1973	I.X	138	170	114	0,28	0,42
"Premières Côtes de Bordeaux" vineyard						
1969	29.IX	107	172	153	0,22	0,36
1970	28.IX	115	200	123	0,41	0,72
1971	27.IX	120	170	140	0,24	0,57
1972	9.X	116	164	194	0,22	0,43
1973	I.X	140	183	154	0,32	0,42

particularthe quantity of phenolic compounds (tannins and anthocyanins), fundamental to the taste of red wines, can double from one year to another.

Another consequence of the influence of variable climatic conditions is the frequent use of several different varieties having complementary characteristics in order to obtain maximum quality. Each vine variety has its own physiological characteristics, particularly in terms of acidity. (table XII); traditionally in the Bordeaux region, Cabernet Sauvignon, the standard variety because of the characteristic wine aroma it produces, is supplemented by a variety with low acidity (Merlot) which gives suppleness in years of insufficient maturity, and by an acidic variety (Malbec or Petit Verdot) which adds freshness in years of full maturity; the modern tendency to produce less acidic, more supple wines has caused an increase in the plantation of Merlot during the last 20 years, to the detriment of the more acidic varieties which are almost nonexistent today.

The conditions of vine cultivation also influence grape composition, and thus influence the organoleptic characteristics of wine. It is well known that all factors which increase excessively the vegetative development of the plant and crop yield, diminish wine quality by, in a sense, diluting the tastes and odors. The production of quality wines requires the respect of strict conditions concerning the rootstock, pruning, soil preparation, and in general, all cultivation practices that influence yield.

Another important element of wine aroma and taste is related to the grape's sanitary condition. Different vine diseases, by modification of grape composition, can produce a strange disagreeable taste in wine; the progress made in the protection of vines against parasites has been a definite factor in the improvement of quality.

The fungus responsible for the rotting of grapes, Botrytis cinerea, presents a particular case; depending upon the conditions of its development it can cause : a) noble rot which concentrates

TABLE XII

AVERAGE COMPOSITION OF BORDEAUX MUSTS AND VINE VARIETIES (RIBEREAU-GAYON et al., 1975 a)

Figures collected under comparables conditions during several harvests. Red and white vineyards are different.

	Vine variety	Alcohol (°GL)	Acidity (méq/l)	Tartaric acid (méq/l)	Malic acid (méq/l)
Red varieties (cultivated in same soil)	Merlot	12°0	92	110	36
	Cabernet franc	11°3	99	102	40
	Cabernet-Sauvignon	11°4	102	111	50
	Malbec	11°0	108	98	55
	Petit-Verdot	12°5	120	120	60
White varieties (cultivated in same soil)	Sauvignon	13°0	90	80	45
	Semillon	12°4	80	82	32
	Muscadelle	11°2	84	90	38

the must and permits the production of botrytised dessert wines (Sauternes) rich in distinct aromas; b) grey rot, is a deterioration, because the fungus tansmits strange tastes such as those of iodine, phenol, and mold. In Muscat varieties, the growth of *Botrytis cinerea* is coupled with the disappearance of the terpenes responsible for their characteristic aroma (BOIDRON, 1978).

B - OAK BARREL STORAGE AND WINE FLAVOR

Traditionally many types of wines, especially red wines, are stored in oak barrels for one or two years before being bottled. The essential benefit is the addition of various aromatic elements to the wine, among which the vanilla scent of wood is dominant; this character is strongest when the wooden barrels are new. This pratice enhances the complexity of wine aroma, provided that the woody character is not excessive and does not dominate the fruity character of the grape.

However storage in wooden barrels causes numerous other modifications in wine composition (table XIII) (RIBEREAU-GAYON, 1971). First, the wood's permeability allows oxygen to penetrate and intervene in the evolution of the aroma; in addition the oxygen modifies the pigment complex and accounts for the higher color intensity of wine stored in wooden barrels despite the lower anthocyanin content as compared to the same wine stored in steel containers (table XIII). In red wine, the increase of tannins from wood is relatively insignificant.

Storage in wooden barrels can also increase wine acidity resulting in a loss of suppleness and balance. This problem increases with barrel age. The storage of empty barrels requires sterilization with SO_2 vapors which progressively oxidize to H_2SO_4, responsible for the acidification. Te figures in table XIII show clearly the increase of total acidity and sulfates and the decrease of pH in wines that have been stored in used barrels.

TABLE XIII

INFLUENCE OF WOODEN BARREL STORAGE ON WINE COMPOSITION (RIBEREAU-GAYON, 1971)

	New wooden barrel	Used wooden barrel	Stainless steel container
Total acidity (g/l, H_2SO_4)	3,63	4,02	3,43
Volatile acidity (g/l, H_2SO_4)	0,62	0,56	0,46
pH	3,35	3,15	3,35
Sulfate (g/l, K_2SO_4)	0,86	1,26	0,92
Ethyl acetate (mg/l)	121	121	88
Tanin (index)	42	40	40
Anthocyanins (g/l)	0,17	0,16	0,18
Color intensity $D_{420} + D_{520}$	0,84	0,64	0,52

Finally, wood is a porous material, difficult to sterilize, which lends itself to the development of micro-organisms. Storage in wooden barrels is accompanied by variable increases in the levels of volatile acidity and ethyl acetate, which are negative quality factors in wine (table XIII).

These observations show that storage in wooden barrels improves the aroma and taste of certain types of wine, fine red wines in particular, on the condition that the two following principles are taken into consideration : a) the woody character is more favorable if the wine itself is of high initial quality, i.e. if the aromatic elements of the wood harmonize with those of the wine without excessively dominating them; b) very strict technical conditions are required, in particular barrel sanitation and maintenance must be rigorous, including the frequent replacement of barrels; the storage of empty barrels must be avoided.

C - INFLUENCE OF SPOILAGE

Because of its high content of alcohol and acidity, wine can be conserved fairly easily; this results in a sometimes excessive sense of security which can lead to a lack of care and cleanliness in processing conditions. Wine is actually very sensitive to contaminations which are sometimes responsible for bacterial development; lactic bacteria develop under anaerobic conditions in wine, transforming different constituents (sugar, tartaric acid, glycerol) and forming acetic acid. Acetic bacteria develop in aerobic conditions on the surface of partially full containers, oxidizing the alcohol into acetic acid, and producing ethyl acetate. In both cases, there is a large deterioration of organoleptic characteristics. For this reason, the maximum content of volatile acidity (acetic acid) is limited by legislation.

Also, wine easily picks up off-tastes and odors; wines whose aroma and taste have been modified by improperly maintained containers are more frequent than would be expected.

Finally, wine can be altered by chemical spoilage, the most common form of which is probably oxidation. The solubility of oxygen in wine, primarily a function of temperature, is approximately 8 mg/l. Manipulations which expose wine to the air (clarification, filtration, bottling...) can cause oxidations which, even if slight, form sufficient acetaldehyde to damage the bouquet. In traditional conservation, the use of small quantities of sulphur dioxide during different processing operations allows the fixation of the acetaldehyde in the form of a bisulfite complex, and as a consequence this accident is avoided.

V - CONCLUSIONS

Wine odor and taste are difficult to interpret because of the complexity of wine composition as well as the diversity of different wine types. However the continued development of analytical methods together with the increasing objectivity of sensory evaluation techniques have resulted in a more rigorous approach to this problem. Nevertheless new developments are required to further our understanding of the complex subject of wine flavor.

REFERENCES

BERTRAND A., 1975 - *Recherches sur l'analyse des vins par chromatographie en phase gazeuse.* Thèse Dr.Etat ès Sc., Université de Bordeaux II.

BOIDRON J-N., 1966 - *Essai d'identification des constituants de l'arôme des vins de* V.vinifera *L. Premiers résultats.* Thèse 3ème Cycle, Université de Bordeaux.

BOIDRON J-N., 1977 - Relation entre les substances terpéniques et la qualité du raisin (Rôle du *Botrytis cinerea*). *C.R IIIè Symposium International d'OEnologie*, 21-25 juin, Bordeaux.

BOIDRON J-N., 1978 - Travaux non publiés.

GLORIES Y., 1978 - *La matière colorante du vin rouge*. Thèse Dr. Etat ès Sc., Université de Bordeaux II.

PARK Y.H., 1974 - *Contribution à l'étude des levures de Cognac*. Thèse 3ème Cycle, Université de Bordeaux II.

PEYNAUD E., 1970 - *Connaissance et Travail du vin*. Dunod éd., Paris.

RIBEREAU-GAYON P., 1971 - Recherches technologiques sur les composés phénoliques des vins rouges III.- Influence du mode de logement sur les caractères chimiques et organoleptiques des vins rouges, plus particulièrement sur leur couleur. *Conn.Vigne Vin, 5(1)*, 87-97.

RIBEREAU-GAYON P., LAFON-LAFOURCADE S. et BERTRAND A., 1975 a - Le débourbage des moûts de vendange blanche. *Conn.Vigne Vin, 9(2)*,117-139.

RIBEREAU-GAYON J., PEYNAUD E., RIBEREAU-GAYON P. et SUDRAUD P., 1975 b - *Sciences et techniques du Vin. Tome II "Caractères des vins. Maturation du raisin. Levures et bactéries"*. Dunod-Bordas ed., Paris.

RIBEREAU-GAYON P., BOIDRON J-N. et TERRIER A., 1975 c - Aroma of Muscat Grape Varieties. *J.Agric.Food Chem.,23(6)*, 1042-1047.

RIBEREAU-GAYON J., PEYNAUD E., RIBEREAU-GAYON P. et SUDRAUD P., 1976 - *Sciences et Techniques du vin, Tome III "Vinifications, Transformations du vin"*. Dunod-Bordas éd., Paris.

SOUFLEROS A., 1978 - *Les levures de la région viticole de Naoussa (Grèce). Identification et classification,d'étude des produits volatils formés au cours de la fermentation*. Thèse Docteur-Ingénieur, Université de Bordeaux II.

TERRIER A., 1972 - *Les composés terpéniques dans l'arôme des raisins et des vins de certaines variétés de* Vitis vinifera. Thèse Docteur-Ingénieur, Université de Bordeaux II.

ALTERATION IN A WINE DISTILLATE DURING AGEING

S. van Straten, miss G. Jonk, L. van Gemert

Central Institute for Nutrition and Food Research TNO
Zeist, The Netherlands

I. INTRODUCTION

Freshly distilled alcoholic beverages, such as brandy, rum and whisky, are most commonly stored in wooden barrels for one year or longer, before they are consumed. During this ageing or maturation period, flavour changes are obvious. A well-aged beverage has a mellow flavour, whereas the unaged distillated has a harsh flavour.

Assumed changes during maturation are:

- reactions of components of an unaged distillate
- reactions of wood and/or wood components with components of the unaged distillate, and
- extraction of wood components.

For example, acids, aldehydes and esters can be formed by oxidation- and esterification reactions of alcohols. By means of radioactive ethanol-14C added to ageing whisky distillates Reazin et al (1) proved that during ageing acetaldehyde, acetic acid and ethyl acetate are produced from ethanol. The presence of 3-methyl-4-hydroxy-octanoic acid lactone - known as "oaklactone" - in barrel-aged alcoholic beverages (2) and oak species have been reported (3). Guymon and Crowell (4) investigated the changes in brandy aged in American and French oak barrels. Among others, they noticed a large amount of oak lactone in brandy that was aged in American oak barrels. Brandy aged in French oak barrels only contained a small amount of oaklactone. Furthermore, bitter tasting compounds extracted from the wood, and aromatic aldehydes derived

ISBN 0-12-169060-1

from wood-lignine are important components in a well-aged alcoholic beverage.

Although extensive studies have been made on compounds present in aged products, relatively little knowledge exists on the actual changes of compounds present in an unaged distillate, these compounds mostly being secondary fermentation products obtained during ageing.

In this study, these compounds have been determined quantitatively in an unaged wine distillate and in distillates stored in three types of containers. For studying the effect of soluble wood compounds on secondary fermentation products, distillates with an additional amount of a wood extract are stored too. Lactones, representing wood extract compounds have also been determined.

The distillates aged for one year have been evaluated sensorially.

II. DISTILLATION, WAREHOUSING AND SAMPLING

A wine distillate of 70 volume % was produced by distillation of a French fortified wine in a potstill equipment. The distillation conditions were the same as commonly used in a brandy distillery.

Six portions of the distillate have been used for the ageing experiment. Three of them are stored in containers of glass, stainless steel and of oak wood, respectively. The wooden barrels were made of French oak (limousin). They had already been used six years for warehousing of wine distillates. Wood extract, prepared by extracting limousin with wine distillate, was added in a concentration of 0.6 % to the other three portions of distillate which were stored too in glass, stainless steel and wooden barrels respectively.

The storage takes place in the cellar of a distillery at normal conditions.

Samples for analytical examination were drawn from the unaged distillated and from each product aged for 4, 8 and 12 months respectively. To exclude uncontrollable differences of the used barrels, ten casks were used to draw a bulk sample. Samples for the sensory analysis were drawn after one year of ageing.

For the next 3 to 5 years warehousing will be continued and in this period analytical and sensorial evaluations are planned, twice and once a year respectively.

III. METHODS OF ANALYSIS

Pre-treatments

Lower-boiling alcohols (up to C_5) and ethylacetate were determined in aliquots diluted to 35 volume % of ethanol, without further treatments.

Lower-boiling aldehydes (C_2 to C_6) were isolated as semicarbazones, and determined quantitatively after regeneration of the aldehydes, using head space analysis (5). This carbonyl specific isolation method enables the analysis of small quantities of aldehydes, that would be lost if extraction methods should have been applied. Acrolein, which might be an interesting component to determine, as well as other unsaturated compounds, cannot be analyzed by means of this method.

High-boiling esters were determined in CH_2Cl_2-extract of aliquots diluted to 15 volume % of ethanol. Furfural and phenethyl alcohol, being present in comparable amounts as high-boiling esters have also been determined in these extracts.

Furfural was also determined with an U.V.method according to an A.O.A.C.method (6).

Aliphatic acids (C_2 to C_{18}) were isolated by means of a distillation and extraction method. The lower acids (C_2 to C_6) were determined as such, the higher ones as methyl esters.

Aliquots of each sample, diluted to 15 volume % of ethanol, were heated under reflux with sodiumhydroxide to hydrolyze the excessive amounts of esters, which would disturb a gas chromatographic analysis of small quantities of alcohols and lactones. The high-boiling alcohols (C_6 and higher) were determined in methylenechloride-extracts of distillates of the hydrolized samples. After addition of an excess of sulphuric acid to the residues, and boiling under reflux, lactones were extracted with pentane-ether (2 : 1). Free fatty acids were removed by washing the extracts with sodiumbicarbonate solution (5%). During this procedure lactones are hydrolyzed, and formed again under acidic conditions.

Chromatographic methods

All compounds were determined quantitatively by means of gas chromatographic method using glass capillary columns, except for low-boiling alcohols, which were analyzed on a stainless steel capillary column. The stationary phases used, were:

- low-boiling alcohols	Carbowax 400
- aldehydes	Carbowax 20 M
- high-boiling esters, alcohols and lactones	SP 2300
- lower acids	SP 1000
- higher acids (as methyl esters)	PEGA

Reference solutions in 70 volume % of ethanol were made of all compounds determined quantitatively. To determined recovery data, these references were treated and analyzed under the same conditions as the actual samples.

An internal standard was added to each sample (e.g. ethylundecanoate for the high-boiling esters), and all chromatographic peaks were evaluated with an internal standard method. Samples without addition of the standard were also analyzed in order to ensure that no appreciable peaks eluted at the same retention time as the internal standard.

Sensoric Methods

The samples were assessed by a panel of 6 members, all experienced in sensory evaluation of strong alcoholic beverages, including the relevant type of brandy.

Samples were diluted to 40 volume % of ethanol, and served in dark blue glasses to hide colour-differences. The assessments took place in an odour-free room.

Sensory evaluation was performed with graphic scales - a straight line anchored at both ends - and concerned the aspects: quality of odour and flavour, intensity of odour and flavour, and the woody character of the flavour. The evaluation was carried out two times.

The results were statistically tested by means of analysis of variance.

RESULTS AND DISCUSSION

Gas Chromatographic Analyses

Analytical data on more than 50 compounds were gathered. The results of some of these are presented in tables 1 to 3. These data show that, as far as the secondary fermentation products are concerned, the composition of a wine distillate does not change drastically during ageing for one year. Besides, the differences between distillates aged in different types of containers are small or even unnoticeable. Consequently, the effect of addition of wood extract is also too small to be measured.

Anyhow, some changes resulting from ageing have to be memorized:

- amounts of some alcohols, especially those being present in small quantities in the unaged product, seem to diminish. Increasing amounts of acids, corresponding to the alcohols, which might be produced by oxidation, have not been found
- the quantity of acetic acid increases during ageing in wooden barrels
- oaklactone has been found in all distillates aged for one year.

When regarding these results, it seems unlikely that a change in composition of the secondary fermentation products during ageing for one year, is important for the quality of brandy. Whether or not the ageing-effect will be more pronounced after ageing for longer periods, is still under investigation.

It is likely that the unquestioned value of ageing is for a great deal based on compounds extracted from wood. One of these compounds is oaklactone, which has a distinctive aroma. As reported by Salo (7) oaklactone has an odour threshold of 0.05 mg/l (in 9.4 % w/w ethanol). Otsuka (2) found odour thresholds of 0.8 and 0.07 mg/l (in 30 % ethanol) for the cis- and trans-oaklactone respectively. Reported by Otsuka, French brandies contain less than 0.5 mg oaklactone/l. Higher quantities found in our experiments need further investigation. It has been proved by means of mass spectrometric analysis that the lactone extract of the distillate aged in glass, contains oaklactone. However, the methylenechloride extract produced without chemical treatments, does <u>not</u> contain oaklactone. These analyses prove that a 'precursor' of oaklactone is present in an aged wine distillate. Evidence of the identity of this precursor is under investigation.

Sensorial analyses

Individual results were tested statistically, and these showed that two of the assessors were much less able to differentiate between the samples in comparison with the other ones. Furthermore, as the results of another assessor were strongly deviating, it was decided not to take the results of these three panel members into consideration.

The results of the selected panel of three members are shown in table 4. The results of odour and flavour concerning quality and intensity are shown together as the scores of odour and flavour did not differ much. Statistical testing by means of analysis of variance shows a significant result in all three instances.

It can be concluded that the samples can be divided into two different groups. One group consists of distillates stored in glass and stainless steel, without addition of wood extract, the other one concerns the other four distillates.

TABLE I. Secondary fermentation products in an unaged and in a one-year-aged wine distillate. Concentration in mg/100 ml

Compound	Unaged distillate	Aged in Glass	Aged in Stainless steel	Aged in Oak
Acetaldehyde	4.8	5.3	5.3	5.6
Isobutyraldehyde	1.2	1.6	1.7	1.3
Isovaleraldehyde	0.3	0.2	0.2	0.2
Furfural (G.C.)	1.1	0.8	0.8	0.8
Furfural (U.V.)	0.7	0.9	0.9	0.9
1-Propanol	23	20	20	20
Isobutanol	70	63	64	62
1-Hexanol	1.4[x)]	1.1	1.5	1.2
1-Octanol	0.07[x)]	0.08	0.1	0.07
2-Phenylethanol[xx)]	0.8[x)]	0.6	0.5	0.5
Acetic acid	0.7[x)]	0.6	0.8	1.2
Isobutyric acid	0.06[x)]	0.06	0.07	0.06
Isovaleric acid	0.06[x)]	0.05	0.05	0.04
Hexanoic acid	0.1[x)]	0.09	0.09	0.07
Decanoic acid	0.5[x)]	0.4	0.5	0.5
Ethyl acetate	41	36	36	40
Ethyl octanoate	2.1	2.4	2.3	2.6
Ethyl decanoate	1.7	1.7	1.6	1.6
Ethyl dodecanoate	1.3	1.0	1.1	0.9
Diethyl succinate	0.5	0.4	0.4	0.4

x) aged for four months in glass
xx) direct determination, without hydrolysis

TABLE II. Secondary fermentation products in an unaged and a one-year-aged wine distillate (with wood extract).

Concentration in mg/100 ml

Compound	Unaged distillate	Aged in Glass	Aged in Stainless steel	Aged in Oak
Acetaldehyde	4.8	3.9	3.9	4.8
Isobutyraldehyde	1.2	1.0	1.2	1.1
Isovaleraldehyde	0.3	0.2	0.2	0.2
Furfural (G.C.)	1.1	0.7	0.8	0.7
Furfural (U.V.)	0.7	0.9	0.9	0.9
1-Propanol	23	19	19	20
Isobutanol	70	62	61	63
1-Hexanol	1.4[x)]	0.9	0.8	0.9
1-Octanol	<0.1[x)]	<0.1	<0.1	<0.1
2-Phenylethanol[xx)]	0.7[x)]	0.6	0.6	0.5
Acetic acid	0.5[x)]	0.6	0.9	1.6
Isobutyric acid	0.08[x)]	0.07	0.09	0.08
Isovaleric acid	0.07[x)]	0.05	0.06	0.06
Hexanoic acid	0.1[x)]	0.1	0.1	0.1
Decanoic acid	0.5[x)]	0.4	0.5	0.4
Ethyl acetate	41	39	39	39
Ethyl octanoate	2.1	2.4	2.4	2.4
Ethyl decanoate	1.7	1.4	1.5	1.6
Ethyl dodecanoate	1.3	0.7	0.8	0.9
Diethyl succinate	0.5	0.5	0.4	0.4

x aged for four months in glass
xx direct determination, without hydrolysis

TABLE III. Oaklactone in a wine distillate after ageing for one year in different containers

Product	Container	Addition of wood extract	mg oaklactone/ 100 ml
unaged	-	no	-
aged one year	glass	no	1.9
" " "	glass	yes	0.3
" " "	stainless steel	no	<0.1
" " "	stainless steel	yes	0.2
" " "	oak barrels	no	0.3
" " "	oak barrels	yes	0.8

TABLE IV. Sensory evaluation of stored wine distillates (ageing one year)

	Mean scores of a selected panel of three members						F-value	significance
	G(-)[x]	G(+)	S(-)[x]	S(+)	W(-)[x]	W(+)		
Quality	153	177	151	172	173	178	2.87	$P \leq 0.05$
Intensity	144	176	143	163	169	174	5.64	$P \leq 0.01$
Woody flavour	13	75	21	65	70	77	26.08	$P \leq 0.01$

x) G : aged in glass; S: aged in stainless steel; W: aged in oak barrel; (-): without wood extract; (+): wood extract added

REFERENCES

1. Reazin, G.H., Baldwin, S., Scales, H.S., Washington,H.W., and Andreasen, A., J. Assoc. Off. Anal. Chem. 59: 770 (1976).
2. Otsuka, K., Zenibayashi, Y., Itoh, M., and Totsuka, A., Agric. Biol. Chem. 38: 485 (1974).
3. Masuda, M., and Nishimura, K., Phytochemistry 10: 1401 (1971).
4. Guymon, J.F., and Crowell, E.A., Am. J. Enol. Vitic. 23: 114 (1972).
5. Maarse, H., and Schaefer, J., in "Analysis of food and beverages, head space techniques", ed. G. Charalambous, p. 17, Academic Press, New York, 1978.
6. Schoenemann, R.L., J. Assoc. Off. Agric. Chem. 44: 392 (1961).
7. Salo, P., Nykänen, L., and Suomalainen, H., J. Food Sci. 37: 394 (1972).

POSSIBILITIES OF CHARACTERIZING WINE QUALITY AND VINE VARIETIES BY MEANS OF CAPILLARY CHROMATOGRAPHY.

A. Rapp, H. Hastrich, L. Engel and W. Knipser

Bundesforschungsanstalt für Rebenzüchtung Geilweilerhof
6741 Siebeldingen, Federal Republic of Germany

INTRODUCTION

The quality of a wine is determined by a balance of all its components as assessed by sensory evaluation. Because of the strong influence on the sensory receptors of volatile aromatic compounds these components can decisively influence the nature and quality of wines. Accordingly, studies into these volatile components will be of value for characterizing wine quality and vine varieties.

For several years now we have been engaged on studies into the aroma compounds of both grapes and wines as well as the biosynthesis of particular components (1-13, 38,42). Two topics which have been most intensively investigated during this period have been : a) the influence of certain must components on the development of bouquet substances formed during fermentation and b) the possibility of characterizing a variety be means of the aroma composition.

Thus, an early diagnosis for the selection of new grape varieties and an objective evaluation of quality both based on exact analytical data rather than subjective organoleptic evaluations may be realized. Early and accurate identification of the taste characteristics of

ISBN 0-12-169060-1

grape varieties would greatly facilitate the whole vine breeding programme.

This report surveys the results of our studies into varietal characterization.

II CONCENTRATION AND SEPARATION OF THE AROMATIC COMPONENTS

The aroma components of grape and wine consist of several hundred individual constituents and include substances of widely different classes (1,2,14-26). The total content of all aromatic substances in wine amounts to 0.8 - 1.2 g/l (abt. 1 % of the ethanol content). The components formed during fermentation, viz. 2-methyl-propanol-1, 3-methyl-butanol-1, 2-methyl-butanol-1 and 2-phenylethanol account for abt. 50 % of the total, the rest being distributed among 400 - 600 aromatic substances. This means that these are present in concentrations between 10^{-4} and 10^{-9} g/l.

The analysis of these trace components is not only a scientific challenge but is of great inportance since the sence organs are extremely sensitive and can easily detect the presence of minor components. From the taste threshold data reported in the literature it is apparent that there is a wide range of sensory responses for various components i.e. 5×10^{-4}g of ethanol in one litre of water can just be detected whereas the threshold for some terpenes, heterocyclic and sulphur containing compounds is much lower (27-33). For methylmercaptan, e.g. a value of 2×10^{-11} g/l has been determined. Due to this variation in the sensitivity of taste (e.g. threshold values ranging in concentration over seven orders of magnitude) those components present in minute amounts can be of far greater significance than those present in relatively large concentration. This circumstance, together with the fact that quite often quantitative variations only are observable between the aroma composition of different varieties, complicate the evaluation of varietial character. The detection of as many aromatic substances as possible with both high-resolution and high-precision chromatographic techniques is therefore a necessary pre-condition for the characterization of a variety.

In order to analyse these trace aroma components a concentration step is necessary. Several methods are available for this purpose. The concentration procedure for the enrichment of aroma substances has to be conducted as carefully as possible in order to avoid artefact development otherwise en assessment of the authentic aroma composition is impossible.

In the case of wine enrichment can be achieved by various methods: distillation, salting out, freezing out, extraction. When the components are concentrated by distillation, freezing out or salting out, ethanol is also concentrated. We therefore prefer to concentrate the aromatic substances by liquid-liquid-extraction (1,4). By using suitable solvents, the concentration of the aroma substances can be controlled in such a way that very little ethanol is co-extracted.

An appropriate solvent must have a low boiling point, he available in high purity, and allow efficient extraction of aroma components. A solvent satisfying these broad criteria and which has been used with advantage for the enrichment of aroma substances is trichlorofluormethane (Freon 11) (7,34,40,41). The extractability of some alcohols (C_4 to C_{10}) and ethyl esters (C_2 to C_{14}) by Freon 11, from aqueons alcohol is shown in Table 1. This data shows that after 4hrs.a limiting value is reached for the extraction of alcohols with a chain length of greater than six carbon atoms. For alcohols with less than six carbon atoms longer extraction times are necessary for efficient recovery i.e. 10 hrs. for pentan-1-ol. Similarly, esters with increasing carbon chain length requrie less time for efficient extraction. Our investigations have shown that for aroma enrichment

Table 1. Extractability of alcohols and ethyl esters by Freon 11.

	2h	4h	6h	10h	15h	20h	25h	50h	100h
n-Alkohole									
C_4	44	65	76	85	91	94	97	99	
C_5	80	96	98	99					
C_6	93	99							
C_7	94	99							
C_8	95	99							
C_9	96	99							
C_{10}	96	99							
Äthylester									
C_2	91	99							
C_3	96	99							
C_4	95	99							
C_5	94	99							
C_6	91	99							
C_7	94	96	98	98	99				
C_8	67	84	89	92	95	96	96	99	
C_9	49	68	74	78	81	83	86	95	99
C_{10}	54	69	73	75	78	80	82	91	98
C_{12}	32	51	58	61	64	68	70	84	97
C_{14}	37	55	61	63	65	67	71	84	97

of grape berry homogenates and grape juices an extraction period of 14 hrs. is adequate. An extension of the extraction time by a farther 10 hrs. affords, in most cases, a small increase in yield only. For exemple the ethyl ester of caprin acid, which was extracted with difficulty only, was recovered;in 78 % yield in 15 hrs. and this was increased to only 82 % in 25 hrs. (see Table 1).

Since the aroma concentrates are complex mixtures of compounds of diverse polarity (alcohols, esters, hydrocarbons, acids etc.) and boiling points (50 - 350 °C) which are present in concentrations ranging over several powers of ten (10^{-2} to 10^{-9} g/l) their separation is difficult and requires columns with high resolution. To analyse an aroma concentrate of several hundred components in a resonable time at least 2 - 3 constituents must he separated per minute. This has only become possible by means of capillary chromatography and particularly by using suitable glass capillary columns. Figure 1 shows a section of 60 minutes (from the 20th to the 80th minute) of an aromagram obtained by means

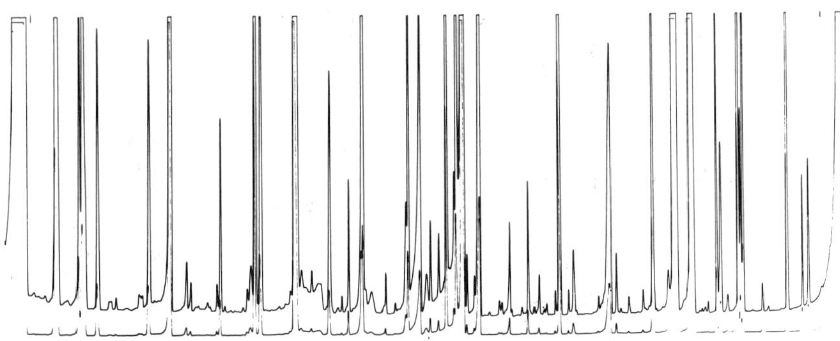

Figure 1. Aromagram of Morio-Muskat. Separation on a 90 m Reoplex glass capillary. Section from the 20th to 80th minute. 50 - 160° C; 1,5° C/min.

of a 90 m long Reoplex 400 glass capillary column. In a total analysis time of 150 minutes, it is possible to separate aroma concentrates into 350 to 400 peaks. This number of separated components should be sufficient for evaluating a vine variety or the wine quality.

III. COMPARISON OF AROMAGRAMS OF DIFFERENT GRAPE VARIETIES

A comparison of the aromagrams of wines and grape juices from different varieties reveals distinct differences in the aroma composition ("fingerprint pattern"). Although there are only quantitative differences in the fingerprint patterns of the varieties, these differences are so distinct as to provide a basis for characterizing a variety. A further problem involves the recognition, from among the several hundred components observed, of those compounds specifically associated with a particular variety. To investigate such variety-specific components (key substances), we compare sections of aromagrams in which distinct differences can be seen. In Figure 2, sections of aromagrams from 4 varieties are compared. It can be seen that these

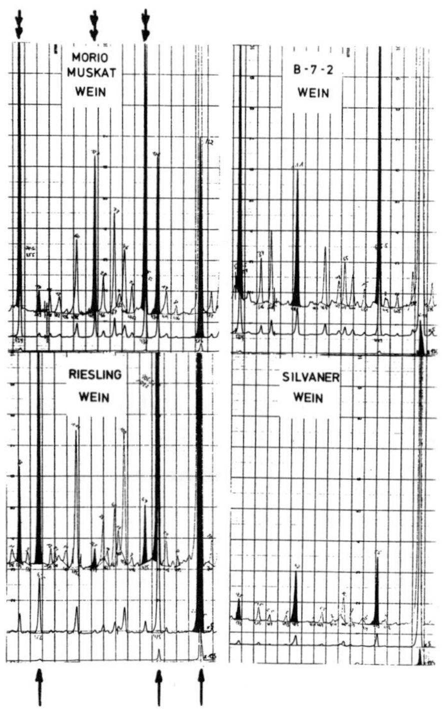

Figure 2. Aromagram sections of wines of different varieties (25th - 40th min).

4 varieties differ greatly in the section shown.

Since these wines were produced from grapes of about the same degree of maturity (70 - 72° Oechsle) and under the same technological conditions (separation, pure culture yeast, duration of fermentation etc.), the observed differences are only due to variety-specific characteristics. The prominent key-substances of Riesling have been marked ⟶, those of Morio-Muskat ⟶⟶. When the peak heights or peak areas of these key-substances are compared, variety - specific relationships are obtained. Such a comparison of the data illustrated in Figure 2 gives the relationships shown in the following table. It can be seen from Table 2 that there are marked differences between the relation-

Table 2. Aroma compounds of wines of different grape varieties (vintage year 1973); relative values

Relation of components min	Riesling 73 - 204	Silvaner 73 - 205	B-7-2 73 - 241	Morio-Muskat 73 - 75
26.1/29.0	93	538	233	3
26.1/31.8	273	1750	3500	6
26.1/35.9	56	97	13	2
28.3/29.0	22	18	23	0,5
28.3/31.8	64	55	343	1
35.9/31.8	5	18	258	3
35.9/29.0	2	6	17	2

ships calculated from the peak sizes of the key-substances of the 4 varieties investigated, (Riesling, Morio-Muskat, Silvaner, B 7-2). In group 1 some relations specific to Morio-Muskat are shown and the last group illustrates those relationships in which B 7-2 differs widely from the other varieties. These results have been obtained from wines of the same vintage year and of similar quality (harvested at about 72° Oechsle and dry enriched to 88° Oechsle). The question then arises as to what extent the year of vintage, the maturity of the grapes and location of vineyard influence these variety - specific relationships.

IV. INFLUENCE OF GRAPE MATURITY ON THE COMPOSITION OF AROMA COMPONENTS

Since the varietial character of a wine might be determined largely by the aromatic components of the grapes from which the wine was produced, further investigations have been conducted to determine whether the variety-specific aroma relationships ca be seen in the aromagrams of grapes also.

Comparable aromagram sections (12th to 82nd minute, range of temperature 70 - 150° C) of different grape

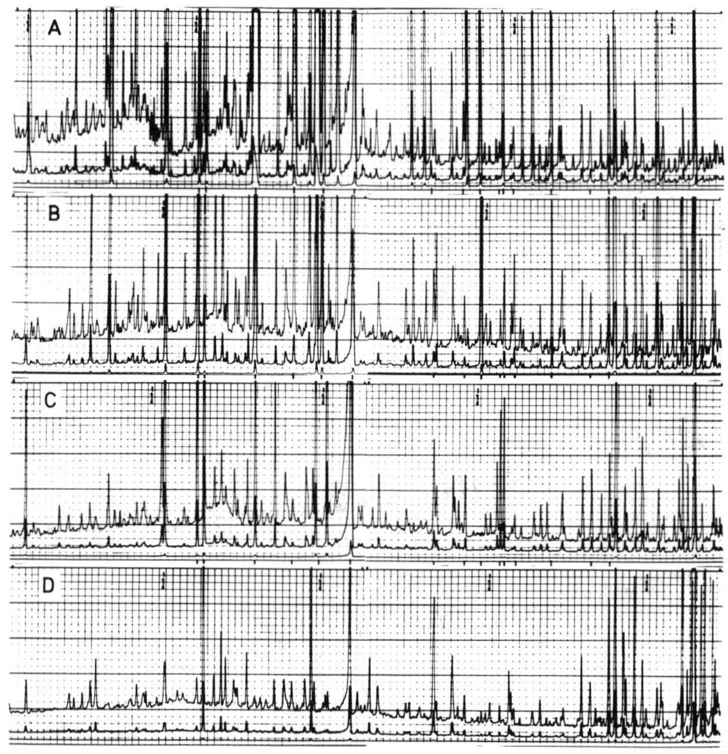

Figure 3. Aromagrams of grapes of different varieties. A = Morio-Muskat, B = Riesling, C = Silvaner, D = B 6-18. Maturity: 71 - 75° Oechsle.

varieties are shown in Figure 3. It is apparent that at approximately the same degree of maturity (weight of the must) there are already quantitative, variety-specific differences in the fingerprint patterns of the grapes. The 4 varieties investigated contain components varying distinctly in quantity. By relating these key-substances of the fruit, values are obtained which distinguish the individual varieties.

The chromatographic data shown in Figure 3 (8) was obtained from fruit prepared under the same conditions (inactivation of enzymes) and gives a comparison of the volatile components of healthy berries at similar maturity (71.5 - 75° Oechsle). The sugar content of the berries was taken as the criterion for the "degree of maturity". To decide whether the variety-specific differences have been superposed by the stage of maturity, the 4 varieties have also been investigated at different phases of maturity.

In Figure 4 aromagram sections (12th - 82nd minute) of Riesling samples at varying degrees of maturity (32.5 - 88.0° Oechsle) are compared. It is apparent that numerous volatile substances are already present in the grapes at the first date of vintage (immediately after sugar accumulation has set in), and that some components (which are also the main components at later dates of maturity) predominate. In spite of a difference in the degree of maturity ranging over 56° Oechsle, the fingerprint pattern of the variety Riesling is clearly recognizable from all aromagrams. In addition, a comparison of the aromagrams in Figure 4 reveals that the content of some aroma components increases with increasing degree of maturity, while the content of a few components decreases. Further investigations in this direction might permit an evaluation of the "aroma maturity" or the optimum aroma quality of a wine variety.

The influence of the berry maturity on aroma composition, as shown in Figure 4 for Riesling, is also

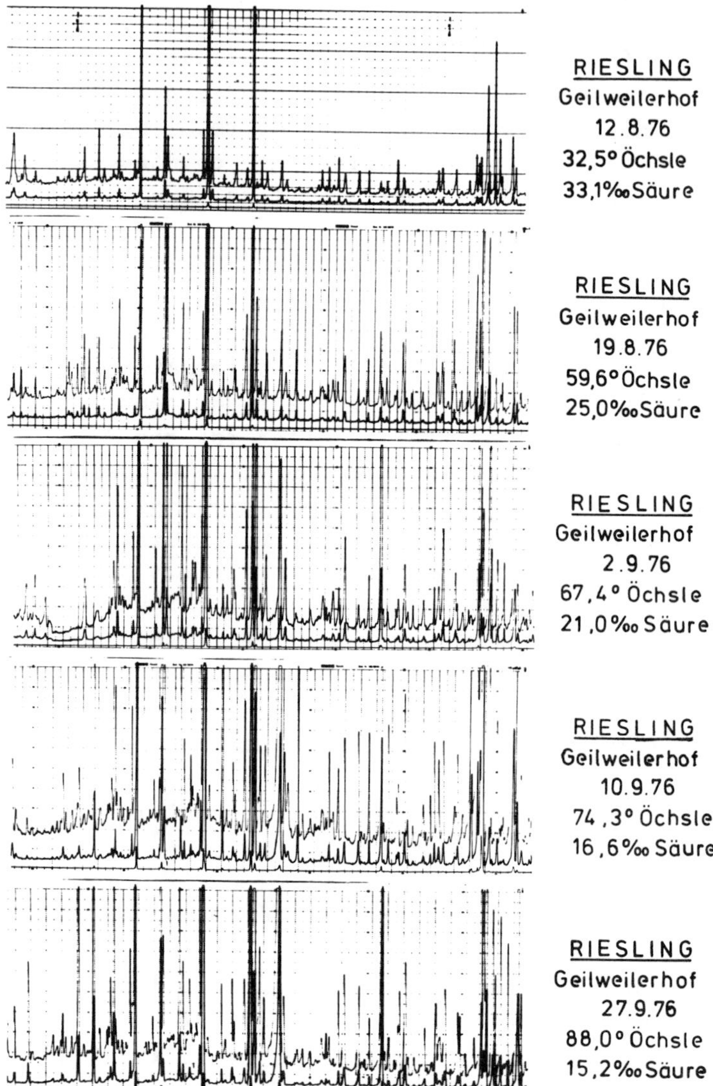

Figure 4. Influence of maturity on the aroma composition of Riesling grapes (vintage year 1976).

recognizable for the other investigated varieties. A similar influence of maturity on aroma composition should also be identifiable in the aroma pattern of wines. From the aromagrams of various Morio-Muskat wines of different quality level (table wines to top quality wines) from the years 1971 - 1974 (Table 3), it can be seen that quantitative changes arise in the finger-

Table 3. Aroma compounds of Morio-Muskat of 4 different vintage years (1971 - 1974) and different degrees of maturity (table wine - top quality wine); relation values.

Relation of components min	123 °Oe 71-374	72° Oe 73-75	62 °Oe 72-87	54 °Oe 74-76
52.5/56.2	319	17	29	3
52.5/51.8	240	14	19	6
52.7/56.2	153	5	22	8
25.7/34.7	8	48	58	188
60.4/50.7	15	2	14	126
56.2/59.7	0,3	2	5	35
35.7/31.7	3	3	1	2
65.3/72.6	2	2	2	2

print pattern typical of Morio-Muskat. Consequently, some relations increase with increasing maturity (table wine to top quality wines), others decrease (middle group) and some remain constant (lower group).

Furthermore, it turns out that those relation values established as typical of Morio-Muskat differ markedly from those of other varieties, in spite of these quantitative changes conditioned by vintage year and quality level (Table 4). Thus, values between 6 and 65 have been found in the relation of components 26.1 and 31.8 in Morio-Muskat (Table 4, middle line), whereas

Table 4. Aroma compounds of wines of different grape varieties; relation values.

Relation of components min	Riesling 73 - 204	Silvaner 73 - 205	B-7-2 73 - 241	Morio-Muskat 1971-1974 Beerenauslese bis Tafelwein
26.1/29.0	93	538	233	3 - 34
26.1/31.8	273	1750	3500	6 - 65
26.1/35.9	56	97	13	2 - 28

this relation is 3500 for B 7-2, 1750 for Silvaner, and 273 for Riesling. However, in considering many vintage years and different quality levels, there are also relations in which the different varieties overlap. These relation values are not suited for a varietal characterization. From these results it is apparent that by investigating a number of key-substances, characterization of all varieties might become possible, quite independent upon vintage year and quality level.

V. INFLUENCE OF THE LOCATION ON THE AROMA COMPOSITION (11)

The influence of the location is another point of great importance. To allow successful characterization of a variety, the fingerprint patterns of the individual varieties must differ significantly and be independent on location.

Figure 5 shows a comparison of aromagram sections (12th to 82nd minute) of different Riesling samples grown in the surroundings of Geilweilerhof. Although the samples have been taken from greatly different sites (south slope, north slope plain), soils, (calcereous,

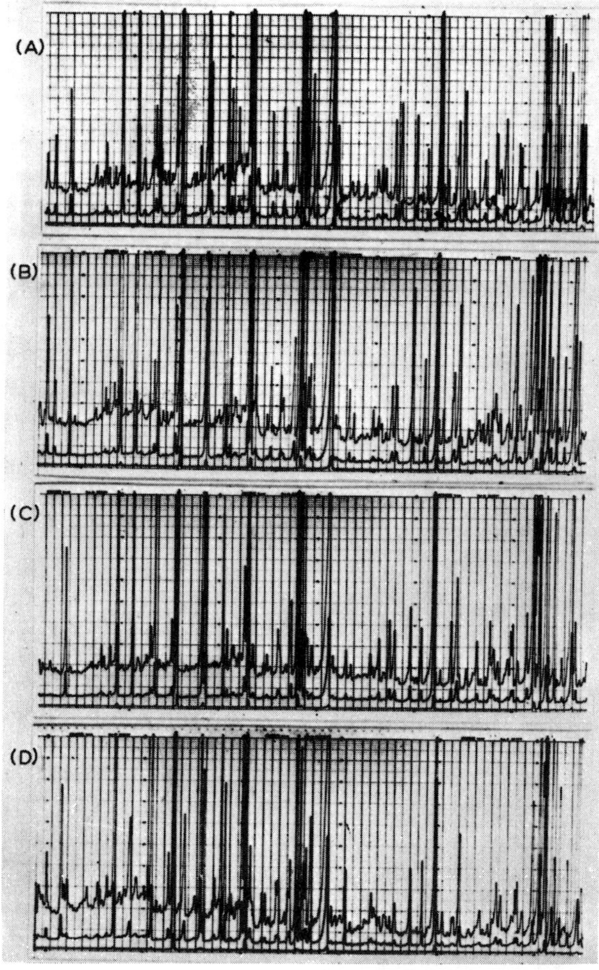

Figure 5. Influence of the location on the aroma composition of Riesling grapes (vintage year 1976).
A = Geilweilerhof, 88° Oechsle; B = Birkweiler, 81° Oechsle; C = Ilbesheim, 95° Oechsle; D = Knöringen, 72° Oechsle.

clay) and degrees of maturity (72 - 95° Oechsle), good agreement can be observed. The fingerprint pattern characteristic of Riesling is clearly visible. The fingerprint pattern shown in the aromagram sections of Figure 6 of Riesling samples from different wine-growing

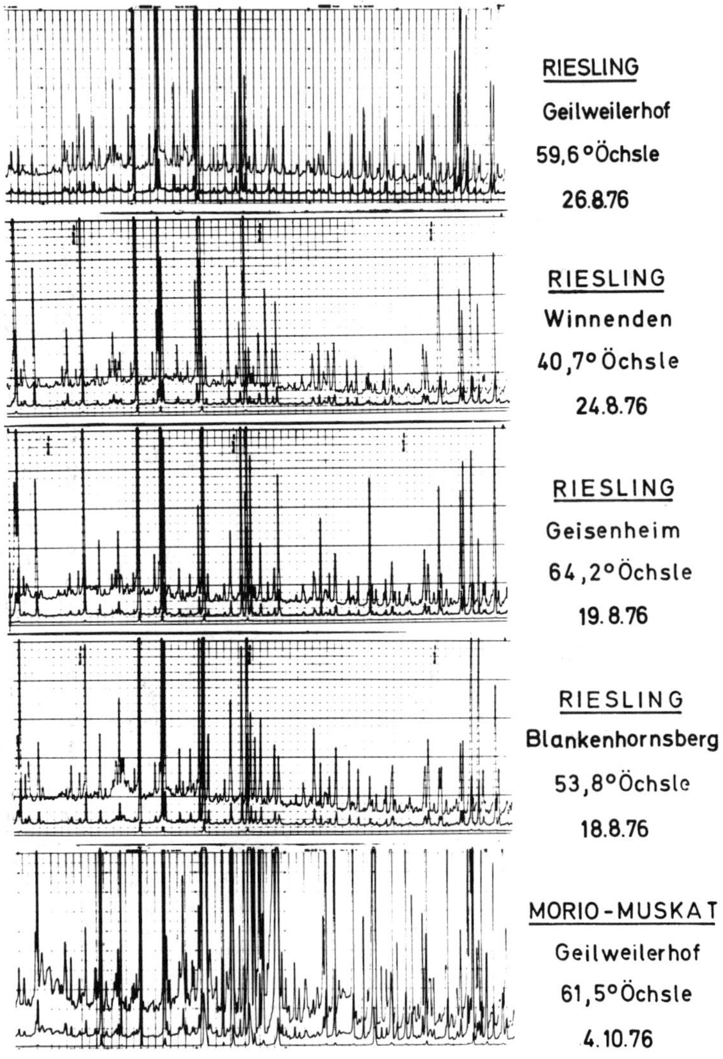

Figure 6. Comparison of fingerprint patterns of 4 Riesling samples from different wine-growing areas of Germany with Morio-Muskat from Geilweilerhof (vintage year 1976).

areas of Germany (Geisenheim/Rheingau, Winnenden/Remstal, Blankenhornsberg/Kaiserstuhl) and different maturity (40° - 64° Oechsle) also exhibit fairly good consistency. Further, the results shown in Figure 7 demonstrate that, even before sugar accumulation begins, numerous

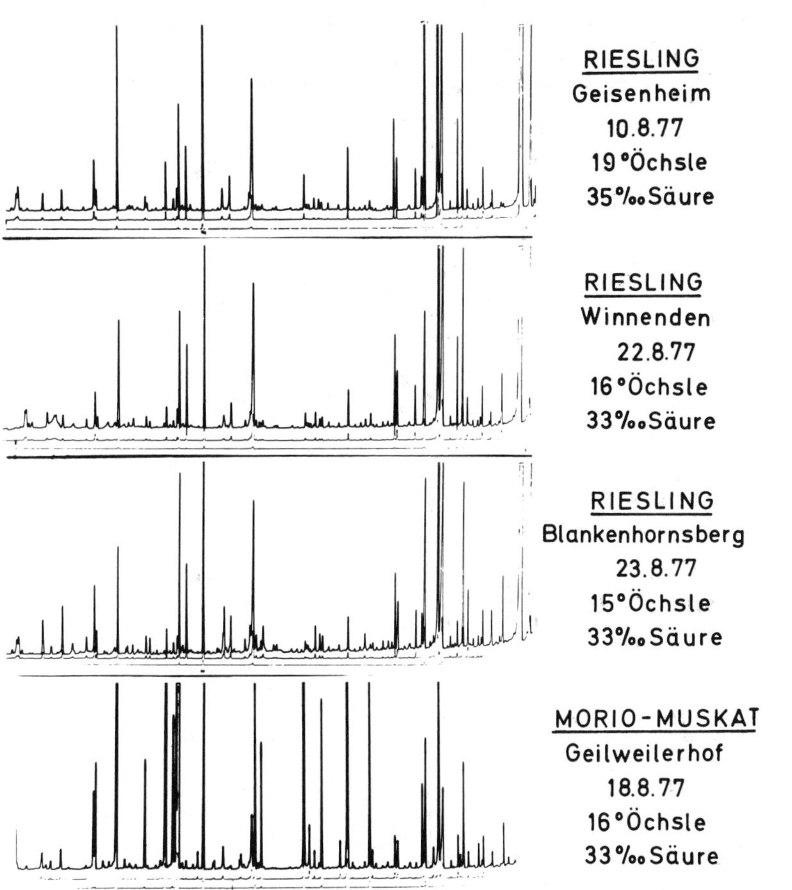

Figure 7. Comparison of fingerprint patterns of unripe Riesling samples from different wine growing areas of Germany with Morio-Muskat from Geilweilerhof (vintage year 1977).

volatile components are present in the grape berries and that these are already sufficient for a differentiation between varieties (here: Riesling and Morio-Muskat).

A peak pattern (fingerprint pattern) can only be used for characterizing a variety if the exogenous influences (location, vintage year, maturity etc.) do not exceed a certain limit. The aromagrams of different varieties illustrate to what extent the pattern of an individual variety may vary and still be clearly recognizable. The fingerprint pattern of Morio-Muskat, shown in Figures 6 and 7 differs markedly from those of various Riesling samples. Some typical key-substances predominate in Morio-Muskat and serve to differentiate this variety from the various Riesling samples of different location and maturity. Our results indicate that the typical "fingerprint pattern" (or"coarse-pattern") of the variety Riesling differs significantly from those of the other investigated varieties (Morio-Muskat, Silvaner etc.), regardless of location and maturity. But some aroma components are also influenced by maturity and location (soil, climate) and so a "fine pattern" in the aromagram is determined by the region of growth. Differences in the variety-specific fingerprints are possible and can be recognized.

VI. CHARACTERIZING DIFFERENT VARIETIES BY SOME KEY-SUBSTANCES

When only the location Geilweilerhof was taken into account, quite a number of components proved to be key-substances which could be used for differentiating the varieties Riesling and Morio-Muskat. In Figure 8 the peak heights of some aroma components from several Riesling samples taken at different locations (Winnenden, Trier, Geisenheim, Würzburg, Geilweilerhof) and varying degrees of maturity are compared with those of Morio-Muskat samples (location Geilweilerhof). The black column represents the peak heights of the selected components at the beginning of berry maturity. The change in peak height during berry maturation is shown by the light column. With increasing maturity, a marked increase in the peak height of component 35 is recogni-

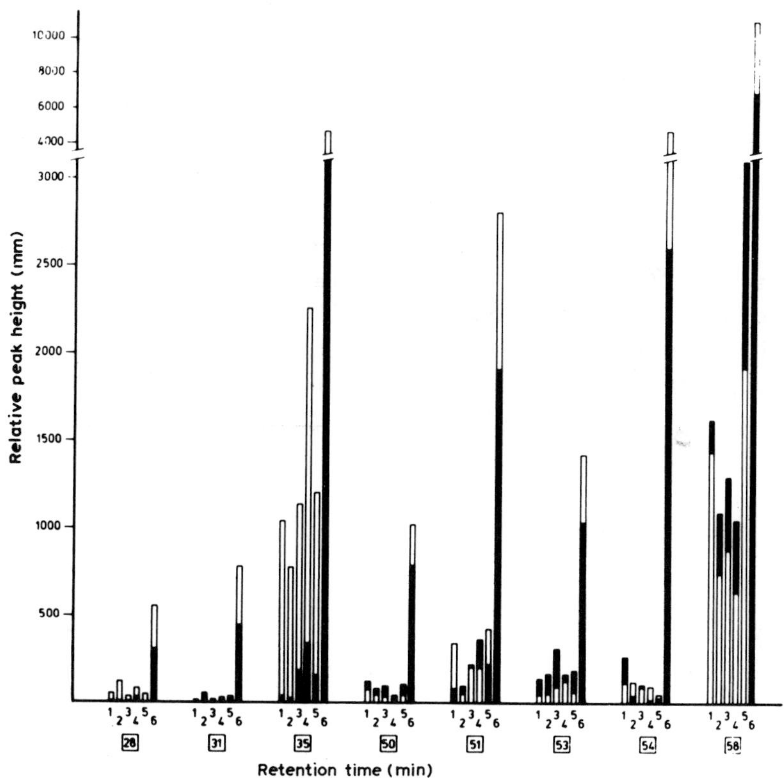

Figure 8. Content of aroma compounds of Riesling and Morio-Muskat; vintage year 1976.
1 = Riesling, Winnenden
2 = Riesling, Trier
3 = Riesling, Geisenheim
4 = Riesling, Würzburg
5 = Riesling, Geilweilerhof
6 = Morio-Muskat, Geilweilerhof

zable in all of the Rieslings as well as the Morio-Muskat. It is apparent that with regard to the peak height of these selected aroma components all of the Riesling samples are in close agreement. When compared, the Riesling and Morio-Muskat show differences which are so pronounced that it is possible to differentiate the 2 varieties even over a range of maturity from 40 - 100° Oechsle and independent of the viticultural district. In addition to those components studied in Figure 8 the total aromagram includes a number of other peaks which might also be useful for characterizing a variety.

By multiple discriminate analysis (11,35) we tested 27 key-components of Riesling and Morio-Muskat to determine if these were suitable for characterizing these varieties. The relative amount of each of the components analysed was estimated by visual comparison of the aromagrams of the two varieties. The results of

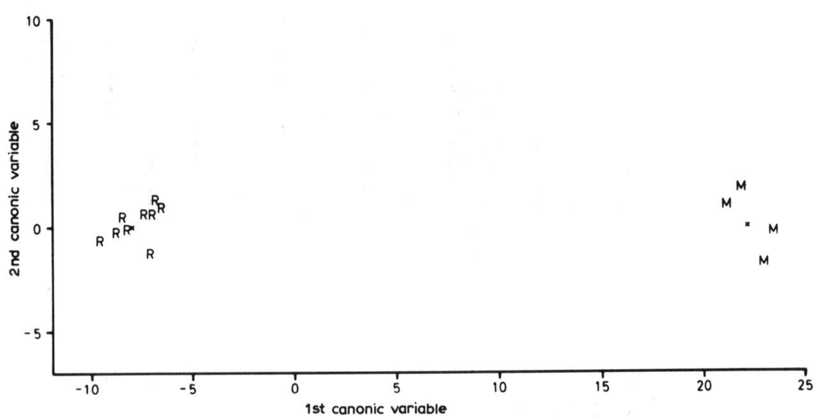

Figure 9. Separation between Riesling (R) and Morio-Muskat (M) by the aid of multiple discriminate analysis.

the multiple discriminate analysis (Figure 9) shows that after 5 steps of calculation a highly significant separation between the cultivars Riesling and Morio-Muskat is apparent (11). This separation was obtained for samples of the two varieties from all German vineyard regions and for berry maturities ranging from 40° to 100° Oechsle. A similar multiple discriminate analysis of these same components also allowed a significant differentiation of Morio-Muskat from several other varieties.

Differentiation of the cultivars Riesling, Silvaner and B 6-18 was also possible after 5 step multiple discriminate analysis of the 27 selected components although differences within this group were not so great as the separation noted above between Morio-Muskat and Riesling. However extension of the analysis from 5 steps to 15 steps led to the development of significant separation between the cultivars Riesling, Silvaner and B 6-18 (see Figure 10).

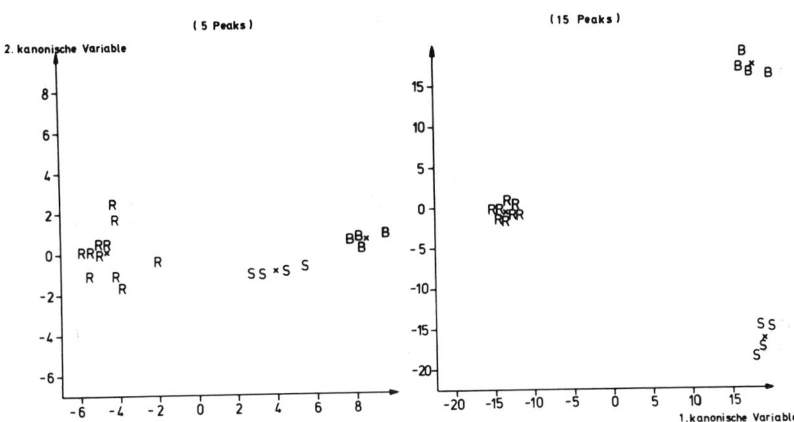

Figure 10. Separation between Riesling (R), Silvaner (S) and B 6-18 (B) by the aid of multiple discriminate analysis.

The question next arises as to what extent these varietal differences, which have been determined up to now, are recognizable in new grape varieties of similar families. Figure 11 shows comparative aromagram

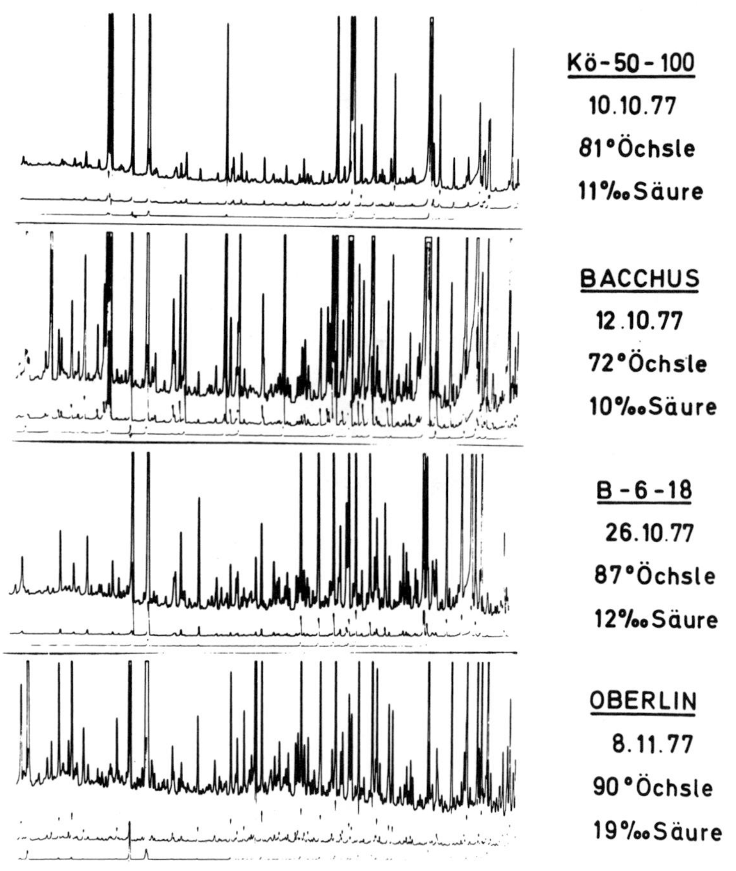

Figure 11. Comparison of fingerprint patterns of different vine varieties; vintage year 1977; 90 m Reoplex.

sections (on Reoplex) of the cvs. Kö 50-100 ((Silvaner x Müller-Thurgau) x Müller-Thurgau), Bacchus ((Silvaner x Riesling) x Müller-Thurgau), B 6-18 ((Vitis riparia x Gamay)F_2 x Fosters White Seedling) and Oberlin 595

(Vitis riparia x Gamay). It can be seen that there are significant differences in the fingerprint patterns between the new cvs. Kö 50-100 and Bacchus on the one hand and Oberlin 595 and B 6-18 on the other hand. Further more it was found that the aroma patterns of related varieties show marked quantitative differences and to such extent that multiple discriminate analysis of the major key-substances allows clear differentiation of the varieties.

From these investigations we expect to establish a quality evaluation in addition to a method of varietal characterization. This means that quality evaluation is based only on analytical data and not on subjective organoleptic assessment. The basis of our technique is the analytical separation of multicomponent mixtures on long glass capillary columns. An essential part of the work involves the recognition of suitable key-substances. Up to now these key-substances have been determined by visual comparison of aromagram sections from different varieties. This purely analytical approach makes use of components which may contribute little to the overall aroma. Accordingly we are now trying to select components for this method of analysis on the basis of their importance to the aroma. The selection can be made after identification of the components by compled g.c. - m.s. and determination of the aroma value of these compounds. A simplier and more rapid selection of those constituents which determine variety and quality can be effected by means of the "sniffing method" (9,36-37). By this technique the type and intensity of odour of every single component arriving at the end of the g.c. column is determined by means of a "sniffing detector". This latter "sniffing detector" is in parallel to the instrument detector. Since our olfactory responses are extremely sensitive, the limit of perception for some components is as low as 10^{-12} g. Figure 12 shows aromagram sections with sensory eva-

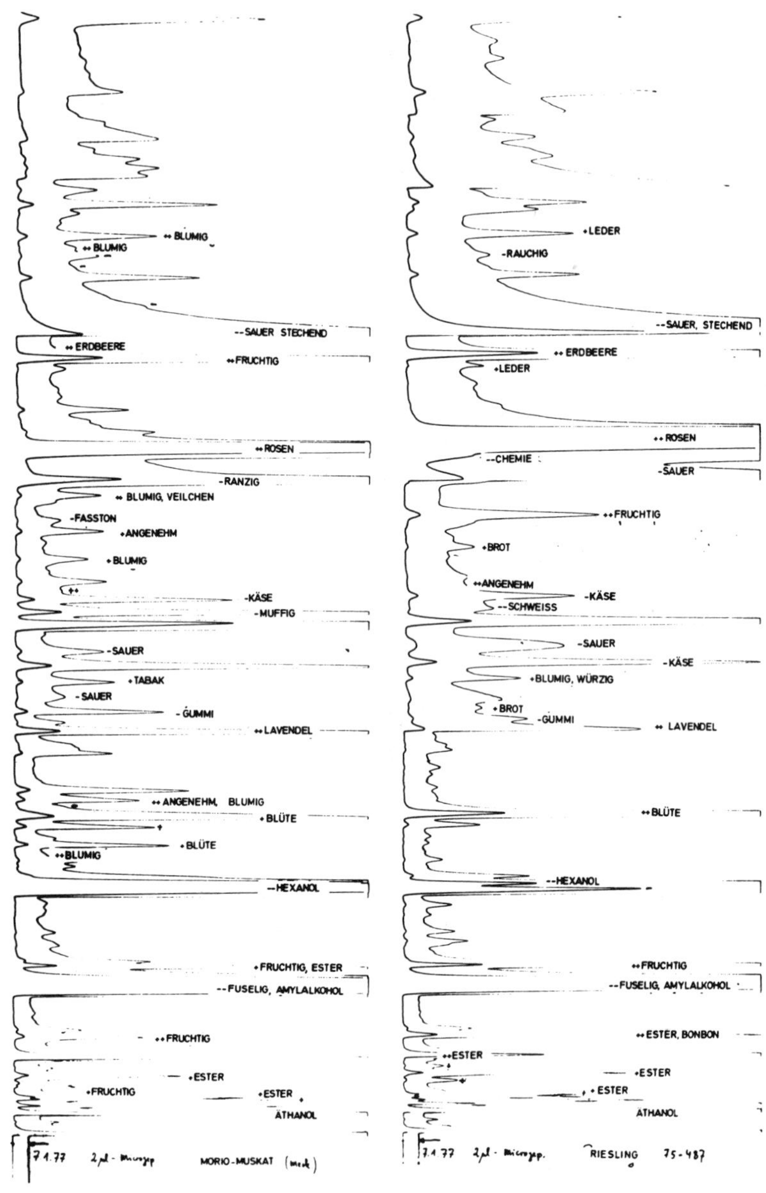

Figure 12. Aromagram of different grape varieties (Riesling and Morio-Muskat) and the quality of aroma ("sniffer method").

luation comments for the varieties Riesling and Morio-Muskat. This combination allows the typical key-substances to be recognized with more accuracy.

VII. SUMMARY

The volatile components of wine can be used to characterize wine quality and vine varieties. Due to the varying sensitivity of taste, components present in minute amounts may be of greater significance to the quality evaluation and the characterization of a variety, than those components present in relatively large amounts.

Using Freon 11 to extract wines and grapes, the volatile aromatic substances have been isolated. These aroma concentrates which are complex mixtures, were separated by g.c. into 350 to 400 components using glass capillary columns. A comparision of the aromagrams of wines from different varieties revealed distinct differences in the quantitative aroma composition ("fingerprint pattern"). These differences provide a basis for characterizing a variety.

In the "fingerprint pattern" of several varieties there are recognizable consistences in the appearance of certain components ("key-substances"). The relationship of these variety-specific "key-substances" permit differentiation of several varieties. A step by step discriminate analysis applied to only a limited number of these key-substances has allowed a definitive analytical differentiation of Riesling and Morio-Muskat grapes which is independent of location, year of harvest and stage of ripening.

VIII. REFERENCES

1. Rapp, A.,Über Inhaltsstoffe von Traubenmosten und Weinen unter besonderer Berücksichtigung der flüchtigen Verbindungen und des stofflichen Geschehens während der Hefegärung. Diss. Univ. Mainz (1965).
2. Drawert, F. und Rapp, A., Gaschromatographische Untersuchung pflanzlicher Aromen. I. Anreicherung, Trennung und Identifizierung von flüchtigen Aromastoffen in Traubenmosten und Weinen. Chromatographia 1, 446-457 (1968).
3. Rapp, A., Franck, H. und Ullemeyer, H., Die Aromastoffe verschiedener Weine. Dt. Lebensmittel-Rundschau 3, 81-85 (1971).
4. Rapp, A., Hövermann, W., Jecht, U., Franck, H. und Ullemeyer, H., Gaschromatographische Untersuchungen an Aromastoffen von Traubenmosten, Weinen und Branntweinen. Chem. Ztg. 97, 29-36 (1973).
5. Rapp, A., Über die Biogenese einiger Aromastoffe des Weines. in "Chemie und Landwirtschaftliche Produktion" Wien, 201-213 (1971).
6. Rapp, A., Steffan, H., Hastrich, H. und Ullemeyer, H., Zur Kenntnis der Biosynthese des Gärungsamylalkohols. Lebensm. Chemie und gerichtl. Chemie 29, 33-38 (1975).
7. Rapp, A., Hastrich, H. und Engel, L., Gaschromatographische Untersuchungen über die Aromastoffe von Weinbeeren. I. Anreicherung und kapillarchromatographische Auftrennung. Vitis 15, 29-36 (1976).
8. Rapp, A. und Hastrich, H., Gaschromatographische Untersuchungen über die Aromastoffe von Weinbeeren. II. Möglichkeiten der Sortencharakterisierung. Vitis 15, 183-192 (1976).
9. Rapp, A., Hastrich, H. und Engel, L., Kapillarchromatographische Untersuchungen über die Aromastoffe von Wein und Weinbeeren. Möglichkeiten zur Sortencharakterisierung. Mitt. Klosterneuburg 27, 74-82 (1977).
10. Rapp, A., Hastrich, H. und Engel, L., Möglichkeiten zur Qualitäts- und Sortencharakterisierung mit Hilfe der Gaschromatographie. in "5th Internat. Oenological Symposium" Auckland, 217-231 (1978).

11. Rapp, A. und Hastrich, H., Gaschromatographische Untersuchungen über die Aromastoffe von Weinbeeren. III. Einfluß des Standortes bei der Rebsorte Riesling. Vitis im Druck.

12. Drawert, F. und Rapp, A., Über Inhaltsstoffe von Mosten und Weinen. Vitis 5, 351-376 (1966).

13. Drawert, F., Rapp, A. und Ullemeyer, H., Radiogaschromatographische Untersuchung der Stoffwechselleistung von Hefen in der Bildung von Aromastoffen. Vitis 6, 177-197 (1967).

14. Bayonove, C. et Cordonnier, R., Recherches sur l'arôme du muscat. Ann. Techn. Agric. 19, 79-93 (1970).

15. Hardy, P.J., Changes in volatiles of muscat grapes during ripening. Phytochemistry 9, 709-715 (1970).

16. Prillinger, F. und Madner, A., Die flüchtigen Inhaltsstoffe von Muskatmosten und -weinen. Mitt. Klosterneuburg 20, 202-205 (1970).

17. Schreier, P. und Drawert, F., Gaschromatographisch-massenspektrometrische Untersuchung flüchtiger Inhaltsstoffe des Weines. Chem. Mikrobiol. Technol. Lebensm. 3, 154-160 (1974).

18. Schreier, P., Drawert, F. und Junker, A., Gaschromatographische-massenspektrometrische Differenzierung der Traubenaromastoffe verschiedener Rebsorten von Vitis vinifera. Chem. Mikrobiol. Technol. Lebensm. 4, 154-157 (1976).

19. Schreier, P., Drawert, F. und Junker, A., Gaschromatographische Bestimmung der Inhaltsstoffe von Gärungsgetränken. Chem. Mikrobiol. Technol. 5, 45-52 (1977).

20. Drawert, F., Schreier, P. und Scherer, W., Gaschromatographische-massenspektrometrische Untersuchung flüchtiger Inhaltsstoffe des Weines. Z. Lebensm. Unters. und Forschung 155, 342-347 (1974).

21. Prillinger, P., Horwatitsch, H. und Madner, A., Versuche zur Charakterisierung von Weinen auf Grund ihrer flüchtigen Inhaltsstoffe. Mitt. Klosterneuburg 17, 271-279 (1967).

22. Cordonnier, R., L'arôme des vins. Progr. Agric. Vitic. 91, 320-329 (1974).
23. Terrier, A. et Boidron, J.N. Identification des derives terpeniques dans les raisins de certaines varietes de Vitis vinifera. Conn. vigne et vin 6, 147-160 (1972).
24. Webb, A.D., Kepner, R.E. and Maggiora, L., Gas-chromatographic comparison of volatile aroma materials extracted from eight different muscat-flavored varieties of Vitis vinifera. Amer. J. Enol. Viticult. 17, 247-254 (1968); 20, 16-24 (1969).
25. van Wyk, C.J., Webb, A.D. and Kepner R.E., Some volatile components of Vitis vinifera variety White Riesling. J. Food Sci. 32, 660-664 (1967).
26. Muller, C.J., Kepner, R.E. and Webb, A.D., Lactones in Wines - A Review. Amer. J. Enol. Viticult. 24, 5-9 (1973).
27. Guadagni, G.D., Buttery, R.G. and Okano, S., Odour thresholds of some organic compounds associated with food flavours. J. Sci. Food Agricult. 14, 761-765 (1963).
28. Guadagni, D.G., Buttery, R.G. and Harris, J., Odour intensities of hop oil components. J. Sci. Food Agricult. 17, 142-144 (1966).
29. Rothe, M. und Thomas, B., Aromastoffe des Brotes. Z. Lebensm. Untersuchung und Forschung 119, 302-310 (1963).
30. Salo, P., Nykänen, L. and Suomalainen, H., Odor thresholdsand relative intensities of volatile aroma components in artificial beverage imitating Whisky. J. Food Sci. 37, 394-397 (1972).
31. Meilgaard, M.C., Aroma volatiles in Beer. in "Geruch- und Geschmacksstoffe".Carl,Nürnberg; 211-254 (1975).
32. Boeckh, J., Die chemischen Sinne Geruch und Geschmack. in "Physiologie des Menschen" Urban und Schwarzenberg, München (1972).

33. Mayer, H.G., Flüchtige Aromastoffe in Lebensmitteln. Angew. Chemie 82, 965-988 (1970).
34. Blakesley, C.N. and Loots, J., A convenient method for multiple extraction of volatile flavor components from Food slurries and pulps using a two-chambered glass-bomb-extractor and dichlorodifluoromethane (Freon 12) solvent. Agric. Food Chem. 25, 961-963 (1977).
35. Schreier, P., Drawert, F., Junker, A. und Reiner,L., Anwendung der multiplen Diskriminanzanalyse für Differenzierung von Rebsorten an Hand der quantitativen Verteilung flüchtiger Weininhaltsstoffe. Mitt. Klosterneuburg 26, 225-234 (1976).
36. Vitzthum, O.G., Flüchtige Inhaltsstoffe des Röstkaffees. in "Geruch- und Geschmacksstoffe" Carl, Nürnberg, 49-64 (1975).
37. Acree, T.E., Butts, R.M., Nelson, R.R. and Lee, C.Y., Sniffer to determine the odor of gas chromatographic effluents. Anal. Chem. 48, 1821-1822 (1977).
38. Rapp, A., Les arômes des vins et des eaux-de-vie. Bull. OIV 45, 151-166 (1972).
39. Chaudhary, S.S., Webb, A.D. and Kepner, R.E., GLC Investigation of the volatile compounds in extracts from Sauvignon blanc wines from normal and botrytised grapes. Amer. Jour. Enol. Viticult. 19, 6-12 (1968).
40. Hardy, P.J., Extraction and concentration of volatiles from dilute aqueous and aqueous-alcoholic solution using Trichlorfluoromethane. J. Agric. Food Chem. 17, 656-658 (1969).
41. Stevens, K.L., Flath, R.A., Lee, A. and Stern, D.J., Volatiles from grapes. Comparison of Grenache juice and Grenache rosé wine. J. agric. Food Chem. 17, 1102-1106 (1969).
42. Rapp, A. und Franck, H., Über die Bildung von Äthanol und einiger Aromastoffe bei Modellgärversuchen in Abhängigkeit von der Aminosäurenkonzentration. Vitis 9, 299-311 (1971).

INDEX

A

Absidia orchidis, stereospecific reduction by, 164
Air odorants, *see also* Large-bore coated columns
  activated charcoal as sampling trap for, 211, 213–215, 217–219
  analysis for, 209–229
  coated supports as sampling traps for, 211, 213–215, 217–219
Alcoholic beverages, *see also* Aroma
  changes during maturation, 381–382
  chemical composition of, 340–347
  major aroma compounds in whisky, cognac, and rum, 341, 343
  phenol concentration in certain whiskies, 340
  sulfur compounds in certain vodkas, 347
Analogs
  meat, 47–55
Aroma, *see also* specific materials
  alcoholic beverages, 339
  beer, 335
  black currant, 6
  bread, 58, 61
  burnt, 7, 35
  cheese, 65
  cocoa, 58, 61
  cooked soybean, 35
  grassy, 34
  meat, 15
  olive oil, virgin, 247
  orange juice, 283
  soy, 34–35
  tea, 305
  wine, 362, 384, 391

B

Barley variety, distinction by principal components analysis, 330
Beer
  classification by tasting, 334
  factorial discriminant analysis of, 336
lactones in, 151, 152
  microbrewed, pilot plant, and commercial beers, physicochemical versus sensory analysis, 330–333
  lactones in, 151, 152
  sensory analysis of, 329
Black caramel color substances isolation and fractionation of pure carbohydrates in, 172–179
  isolation and fractionation of
    breadcrust, 180–182, 185, 189, 191–193
    chicory, 184, 187, 191–193
    coffee, 183, 188, 191
    molasses, 186, 194
Black tea, 134, 312, 316
Bread, *see also* Aroma
  compounds in bread crust, 63
  flavor, 57–58, 62
  preparation of "Troz," 59
Browning reaction products, 8

C

Cat taste, 233
Ceratocystis variospora
  formation of terpenoids by, 159–161
  gas chromatogram of aroma, concentrate from, 158
Cheese
  bacterial surface-ripened, 71
  flavor compounds related to specific microflora, 66–67, 71–72
  internal propionibacteria fermentation, 71–72
  mold-ripened, 66–68
Chocolate
  conching, 77
  flavor precursors, 75
  volatiles characterization, 76

Cocoa
applications, 85–86
"chocless," 89
composition, 83
definition, 82
flavor, 89
substitution, 86–89

E

Enrichment procedures, 212, 392
Esters, contribution to the perceived aroma of whisky, 349

F

Flavor, general
bitter, 9–10
black currant juice, 6–7
changes, heat induced, 5–7
compounds, synthesis *de novo,* 2–5
modification of, 1
retort, 7
Food flavors
characterization of synthetics by isotopic analysis, 202
classification—natural, artificial, 199–200
determination of the ^{13}C-^{12}C ratio of
citral, 202–203
experimental systems, 200
vanillin, 204–205
determination of the ^{2}H-^{1}H ratio of
citral, 205
experimental systems, 205
menthol, 205
trans-anethole, 206

G

Glass capillary columns, 305, 392–395

H

Headspace techniques
flavor interactions with food ingredients, 116–118
olive oil, 249–255
orange juice, 284–286
Heyns rearrangements, 131–132
Human taste measures, 232
Hydrophobicity, 11

L

Lactones
in beer, 151, 152
in fermentation products, 146
formation by microorganisms, 145 Lipid oxidation, 8, 11
Large-bore coated columns (LBC)
analysis for breath odorants by sampling on, 227–228
coating of, 214
large porous polymer columns versus LBC, 221–227
sample desorption of, 215
sampling for air odorants by, 209

M

Meat
analogs, 43–44, 47, 52
aroma analysis, 16, 23
aroma compounds structure, 18–22, 24–28
flavor development, 40, 44–47
flavor precursors, 17
processing, 5, 54

O

Olive oil
aroma, 247
phenolic compounds, 248, 255
biogenesis, 270
content, 255
localization in fruit, 270
mechanical extraction process effects, 258
organoleptic significance, 275
pigmentation rate of the influence of fruits, 258
qualitative and quantitative analysis, 259
seasonal variations, 262
storage of fruits effect, 274
varietal differences, 264
volatile constituents
biogenesis, 267
headspace, 252, 254
localization in fruit, 267
mechanical extraction process effects, 271
organoleptic significance, 258
qualitative and quantitative analysis, 248
seasonal variations, 262
storage of fruit effects, 274
varietal differences, 264
Orange juice
aroma, 283

compounds identified in, 291–298
degradation compounds identified in off-flavor, 299–300
headspace analysis of volatiles in 284–286
off-flavor, 298–299
total volatile analysis, 286–291

P

Protein hydrolysates, bitterness of, 9

R

Rearrangement products
identification by proton magnetic resonance, 136–138
isolation from
black tea, 134
roast meat, 135
tomato powder, 133
thermal degradation, 138–143

S

Soy
flour composition, 48–49
proteins
headspace gas volatile composition, 51–52
interaction and release of flavors, 37–39
off-flavors, 34
prevention of off-flavor formation, 36
removal of off-flavors, 36
Sporobolomyces odorus
conversion of (1-^{14}C)-deca-and dodecanoic acids by, 154
conversion of (U-^{14}C)-linoleic acid by, 155
formation of lactones, 151, 158
gas chromatogram of aroma concentrate from, 153
lactones produced by, 157
Stereospecific reduction by
Absidia orchids, 164
Pseudomonas ovalis, 165
Sweet compounds
biomolecular studies, 91–96
glycyrrhizin, 99–102
Katemfe, 107–109
Lo Han fruit, 102
miracle fruit, 104
monellin, 107
natural origin, 97
nitroanilines, 93–95
osladin, 103
serendipity berries, 106
stevioside, 98

T

Tea
aroma constituents, 306–311
contributing fraction composition
black tea, 312
green tea, 319
manufacturing process
black tea, 312
green tea, 312
Pouchong tea, 313
sweet tea, 325
topnote aroma composition
Ceylon black tea, 313–317
green tea, 318–319
Keymun black tea, 317–318
Lotus tea, 325
Pouchong and Jasmine teas, 322–324
Tenax GC trapping of
air volatile organic, 212
virgin olive oil volatiles, 249
Terpenoids
formation by microorganisms, 156
transformation by microorganisms, 161

V

Volatile organic compounds in equilibrium with food ingredients
headspace analysis, 116–118
heats of sorption, determination, 119, 124, 128
interactions with liquid and solid substrates, 121, 123

W

Wine
aroma components of
concentration and separation, 392–395
different grape varieties, 395–397
grape maturity influence on composition of, 398–402
location influence on composition of, 402–406
primary, from the grape, 362–364
secondary, from fermentation, 364–371, 387, 388
characterization by key substances, 406–413
distillate ageing experiment
chromatographic analysis, 384–385
oaklactone in, 389
pretreatments, 383
sensory evaluation, 384, 386, 389

quality factors
  natural, 371–376
  spoilage, 379
  storage in wooden barrels, 376–379
suppleness index, 357–361
taste, 356

Whisky, perceived aroma of
  aroma compounds, 347
  core of, 348
  ethyl esters of C_6-C_{12} acids, 351
  strength and quality of, 350–353

A
B
C 8
D 9
E 0
F 1
G 2
H 3
I 4
J 5